Forest Recreation and Landscape Protection

Forest Recreation and Landscape Protection

Editor

Radu-Daniel Pintilii

MDPI • Basel • Beijing • Wuhan • Barcelona • Belgrade • Manchester • Tokyo • Cluj • Tianjin

Editor
Radu-Daniel Pintilii
University of Bucharest
Romania

Editorial Office
MDPI
St. Alban-Anlage 66
4052 Basel, Switzerland

This is a reprint of articles from the Special Issue published online in the open access journal *Forests* (ISSN 1999-4907) (available at: https://www.mdpi.com/journal/forests/special_issues/Recreation_Landscape).

For citation purposes, cite each article independently as indicated on the article page online and as indicated below:

LastName, A.A.; LastName, B.B.; LastName, C.C. Article Title. *Journal Name* **Year**, *Volume Number*, Page Range.

ISBN 978-3-0365-6493-7 (Hbk)
ISBN 978-3-0365-6494-4 (PDF)

Cover image courtesy of Radu-Daniel Pintilii

Contents

About the Editor

Radu-Daniel Pintilii

Radu-Daniel Pintilii (Associate Professor) is a geographer who achieved a Ph.D. in the field of Geography from the University of Bucharest in 2008. He is an employee of the Faculty of Geography (Department of Human and Economic Geography) and a Ph.D. advisor of two Doctoral Schools (The Faculty of Geography and the Interdisciplinary School of Doctoral Studies) in the same University. He is also the Scientific Manager of the Research Center for Integrated Analysis and Territorial Management (CAIMT), with main interests in Natural Resources, Forestry, GIS, Spatial Planning and Statistics. He has attended more than 50 International Scientific Conferences and published 45 ISI-indexed Articles, scientific reviewer of more prestigious journals and the H-index, according to WoS, is 13. He developed important collaborations with prestigious European higher education institutions, via the CIVIS Network (Austria, Belgium, France, Germany, Great Britain, Greece, Italy, Spain, Sweden and Switzerland).

forests

MDPI

Editorial

Forest Recreation and Landscape Protection

Radu-Daniel Pintilii [1,2,3]

[1] Department of Human and Economic Geography, University of Bucharest, 1, Nicolae Bălcescu Avenue, 010041 Bucharest, Romania; pinty_ro@yahoo.com
[2] Research Center for Integrated Analysis and Territorial Management, University of Bucharest, 4-12, Regina Elisabeta Avenue, 030018 Bucharest, Romania
[3] Section of Life, Environmental and Earth Sciences, Research Institute of the University of Bucharest (ICUB), 90, Panduri Street, 050663 Bucharest, Romania

1. Introduction

Forests, which are viewed as an important natural resource for humanity, have been constantly threatened with degradation or even extinction since the dawn of the modern period. Landscape protection, especially forest conservation, represents an important research topic of the last century. While forest recreation focuses on forests as the main destinations for public recreational activities, landscape protection refers to the conservation, development, and restoration of natural beauties. Urban and peri-urban forests have become places for recreational activities of populations living in big cities all over the world. The lack of green spaces is always felt by modern society within its current built environments. In recent years, demands for forest recreation activities have been increasing both in volume and in diversity. In some areas, predominantly rural ones, the forest acts as a site for the seasonal collection of berries and mushrooms (non-timber forest products, NTFPs), which are integrated into rural life as key social activities. Additionally, landscape protection measures have been imposed during the last two or three decades due to the immense, adverse human effects on all the landscape components. Landscape protection is imperatively required due to the development of human settlements over the last few decades, which may be viewed as a forceful and constant expansion. It is imposed not only for scientific and ecologic reasons but also for social and cultural ones, and this is why the driving mechanisms for managing forests in the future should be better understood.

Citation: Pintilii, R.-D. Forest Recreation and Landscape Protection. *Forests* **2022**, *13*, 1440. https://doi.org/10.3390/f13091440

Received: 5 September 2022
Accepted: 7 September 2022
Published: 8 September 2022

Publisher's Note: MDPI stays neutral with regard to jurisdictional claims in published maps and institutional affiliations.

2. The Methodology Used in the Papers of the Special Issue

2.1. Study Areas

The origins of the studies presented the papers of this Special Issue cover a large selection of countries (Figure 1), which encompass the three continents (Asia, Europe, and Oceania) and have different natural and socio-economic conditions, as reflected in their climates and microclimates, vegetation (a repartition of plant species), air quality (pollution), and different profiles of economic activities (agricultural, industrial, or services, including tourism and recreational activities). There are two papers that focus on large areas, encompassing the countries around the Mediterranean Sea [1,2]. The other papers cover smaller areas, such as islands (Sulawesi in Indonesia [3], New Zealand in Oceania [4]), countries (China, South Korea, and Taiwan [5,6]), or even smaller areas, such as mountains (Moravian Karst Plateau in the Czech Republic and Plitvice Lakes National Park in Croatia) [7,8], or cities (Warsaw in Poland and Xi'an in China) [9,10].

Figure 1. Study areas of the papers in this Special Issue (left: paper number, according to the references; middle: study area; and right: continents). Data source: [1–10].

2.2. Data Acquisition and Processing

Authors used data obtained from diverse sources, with the data providers including official authorities [3,9,10], satellite images or GIS [3], other platforms [1,6,8], and questionnaires [4,5]. The methodologies applied are diverse and complex, from simple monitoring and descriptive statistics [1,2,7] to advanced models (Shapiro–Wilk and Kappa tests [1,3], social media content, cluster analysis, natural language processing [8], and sentiment analysis [6,8]). The processing of the data was performed using specialized software, including R Software [1,10], SPSS [4,5,9], KH Coder Software [8], and GIS [1–10].

3. Summary of the Papers in This Special Issue (Forest Recreation and Landscape Protection)

The papers published in this Special Issue, "Forest Recreation and Landscape Protection", cover a wide range of topics that can be divided into two main categories: studies focusing on forest recreation and those focusing on landscape protection. The papers investigating forest recreation refer to the tourism activities in forests, which are regarded as leisure activities (e.g., mountain biking). The others, focusing on landscape protection, concentrate on the impact of forest (in)stability on human health (air quality) and, especially, the lack of forests and forest degradation as direct consequences of human economic activity.

In the first paper, Ciobotaru et al. [1] present the environmental issues related to forest loss in one of the important areas of the Earth, the Mediterranean region. The paper analyzes the tree cover loss and cumulative tree cover loss in each country, acting as an important index of the tendency towards tree cover loss in each country, over a total area of 581 Mha. The forest loss is determined by certain natural and anthropogenic activities, such as intense forest fires, overgrazing, agricultural expansion, and illegal logging.

In the second paper, Hrůza et al. [7] underline the important role of "controlled recreation" by building and tracing certain bike routes in the mountainous area of the Moravian Karst in Poland. While cycling has become one of the popular recreational activities of today's world, it is also important for the distribution of groups of visitors in the area, especially the bikers, who use only the designed routes.

Kim et al. [5] offer an international comparative analysis of the perception of the natural environment in South Korea, Taiwan, and Indonesia. Using a specific methodology, the perceptions regarding environmental conservation and the use of urban forests were analyzed, and some significant differences between the three countries, which have different histories, cultures, and economic developmental levels, were identified. Among all the countries, Taiwan showed the highest positive results, while concerns regarding the "limits of growth" were significant in South Korea and those regarding the "ecological crisis" more pressing in Taiwan and Indonesia.

Jiang et al. [9] show that, as the urbanization and industrialization processes in China have accelerated, pollution has become one of the most dangerous issues in the cities. The urban green spaces, as "green infrastructure", are an important part of the urban ecosystems, with the purposes of air purification and refreshing. By considering the meteorological factors over time, the effects of green spaces with different vegetation structures and effects on air purification were determined. The results underline the positive effect of urban green spaces on air quality, and this could be used as an important theoretical basis for optimizing the structure of urban green spaces.

Another paper focusing on air quality, Hoppa et al. [10], studies the impact of air pollution on people's health, which presents as one of the world's leading environmental health threats. The authors identify children as the most vulnerable social group, who are most exposed to the harmful effects of air pollution. With greater exposure time, more negative effects are observed. The study investigates the role of trees in air purification during winter, as a leafless period, on children's school routes. The study was conducted in Warsaw, and the results obtained demonstrate the weak impact of the tree canopy in the winter season.

Based on habitat preferences, Kinho et al. [3] investigate the spatial distribution of the *Magnolia* genus in the northern part of Sulawesi using 786 waypoints. They confirmed that the *Magnolia* spp. distribution is influenced by factors such as the annual temperature range, precipitation seasonality, and elevation. Endemic and endangered species, such as *Magnolia sulawesiana* Brambach, Noot., and Culmsee, were identified in the central part of Sulawesi. This study could provide a scientific basis for efficient forest management and the development of conservation strategies and landscape protection measures.

Sergiacomi et al. [8] conducted a survey, using the social media TripAdvisor platform, of 15,673 visitor reviews of the Plitvice Lakes National Park in Croatia. Their aim was to detect the strengths and weaknesses of the National Park. The methods of analyses that the authors used could be applied in different countries with similar natural contexts and protected areas. The results obtained show that visitors were both interested in naturalistic and landscape aspects and the accessibility and management of routes and visits to the park.

Zeng et al. [6] used deep learning technology to select six case studies in China and to analyze geotagged photos of forest landscapes posted by forest recreationists in a specific application. The paper also used Deepsentibank to perform a sentiment analysis of forest landscape photos. This facilitates a better understanding of Chinese forest recreationists' forest landscape preferences. The authors concluded that people preferred flat views and had a strong preference for forest trail landscapes, which plays an important role in forest landscape recreation, and that the photos evoked both positive and negative emotions.

Sancho-Knapic et al. [2] also underline the impact of air quality on human health. The study was conducted in Mediterranean forests, examining the variations in air quality under the influence of different factors (temperature and leaf development). The six most abundant monoterpenes were detected, and it was found that their abundance in the air increased after the drought period. The study results show that people who enjoy forest-based activities in Mediterranean conifer areas are more exposed to air monoterpenes when the temperature increases. This happens when the leafy canopy is well developed and after a drought period.

Bayne et al. [4] refer to recreational activities in the production forests of New Zealand. These kinds of forests offer many opportunities for mountain biking, particularly in smaller commercial peri-urban plantations and forest parks maintained as dedicated mountain bike parks. The paper explains the impacts of, and the responses of forest owners and managers to, increased mountain biking activities in commercial forest estates. This paper is useful for urban and peri-urban forest planners (managers) who accommodate recreational activities such as mountain biking in production forests. It suggests policies and best practices that can help to protect commercial interests and forest ecology.

4. Conclusions

This Special Issue, entitled "Forest Recreation and Landscape Protection", comprises 10 papers by 59 authors from 12 countries on three continents: Asia (Indonesia, China, South Korea, India, Japan), Europe (Spain, Italy, Poland, Czech Republic, Croatia, Romania), and Oceania (New Zealand). Their multiple affiliations permitted the authors to study and research a great expanse of topics and areas related to forests, as an important natural resource and forest ecosystem service provider, as well as other elements with direct or indirect impacts on the forests (Figure 2).

Figure 2. Authors' affiliation(s) and country (or countries) (right) by paper, with references (left). Data source: [1–10].

Another riveting aspect is the diversity of the keywords used in the papers (Figure 3). All the papers are centered around the *forest* as the main keyword, appearing in diverse combinations: forest management [4,7], forest recreation [7,8], forest landscape [6], and forest air [2], plantation forest [4]. Others relate to the tree species (*Magnolia sulawesiana* [3], *Pinus pinaster* [2], *Pinus radiata* [4]). Other keywords relate to the activities linked to forests, such as recreational activities (visitor monitoring [7], nature experience [5], single-track bike trails [7], TripAdvisor reviews [8], geotagged photos [6], path analysis [2]) and economic activities (commercial forestry, trade-off [4]). Other keywords center on the landscape, the other key focus of this Special Issue, and associated terms such as environment [1], landscape protection, and conservation [3].

Figure 3. Keywords of the papers in this Special Issue. Data source: [1–10].

Figure 4 offers an illustration of this multitude of 58 keywords grouped into 11 families, from all 10 papers (economic activity, environment, impact, index, landscape, management, methodology, pollution, region, resources, and tourism). The family of the key words is placed in the middle of the Figure, the paper numbers are on the left side, and the keywords are on the right side of Figure 4.

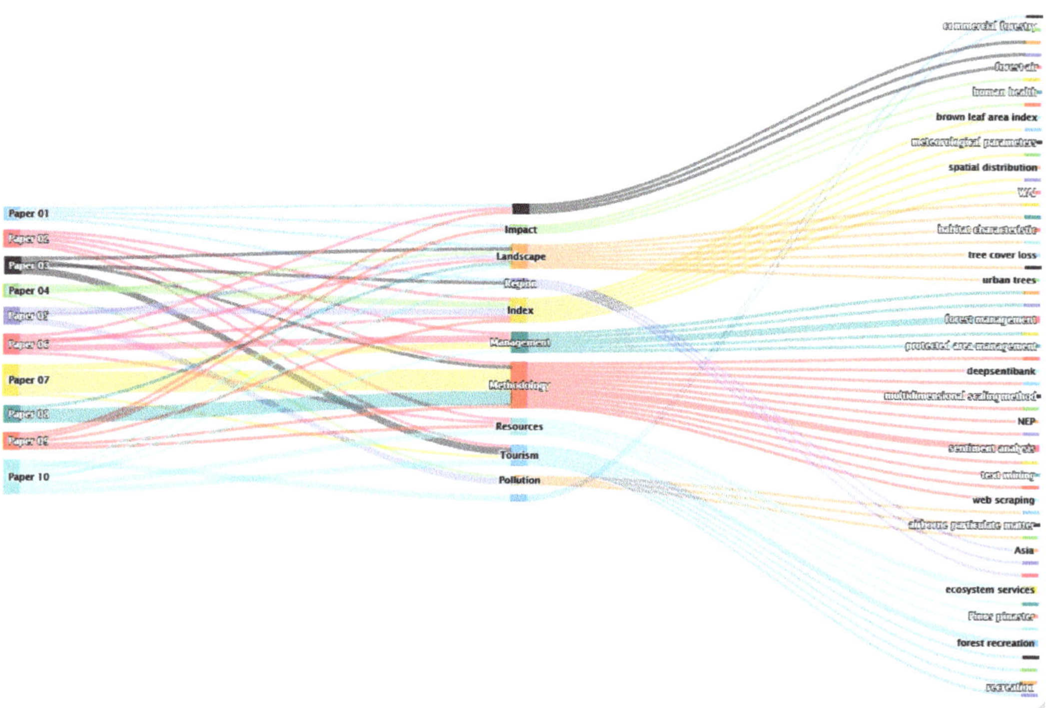

Figure 4. Papers in the Special Issue/family of keywords/keywords. Data source: [1–10].

The papers from this Special Issue adopt an interdisciplinary approach that considers the aspects affecting forest degradation, including tree cover loss and deforestation, on the one hand, and those that lead to landscape conservation and protection, on the other. All the papers use an adequate and modern variety of methodologies, providing a real source of inspiration for further research papers and Ph.D. theses.

Funding: This research received no external funding.

Conflicts of Interest: The author declares no conflict of interest.

References

1. Ciobotaru, A.-M.; Patel, N.; Pintilii, R.-D. Tree Cover Loss in the Mediterranean Region—An Increasingly Serious Environmental Issue. *Forests* **2021**, *12*, 1341. [CrossRef]
2. Sancho-Knapik, D.; Gil-Pelegrín, E.; Ferrio, J.P.; Alonso-Forn, D.; Martín-Sánchez, R.; dos Santos Silva, J.V.; Imanishi, J.; Peguero-Pina, J.J.; Sanz, M.Á. Changes in the Abundance of Monoterpenes from Breathable Air of a Mediterranean Conifer Forest: When Is the Best Time for a Human Healthy Leisure Activity? *Forests* **2022**, *13*, 965. [CrossRef]
3. Kinho, J.; Arini, D.I.D.; Abdulah, L.; Susanti, R.; Irawan, A.; Yulianti, M.; Subarudi, S.; Imanuddin, R.; Wardani, M.; Denny, D.; et al. Habitat Characteristics of *Magnolia* Based on Spatial Analysis: Landscape Protection to Conserve Endemic and Endangered *Magnolia sulawesiana* Brambach, Noot., and Culmsee. *Forests* **2022**, *13*, 802. [CrossRef]
4. Bayne, K.M.; Scott, M.B.; Yao, R.T. Getting Flow: The Place of Production Forests in the Rise of Mountain Biking. *Forests* **2022**, *13*, 1326. [CrossRef]
5. Kim, D.; Avenzora, R.; Lee, J.-h. Exploring the Outdoor Recreational Behavior and New Environmental Paradigm among Urban Forest Visitors in Korea, Taiwan and Indonesia. *Forests* **2021**, *12*, 1651. [CrossRef]
6. Zeng, X.; Zhong, Y.; Yang, L.; Wei, J.; Tang, X. Analysis of Forest Landscape Preferences and Emotional Features of Chinese Forest Recreationists Based on Deep Learning of Geotagged Photos. *Forests* **2022**, *13*, 892. [CrossRef]
7. Hrůza, P.; Pelikaán, P.; Olišarová, L. Single-Track Bike Trails in the Moravian Karst as Part of Forest Recreation. *Forests* **2021**, *12*, 1601. [CrossRef]

8. Sergiacomi, C.; Vuletić, D.; Paletto, A.; Barbierato, E.; Fagarazzi, C. Exploring National Park Visitors' Judgements from Social Media: The Case Study of Plitvice Lakes National Park. *Forests* **2022**, *13*, 717. [CrossRef]
9. Jiang, B.; Sun, C.; Mu, S.; Zhao, Z.; Chen, Y.; Lin, Y.; Qiu, L.; Gao, T. Differences in Airborne Particulate Matter Concentration in Urban Green Spaces with Different Spatial Structures in Xi'an, China. *Forests* **2022**, *13*, 14. [CrossRef]
10. Hoppa, A.; Sikorska, D.; Przybysz, A.; Melon, M.; Sikorski, P. The Role of Trees in Winter Air Purification on Children's Routes to School. *Forests* **2022**, *13*, 40. [CrossRef]

forests

MDPI

Article

Tree Cover Loss in the Mediterranean Region—An Increasingly Serious Environmental Issue

Ana-Maria Ciobotaru [1], Nilanchal Patel [2,†] and Radu-Daniel Pintilii [3,4,5,*,†]

1 "Gheorghe Balș" Technical College, 107 Republicii Street, 625100 Adjud, Romania; ciobotaruanamaria@yahoo.com
2 Department of Remote Sensing, Birla Institute of Technology Mesra, Ranchi 835215, India; nilanchal.patel@gmail.com
3 Research Center for Integrated Analysis and Territorial Management, University of Bucharest, 4-12, Regina Elisabeta Avenue, 030018 Bucharest, Romania
4 Research Institute of the University of Bucharest (ICUB), 90 Soseaua Panduri, 050663 Bucharest, Romania
5 Faculty of Geography, University of Bucharest, 1, Nicolae Bălcescu Avenue, 010041 Bucharest, Romania
* Correspondence: radu.pintilii@geo.unibuc.ro; Tel.: +40-788-990060
† Contribution equal to that of the first author.

Abstract: The Mediterranean Region currently faces major environmental issues that require constant analysis and monitoring. This study presents a thorough approach based on the application of Landsat imagery from Global Forest Change during 2001–2019. Spatial distribution mapping was one of the objectives of the study. We approached the analysis of tree cover loss areas by analyzing the cumulative tree cover loss and Tree Cover Loss Rate. This indicator offers information about the trend of tree cover loss in each Mediterranean country. A total of 581 Mha of deforested area was mapped during the analyzed period. Analysis was further supplemented by some statistical operations (distributions shown via histograms, validation via Shapiro–Wilk normality test, and testing via one-sample *t*-test). Agricultural expansion, intense forest fires, illegal logging, overgrazing (especially in the northern part of Africa), and extensive livestock farming have influenced the Mediterranean forest ecosystem's stability. The continuation of these activities could cause extreme climatic events, severe degradation, and desertification.

Keywords: tree cover loss; environment; degradation; Mediterranean Region

Citation: Ciobotaru, A.-M.; Patel, N.; Pintilii, R.-D. Tree Cover Loss in the Mediterranean Region—An Increasingly Serious Environmental Issue. *Forests* **2021**, *12*, 1341. https://doi.org/10.3390/f12101341

Academic Editor: John Innes

Received: 10 August 2021
Accepted: 28 September 2021
Published: 30 September 2021

1. Introduction

The effects of tree cover loss and forest degradation are critical environmental problems [1]. The fragility of forest ecosystems is an aspect intensely present in research in recent years, whether tropical forests, mangrove forests, or temperate forests [2–5].

Forest ecosystems are a critical component of the world's biodiversity, characterized more by diversity and unicity than other ecosystems of the world [6,7]. Forests cover 31% of the global land area, 4.06 billion hectares are natural forests or plantations, and approximately 15% are compact forest areas [8].

FAO estimated that, in 2015, across all Mediterranean countries, there were 88 million hectares of forest [9]. These forests represent biodiversity hotspots with around 60% endemic plant species from 25,000 species, but they are also very fragile ecosystems depending on variations in environmental conditions [10–12].

Several drivers determine tree cover loss: (i) commodity-driven deforestation, (ii) urbanization and demographic changes, (iii) shifting agriculture due to small- or medium-scale agriculture, (iv) forestry activities that through forest harvesting affect the forest stability, and (v) wildfires that determine the temporary loss of forests [13]. Anthropic activities have significantly altered biodiversity, and conservation efforts are needed to protect the mountains and coastal areas as a network of protected areas [14–16]. The long period of forest exploitation, significant socio-economic changes, rapid urbanization, and severe climatic events are important threats to the environmental stability of Mediterranean forests; a conservation strategy and sustainable forest management are thus required [17–19].

Human activities in forests also have a significant impact on the biodiversity of the Mediterranean Region. As a result, World Heritage Natural Sites have experienced a decrease in their degree of uniqueness and are subject to threats driven by climate change and increasing population in littoral areas [20–22].

Agricultural activities have influenced Europe's landscape through new crops and pastures, forest exploitation for fuel and wood processing, changes in soil properties, increasing temperatures in urban areas, and increased fire risk [23–27].

Additionally, natural hazards (storms, forest fires, strong winds) are strongly related to changes in land use and land cover, which fragment the landscape and result in environmental damage [28–31]. Large fires (>1000 ha) often claim human victims and cause greater burned forest areas and property damage [24].

Forest ecosystems are crucial in the fight against climate change, and reforestation can contribute to reducing the concentrations of greenhouse gases in the atmosphere [32–35].

For these major causes of tree cover loss, quantifying and documenting the extent of tree cover loss is a priority activity for environmental stability. The objectives of this study were (1) to explore tree cover loss rates in the Mediterranean Region; (2) to illustrate the spatial distribution of deforested areas; and (3) to show the evolution of the Tree Cover Loss Rate (*TCLR*) for the period 2001–2019 to illustrate the situation for this fragile environment.

The Mediterranean environment's stability is influenced by the tree cover loss in the context of actual global climate changes (increasing about 0.85 °C globally and 1.3 °C in the last century) [9]. That is why it is important to constantly monitor the evolution of tree cover loss, to observe the general trend. The tree cover loss is responsible for the present situation of flash floods and fires [36–38]. With the results presented in this paper, we aim to develop a better understanding of the impact of tree cover loss on the environment in the Mediterranean Region.

2. Materials and Methods

2.1. Study Area

As the study area, we focused on Mediterranean countries; these are characterized by certain patterns of the climate, such as dry and hot summers and moist and cool autumns and winters. Sometimes, extreme climatic events can influence the forests [39,40]. The natural vegetation of the Mediterranean region is related to the Mediterranean climate, but it is influenced by the presence of mountainous areas [17]. The Mediterranean vegetation is adapted to its environmental conditions with a deficit of precipitation during the warm season. It comprises predominately xerophilous vegetation, shrublands, broadleaf forests (60%, with species *Castanea sativa*, *Quercus suber*, and *Quercus ilex*), and coniferous forests (*Pinus pinea*, *Cupressus sempervirens*, and *Castanea sativa*), varying in proportion from Italy (76%) to Portugal (49%) [9,41,42]. The cork oak (*Quercus suber*) savannas in southwestern Europe and northwestern Africa have great conservation value and are characterized by shrub formations to grasslands with high biodiversity [43,44]. These countries are across three continents, most of them covering the European (13 countries), African (4 countries), and Asian (4 countries) territories (Figure 1).

2.2. Data Acquisition and Processing

The analysis of tree cover loss in the Mediterranean Region began with the collection of satellite images with a spatial resolution of 30 m from the Global Forest Change (GFC) dataset, courtesy of the Department of Geographical Sciences, University of Maryland (UMD), GLAD Laboratories in partnership with Google, United States of America from the Global Forest Watch, for a period of 19 years (2001–2019) [45]. The most suitable data to monitor forest cover changes are satellite images that can provide, through post-analysis, information about land use and forest changes. These images were used to extract the tree cover loss areas (annually from 2001 to 2019). The first step consisted of downloading the raster data. They were downloaded as individual, 21 (10 × 10 degree) granules, for the entire Mediterranean Region considering the geographical coordinates in Table 1. Using the function "extract by mask", the resulting merged raster image was cut according to the Mediterranean Region border (vector).

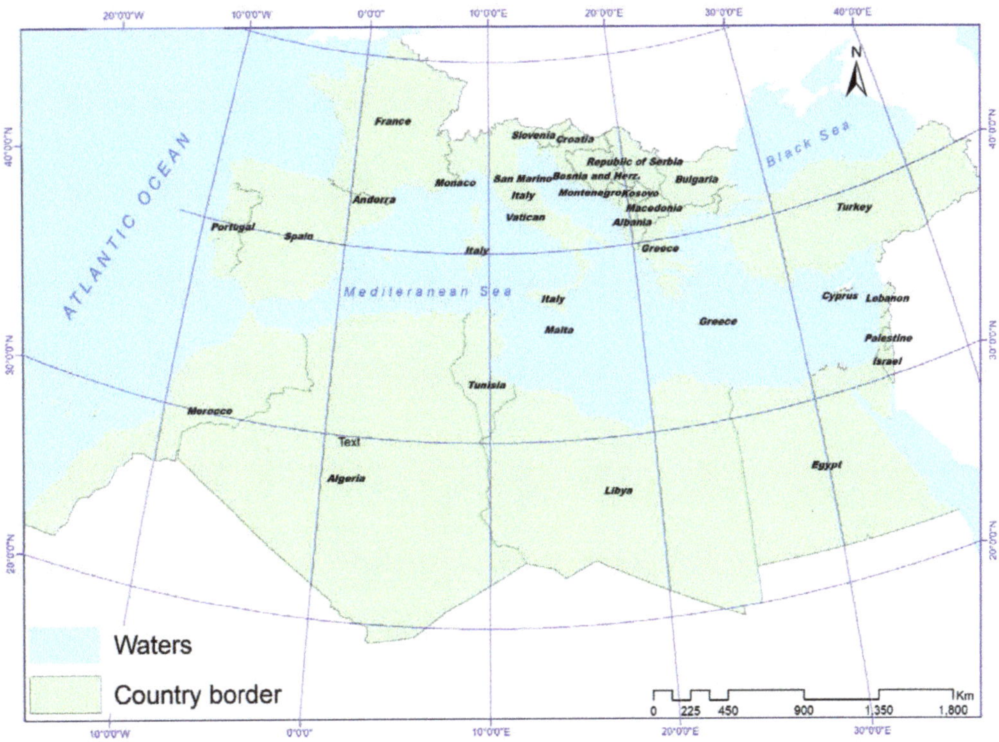

Figure 1. The geographical positions of countries in the Mediterranean Region.

Then, the data were merged to produce a single raster image. The conversion into points of the raster image was performed using the function "raster to points" to extract the annual tree cover loss for each country (Figure 2). Each pixel had a different color representing a different year (e.g., for the period 2001–2019, 19 different colors were used for each year).

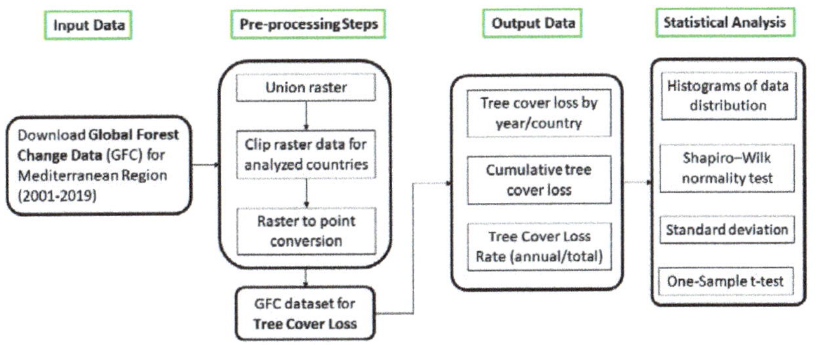

Figure 2. Flowchart including input data, pre-processing steps, statistical analysis, and output data to determine the tree cover loss in the Mediterranean Region.

Table 1. Landsat-7 ETM+ geographical coordinates covering the study area (Data obtained from the Global Forest Change platform [45]).

No.	Satellite Images and Spatial Resolution	Longitude	Latitude	Data Source
1	LANDSAT 7 ETM+, 30 m	10°–20° N	0°–10° E	GFC
2		10°–20° N	20°–30° E	
3		20°–30° N	0°–10° E	
4		20°–30° N	10°–20° E	
5		20°–30° N	0°–10° W	
6		20°–30° N	20°–30° E	
7		20°–30° N	10°–20° W	
8		20°–30° N	30°–40° E	
9		30°–40° N	0°–10° E	
10		30°–40° N	10°–20° E	
11		30°–40° N	0°–10° W	
12		30°–40° N	20°–30° E	
13		30°–40° N	30°–40° E	
14		30°–40° N	40°–50° E	
15		40°–50° N	0°–10° E	
16		40°–50° N	10°–20° E	
17		40°–50° N	0°–10° W	
18		40°–50° N	20°–30° E	
19		40°–50° N	30°–40° E	
20		40°–50° N	40°–50° E	
21		50°–60° N	0°–10° E	

The data are expressed in hectares, and for our analyses, the tree cover loss by year from 2001 to 2019 and the canopy cover levels from ≥10 to ≥75 (≥10, ≥15, ≥20, ≥25, ≥30, ≥50, ≥75) seemed to be remarkably interesting. The recommendation is to select the desired percent canopy cover level and use it consistently throughout any analysis, but the Global Forest Watch website uses a ≥30% canopy cover threshold as a default for all statistics, the same canopy level used by us in the present study. The processing of the data was performed by using descriptive statistics and was materialized in graphic and cartographic materials and boxplots, made using the ArcGIS (ESRI, Redlands, CA, USA), Microsoft Excel, and R Software platforms.

2.3. Methodology

To observe the distribution of data, histograms by country were made with all the tree cover loss. Then, some specific operations and tests were applied: the standard deviation for each country and one-sample *t*-test, also by country. To check the normality of the tree cover loss distribution, the Shapiro–Wilk statistical test was applied, due to the small amount of data (2001–2019). The data displayed in Table 3 indicate, for most of the countries analyzed, a normal distribution (*W* values are over 0.75). This hypothesis is also sustained by the *p*-values, which are above 0.05 in most cases. Among all the 30 countries, there were 9 that did not report any information about tree cover and tree cover loss areas, by year or by canopy cover level. That is why they were excluded from our analyses (Figure 2).

The *TCLR* was calculated to observe the trend in the tree cover loss evolution by year and for the entire period. The annual evolution was marked in blue, and the upward trend, for the entire period, was marked in red (meaning a negative aspect—a high tree cover loss rate) or green (meaning a positive aspect—a low tree cover loss rate).

It was calculated by using the following equation:

$$TCLR\ (\text{annual}) = \left(\frac{TCL\,n - TCL\,n-1}{TCL\,n-1}\right) \times 100.$$

where $TCL \rightarrow$ tree cover loss; $n \rightarrow$ year.

$$TCLR\ (\text{entire period}) = (\frac{TCL\ n - TCL\ n - 19}{TCL\ n - 19}) \times 100$$

where $TCL \rightarrow$ tree cover loss; $n \rightarrow$ year.

3. Results

Tree Cover Loss in the Mediterranean Region

In Figure 3, showing the tree cover loss by country in the 2001–2019 period, there are differences from one country to another; these are due, on the one hand, to their total areas, tree cover areas, and timber needs and, on the other hand, to other natural or anthropogenic factors such as wildfires, diseases, the expansion of urban areas, and so on. The biggest tree cover loss areas were registered in Spain (1,231,065 ha), France (1,142,699 ha), Portugal (1,027,175 ha), and Turkey (499,959 ha). On the opposite side, the countries with low tree cover loss included Palestine (18 ha), Malta (13 ha), and Jordan (7 ha). These countries have forest areas that are large but fragile to main threats: climate change, extreme temperature, human activities, urbanization, and the need for agricultural land. The smallest values were recorded in countries with small administrative territories.

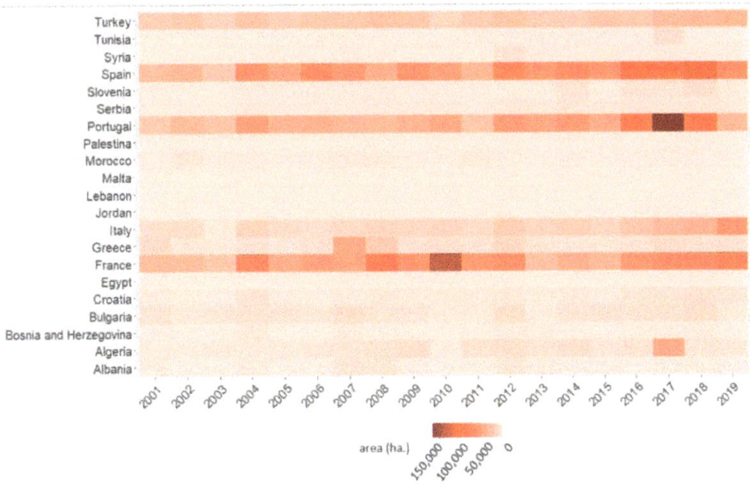

Figure 3. Tree cover loss by country (2001–2019).

To see the importance of the tree cover loss, it is particularly useful to analyze the distribution of the tree cover loss. By ordering all the countries alphabetically, we obtain the most relevant and non-discriminatory view of all the countries in the whole territory considering a canopy level of $\geq$30, the same level taken in all analyses [27] (Table 2).

To identify the shape of the tree cover loss data and see whether tree cover loss process changes occurred from one year to another, some descriptive statistic operations were applied. The first step was to determine the simple distribution of them by constructing a histogram for each country of the tree cover loss in the period 2001–2019. The histograms in Figure 4 show a very different distribution of the data for each country, and they can be grouped into only two main categories. Most of the countries (15 countries) have a skewed-right distribution; among these, Albania, Algeria, Egypt, Greece, Portugal, and Tunisia can be mentioned as representatives. In the second category (symmetric/normal distribution), there are only six countries (Bulgaria, Lebanon, Morocco, Serbia, Spain, and Turkey).

The normality of the tree cover loss data distribution for each country was checked using the Shapiro–Wilk test in the present study. The data in Table 3 indicate a normal distribution for most of the countries (W values over 0.75), but not in some cases (Algeria, 0.48; Egypt, 0.56; Greece, 0.59; Jordan, 0.65; and Portugal, 0.71). This hypothesis is also sustained by the

p-values, which are above 0.05 in most cases (Lebanon, 0.1503; Morocco, 0.1335; and Serbia, 0.9262). The standard deviation exhibits significant variation, ranging from 0.60 (Jordan) and 0.91 (Palestine) to 35,264.54 (Portugal), 26,971.25 (France), and 13,047.56 (Italy).

One-sample t-tests were applied for the sample of 21 countries to test whether the mean tree cover loss in each country was different from a specific value; statistical significance was observed for all countries as indicated by the t, df and p values registered. In all cases, the degree of freedom was 18, which means that the power of the test is high. The obtained values of t varied from 2.689 (Jordan) to 20.211 (Turkey) and the probability level p in all cases is under 0.05 (from 0.00001 for Bosnia and Herzegovina, Bulgaria, Croatia, Italy, Lebanon, Morocco, Portugal, Spain, and Turkey to 0.01500 for Jordan) (Figure A1). The evolution of the *TCLR* for the 21 countries shows three situations: one where the tree cover loss increases (represented in red color—negative aspect), one where the surfaces decrease (green color—positive aspect), and one country with no evolution (0%, Jordan) (Figure 5). For most countries, the tree cover loss area as a percentage for the period 2001 to 2019 increased; the highest values (more than 100%) were registered in Italy (504.85%), Slovenia (461.95%), Syria (340.82%), Tunisia (148.05%), France (136.2%), and Spain (116.46%). Other countries encountered negative evolution rates for the same period; among them were Palestine and Malta (−100% for each), Egypt (−85.81%), and Albania (−71.20%).

Table 2. Tree cover loss areas in countries in the Mediterranean Region [Data obtained from the Global Forest Change platform and online documentation [45].

Country	Country Area	Tree Cover Loss 2001	Tree Cover Loss 2005	Tree Cover Loss 2010	Tree Cover Loss 2015	Tree Cover Loss 2019	Cumulative Tree Cover Loss (2001–2019)
				(ha)			
Albania	2,873,537	3729	695	656	284	1074	39,047
Algeria	124,831,323	3469	4470	2606	5707	6291	158,275
Andorra	46,800	*NA* *	*NA* *	*NA* *	*NA* *	*NA* *	*NA* *
Bosnia and Herzegovina	5,106,883	1496	574	911	581	1232	30,073
Bulgaria	11,158,731	8202	6567	3569	7293	4125	123,569
Croatia	5,707,840	3378	2747	3499	2669	5978	76,297
Cyprus	925,100	*NA* *	*NA* *	*NA* *	*NA* *	*NA* *	*NA* *
Egypt	98,376,439	148	53	98	32	21	1732
France	54,951,498	34,421	46,391	143,265	37,293	81,304	1,142,699
Greece	13,257,505	12,617	4181	7785	3840	7604	182,913
Israel	2,077,000	*NA* *	*NA* *	*NA* *	*NA* *	*NA* *	*NA* *
Italy	30,075,443	9871	10,896	15,806	14,124	59,705	358,569
Jordan	8,911,879	1	0	0	0	1	7
Kosovo	1,088,700	*NA* *	*NA* *	*NA* *	*NA* *	*NA* *	*NA* *
Lebanon	1,023,804	299	300	115	174	390	4147
Libya	176,000,000	*NA* *	*NA* *	*NA* *	*NA* *	*NA* *	*NA**
Macedonia	6,700,000	*NA* *	*NA* *	*NA* *	*NA* *	*NA* *	*NA* *
Malta	32,335	1	0	1	0	0	13
Monaco	202	*NA* *	*NA* *	*NA* *	*NA* *	*NA* *	*NA* *
Montenegro	1,381,200	*NA* *	*NA* *	*NA* *	*NA* *	*NA* *	*NA* *
Morocco	41,348,767	1868	1526	988	952	1250	37,419
Palestine	621,997	2	0	1	0	0	18
Portugal	8,955,506	24,405	38,934	48,401	46,097	36,886	1,027,175
San Marino	6120	*NA* *	*NA* *	*NA* *	*NA* *	*NA* *	*NA* *
Serbia	7,823,140	2380	1551	2462	2815	3962	52,807
Slovenia	1,998,091	615	1951	1434	1025	3456	45,059
Spain	50,604,279	32,811	48,955	56,093	65,414	71,022	1,231,065
Syria	18,691,800	267	792	241	2393	1177	20,681
Tunisia	15,486,458	231	633	398	1880	573	26,937
Turkey	78,070,341	22,910	20,411	17,008	27,735	39,118	499,959
Vatican	44	*NA* *	*NA* *	*NA* *	*NA* *	*NA* *	*NA* *

* *NA*—countries with no tree cover loss data.

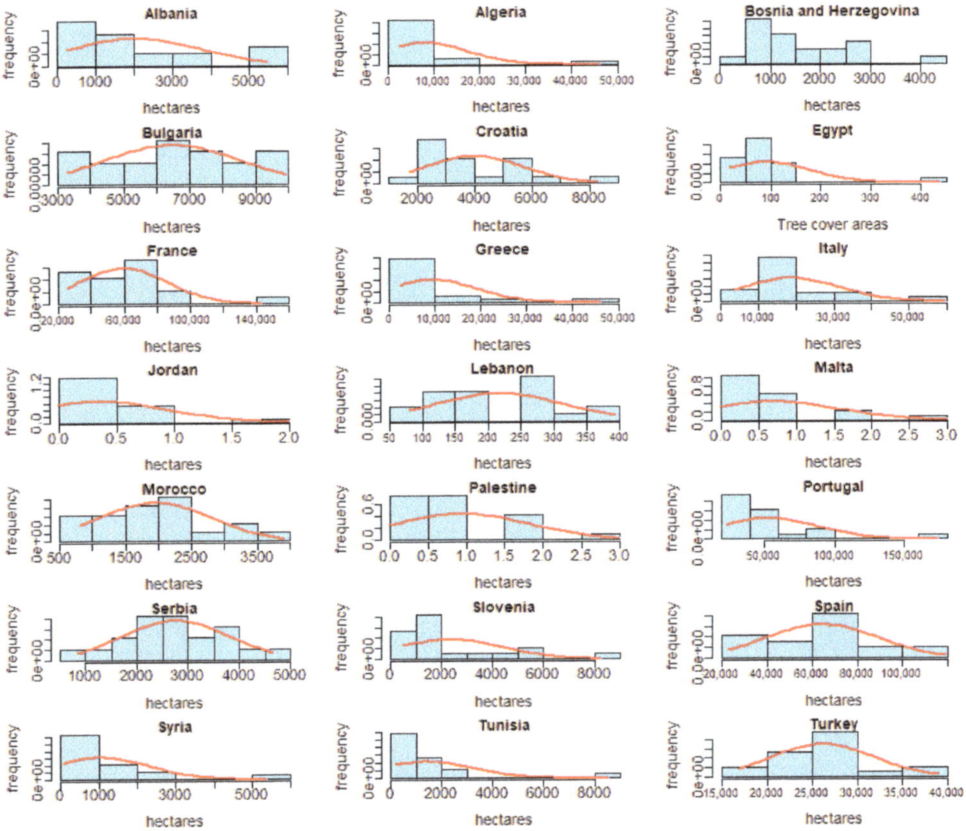

Figure 4. Histograms of tree cover loss by country (2001–2019).

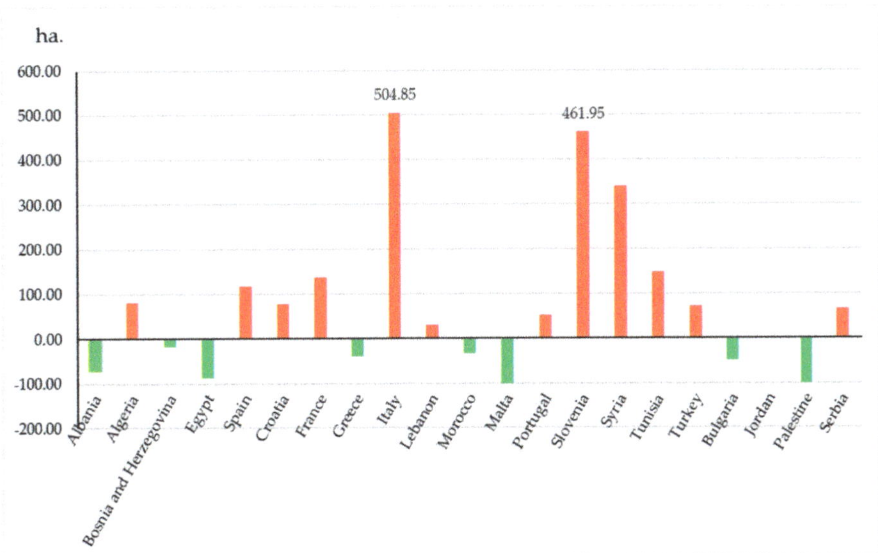

Figure 5. Evolution of the Tree Cover Loss Rate (2019 vs. 2001).

Table 3. Shapiro–Wilk normality test for data validation.

No.	Country	W	*p*-Value	Standard Deviation
1	Albania	0.81	0.001462	1724.88
2	Algeria	0.48	3.579×10^{-7}	9438.98
3	Bosnia and Herzegovina	0.92	0.1103	993.18
4	Bulgaria	0.95	0.4318	2105.93
5	Croatia	0.91	0.08276	1753.81
6	Egypt	0.56	2.787×10^{-6}	90.24
7	France	0.87	0.01634	26,971.25
8	Greece	0.59	3.138×10^{-6}	9791.96
9	Italy	0.79	0.0007701	13,047.56
10	Jordan	0.65	1.374×10^{-5}	0.60
11	Lebanon	0.93	0.1503	100.44
12	Malta	0.76	0.0003579	0.89
13	Morocco	0.92	0.1335	866.03
14	Palestine	0.85	0.006288	0.91
15	Portugal	0.71	7.214×10^{-5}	35,256.54
16	Serbia	0.98	0.9262	1053.05
17	Slovenia	0.79	0.0007622	2124.55
18	Spain	0.96	0.6258	25,074.63
19	Syria	0.74	0.0001789	1264.86
20	Tunisia	0.57	2.128×10^{-6}	1841.72
21	Turkey	0.96	0.602	5675.02

Analyzing the situation by country, Albania—although it presented a generally decreasing tree cover loss percentage—in many years showed increasing values over 400% (2004 vs. 2003 and 2007 vs. 2006). High values were also found in 2012 vs. 2011 (over 300%) and 2016 vs. 2015 (over 200%). Algeria also exhibited two periods with high values (2011 vs. 2010, over 200%, and 2017 vs. 2016, up to 500%). For Bosnia and Herzegovina, in 2004 vs. 2003, the tree cover loss rate was up to 1300%. For Bulgaria, the values were low (only 200% in 2012 vs. 2011). In Croatia and Egypt, the highest values oscillated around 300% (2004 vs. 2003 for Croatia and 2012 vs. 2011 for Egypt). For Greece and Italy, the values were up to 600%, and the latter country also showed a considerable increase for the entire period. Jordan and Palestine registered negative evolution for the entire period, but for Palestine, in 2008 vs. 2007, the tree cover loss rate was 200%. Slovenia and Syria had the highest values, with 800% for Slovenia in 2014 vs. 2013 and 3400% for Syria in 2012 vs. 2011. Very different was the situation of Turkey: for this country, the value was up to 75% each year—a moderate rate (Figure A2).

The tree cover loss in the different countries during the period of analysis (2001–2019) is depicted in Figure 6. It can be observed that in the upper part of the Mediterranean region (characterized by large areas of tree cover and situated closer to the temperate climate area—temperate forests), in countries such as Portugal, France, and Spain, forest loss is predominant. On the other side, in countries close to the desert areas and where the tree cover is not so dense, their loss is moderate to low. This image exhibits a significant amount of correlation with the two previous figures, which present the forest loss in comparison to the tree cover extent in 2000 and 2010.

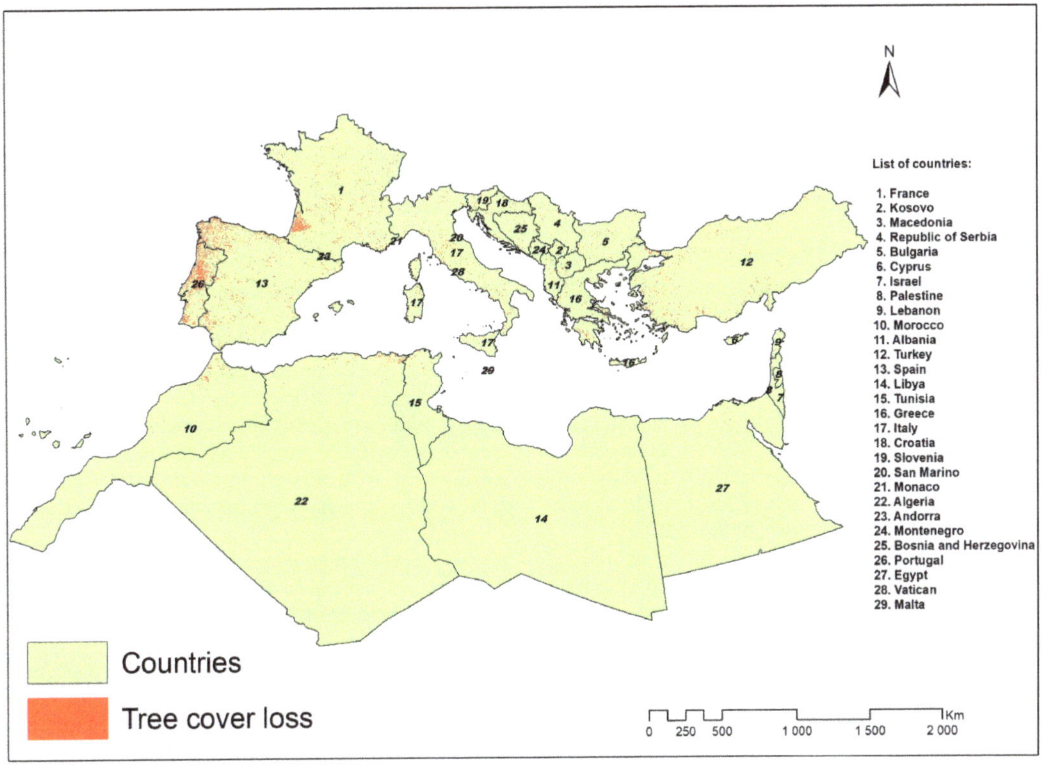

Figure 6. Spatial distribution of the tree cover loss areas in the Mediterranean Region (2001–2019).

4. Discussion

This article describes a methodology for studying the tree cover in the Mediterranean Region, providing a consistent spatial distribution of tree cover loss changes at the national scale for 2001–2019. The methodology used in this study based on the analysis of tree cover loss data is shown to accurately monitor the tree cover loss.

The vegetation is in strong correlation with the climate, characterized by a deficit of precipitation in the warm season and moist and cool winters [17].

In the last century, in the Mediterranean Region, the temperature has increased 1.3 °C [9]. The stability of forests is affected by both climatic factors and socioeconomic pressure. Therefore, the forest ecosystems are influenced by intense wildfires [20,24], loss of biodiversity [2,12,41], land degradation and fragmentation [9,15,46], and traditional agricultural techniques [16,23,46]. However, the drivers that modify the distribution of forest areas vary from one country to another.

Analysis of the forest loss data shows that the forests in the Mediterranean Region represent an ecosystem that is vulnerable to external threats and has experienced intense tree cover loss (a total of 581 Mha of tree cover loss area) [9,45].

In European countries, the major activity that affects forests is harvesting done to provide wood for industry [19,23,46]. On the other hand, in Africa, the dominant driver of tree cover loss is shifting agriculture, which determines temporary loss of forests, especially in Morocco, Algeria, and Tunisia [46,47]. Of the dominant threats to tree cover loss, including agricultural expansion, the underlying causes of tree cover loss are the global markets for cork oak, timber, and pulp [11,27,39,48]. In addition to this, another direct threat to tree cover loss includes intense forest fires [9,13,24,49], illegal logging [13,17], and some traditional tools for creating grasslands for extensive livestock farming and overgrazing in states such as Algeria, Lebanon, Morocco, Tunisia, and Turkey [9,16].

The utility of this study for the Mediterranean Region is to encourage continuous monitoring of tree cover loss evolution, as has already been conducted in countries in South America [50,51], Europe [52], and Africa [53,54].

We evaluated the tree cover loss in the Mediterranean Region and the significant drivers that affect the forest area. Second, analysis of tree cover loss by canopy cover signifies that the highest loss areas are specific to the Mediterranean countries.

Third, the highest tree cover loss areas have specific environmental conditions (high temperatures, dry periods), severe climatic events, frequent forest fires, and changes in land use among the principal causes of tree cover loss. Portugal and Spain recorded the highest rates of tree cover loss (1,231,065 ha in Spain and 1,027,175 in Portugal) [16,46].

Some sustainable measures of specific vegetation for the Mediterranean cork oak savanna are suitable to reduce the effects of tree cover loss and degradation [43–45].

Fourth, tree cover loss is strongly correlated with environmental conditions, the intensity of human activities such as intense urbanization, and the need for agricultural land.

The histograms show a normal distribution of data for most of the countries. This allowed us to conduct the analysis. Shapiro–Wilk normality tests and *t*-tests showed the suitability of the data for the analysis, with their significance considered to subscribe to the normal interval.

The scatterplot in Figure 7 shows the relation between country areas and tree cover loss; the dispersion of the points allows us to conclude that the disturbances in the tree cover loss are from natural causes. The small value of R^2, at about only 0.0399, illustrates that there is no correlation to the analyzed data.

We infer that the tree cover loss is influenced by a complex category of drivers (intense fires, increased temperatures, agricultural activities, and illegal logging) that determine the fragmentation of forests and the fragility of the Mediterranean environment.

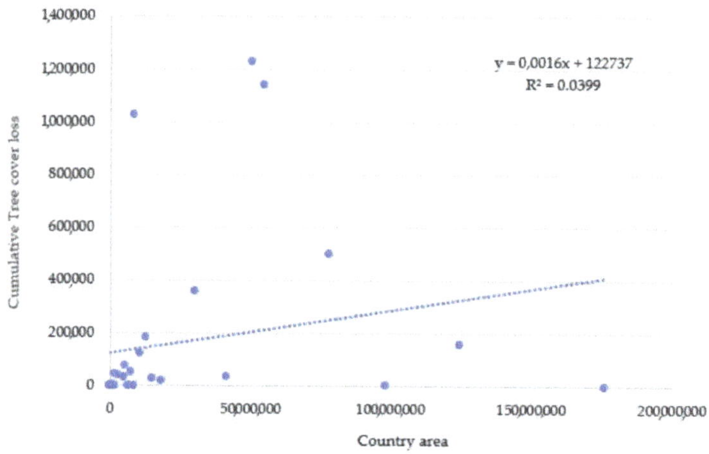

Figure 7. Relation between country area and tree cover loss area.

5. Conclusions

Tree cover loss in the Mediterranean Region is induced by both direct and indirect causes: local policies that need improvements for sustainable forest management; demographic changes and urbanization, which can cause degradation; and desertification, especially in the northern part of Africa.

Analysis of tree cover loss changes using Landsat imagery in the Mediterranean Region represents the focus of much research over the years. The current study presents a tree cover loss mapping and evaluation of the *TCLR* for the period 2001–2019 in the context of significant threats in southern Europe, North Africa, and southwest Asia, such as intense forest fires, overgrazing, development of urban areas, illegal logging, intense agriculture, and changes in land use and land cover. Improved policies regarding management and maintenance of land use are important for the constant monitoring of forest ecosystems in the Mediterranean Region.

Author Contributions: Conceptualization, A.-M.C., N.P., R.-D.P.; methodology, R.-D.P.; software, R.-D.P.; validation, A.-M.C. and N.P.; formal analysis, A.-M.C., N.P., R.-D.P.; investigation, A.-M.C.; resources, A.-M.C. and R.-D.P.; data curation, N.P.; writing—original draft preparation, A.-M.C., N.P., R.-D.P.; writing—review and editing, A.-M.C. and N.P.; visualization, R.-D.P.; supervision, A.-M.C.; project administration, A.-M.C.; funding acquisition, A.-M.C. and R.-D.P. All authors have read and agreed to the published version of the manuscript. All authors made equal contributions to the preparation of this scientific paper.

Funding: This research received no external funding.

Data Availability Statement: We choose to exclude this statement because the study did not report any data.

Conflicts of Interest: The authors declare no conflict of interest.

Appendix A

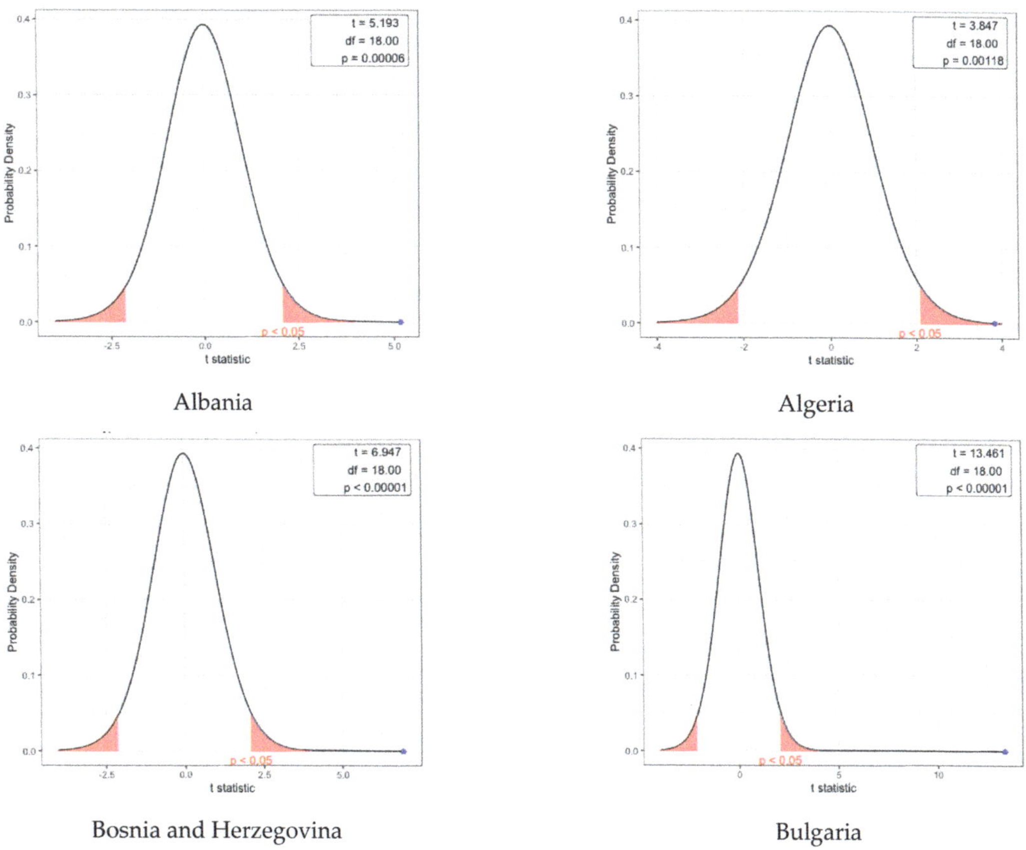

Figure A1. *Cont.*

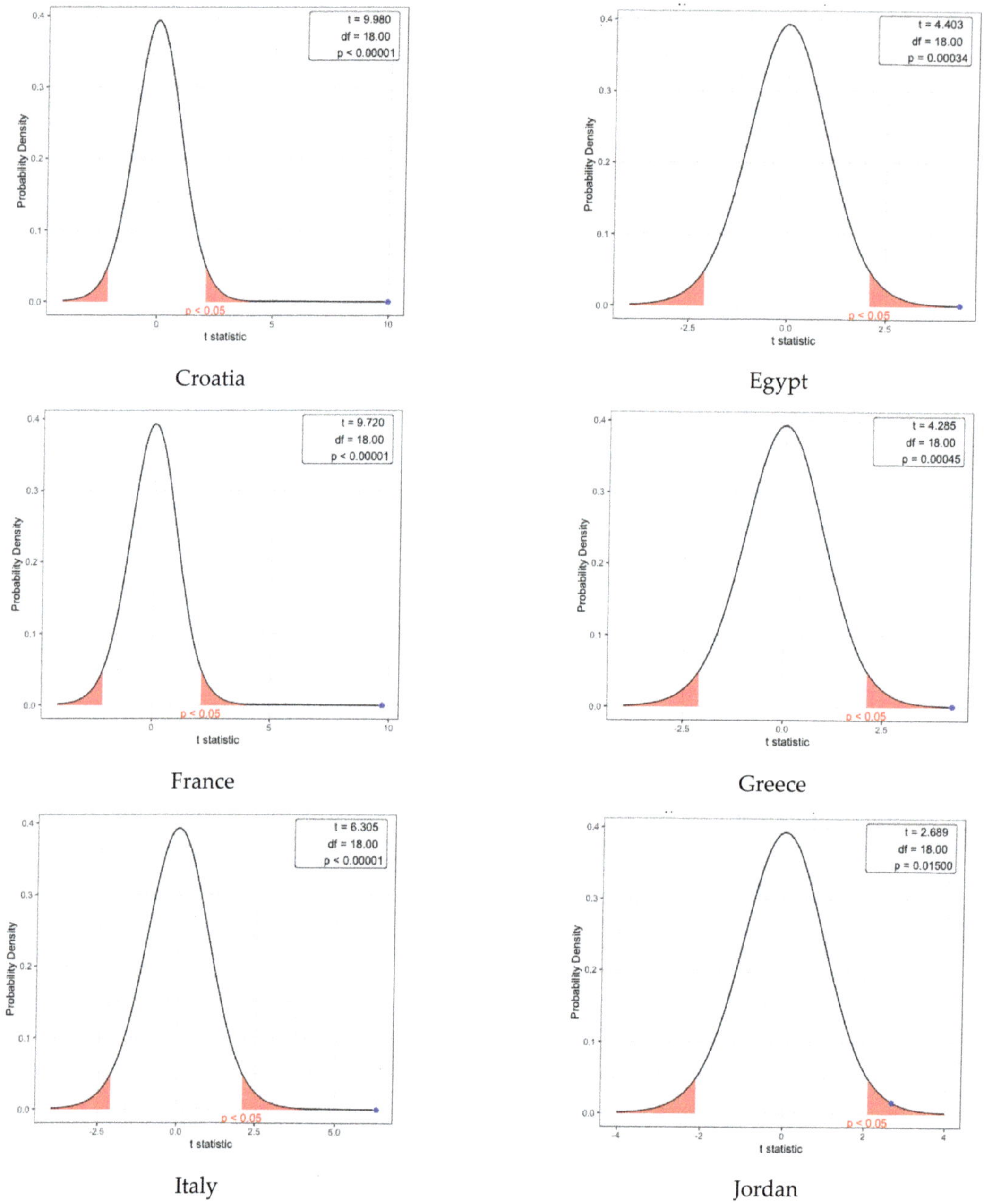

Figure A1. *Cont.*

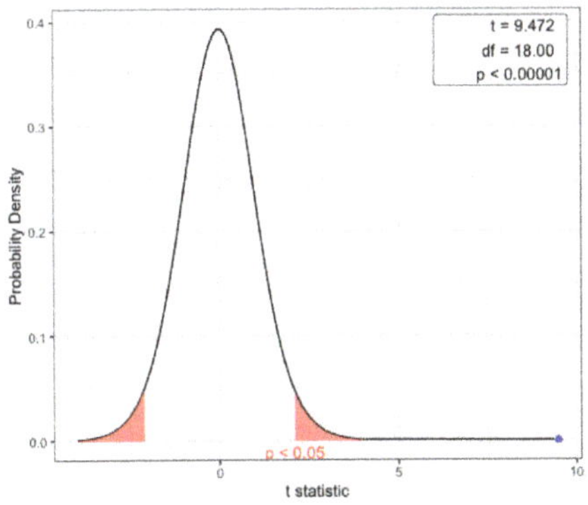

Lebanon

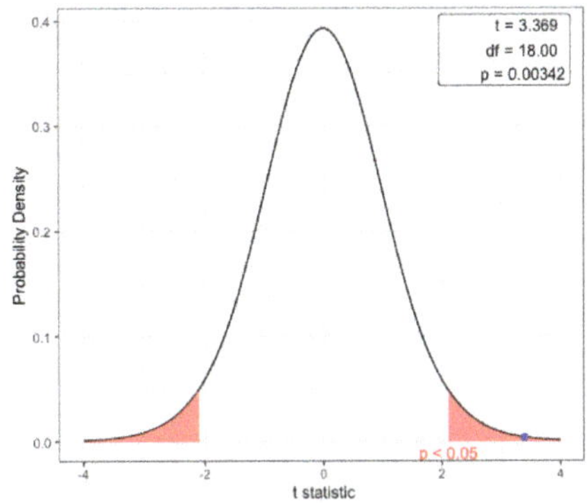

Malta

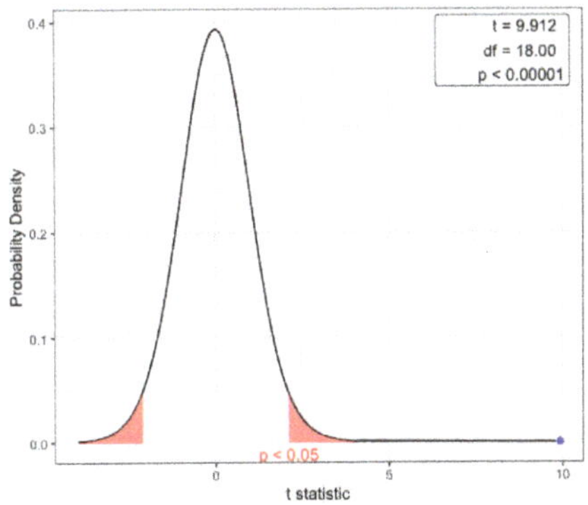

Morocco

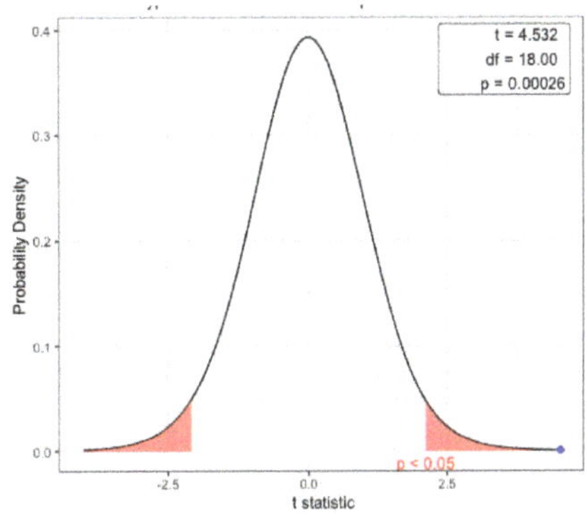

Palestine

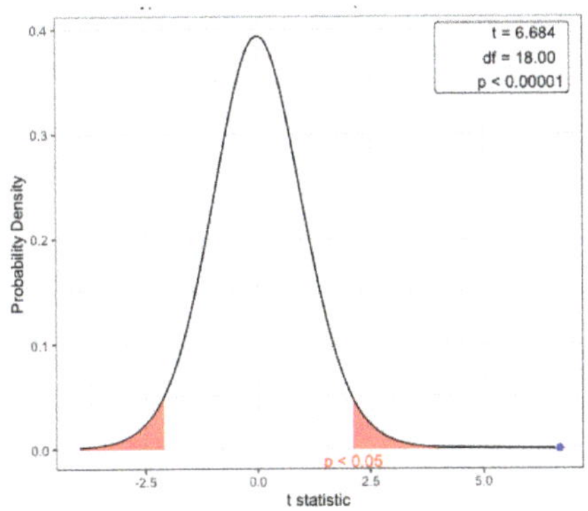

Portugal

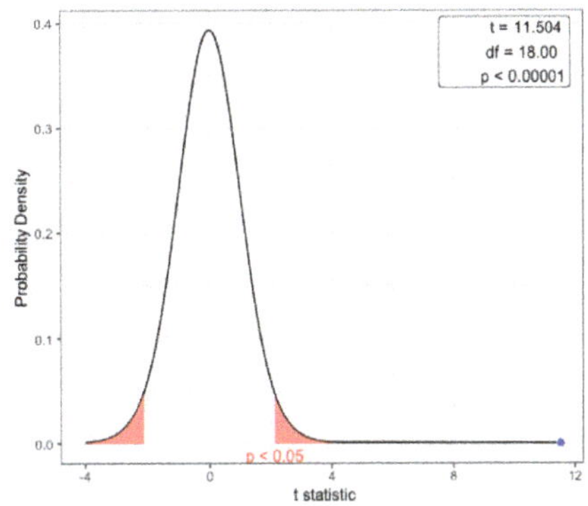

Serbia

Figure A1. *Cont.*

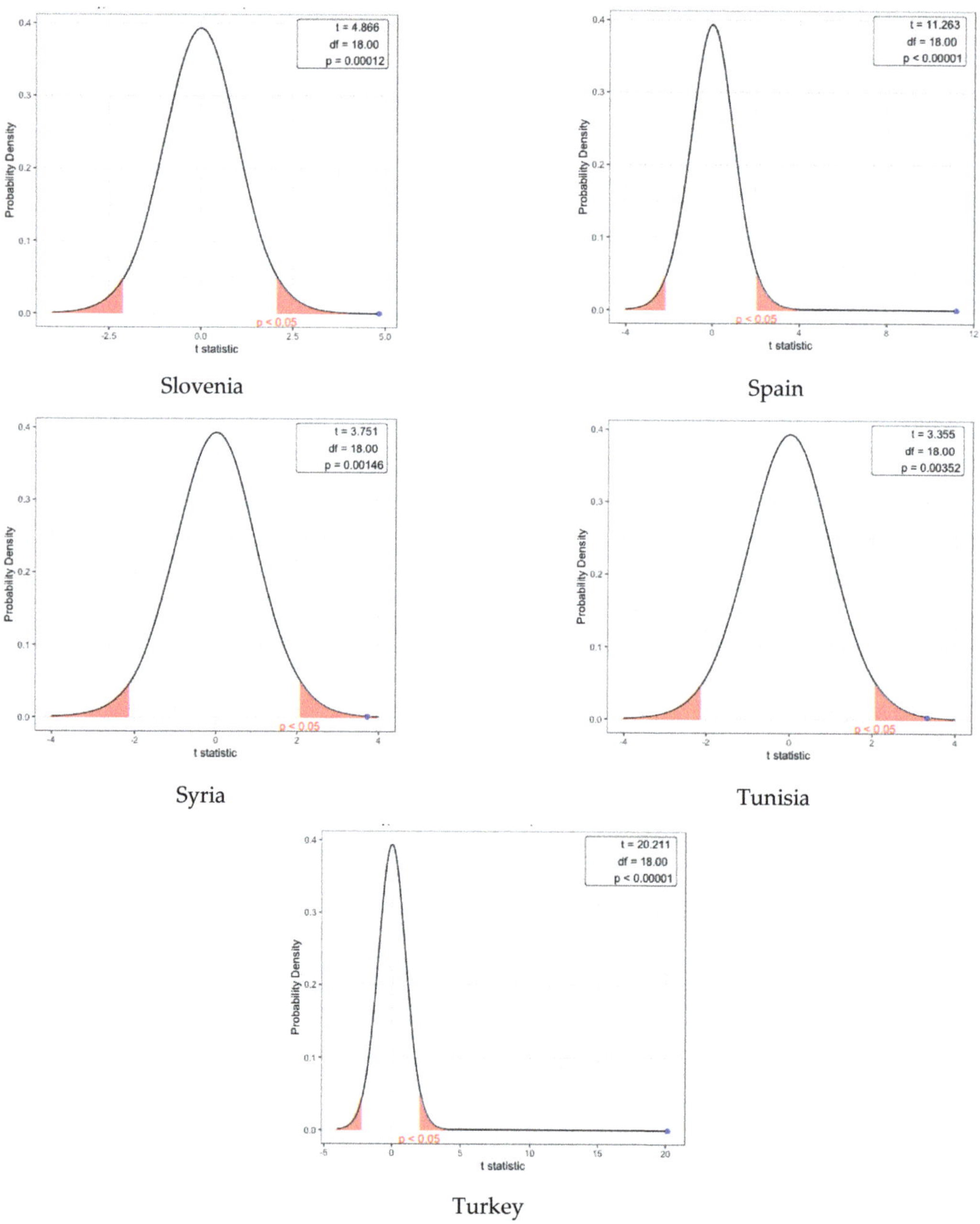

Figure A1. One-sample *t*-test results on tree cover loss data by country.

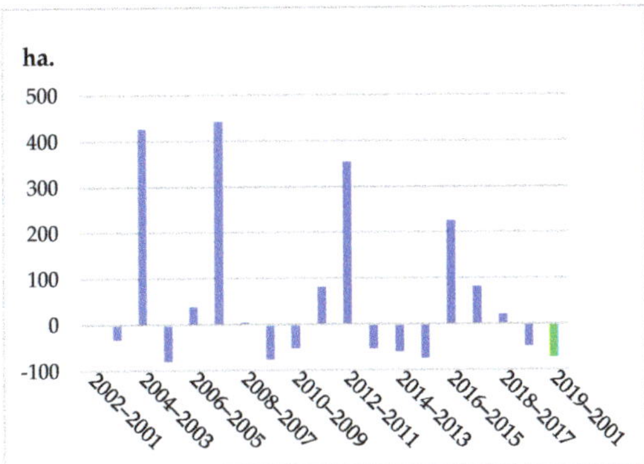

Albania

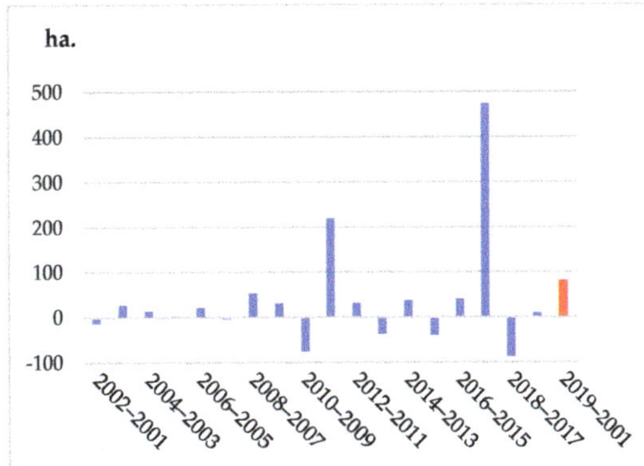

Algeria

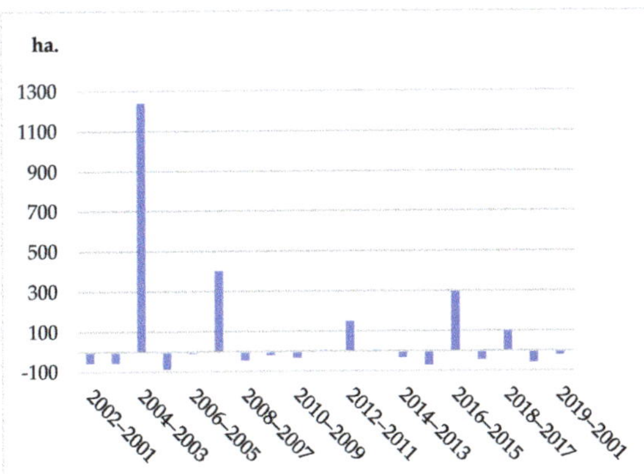

Bosnia and Herzegovina

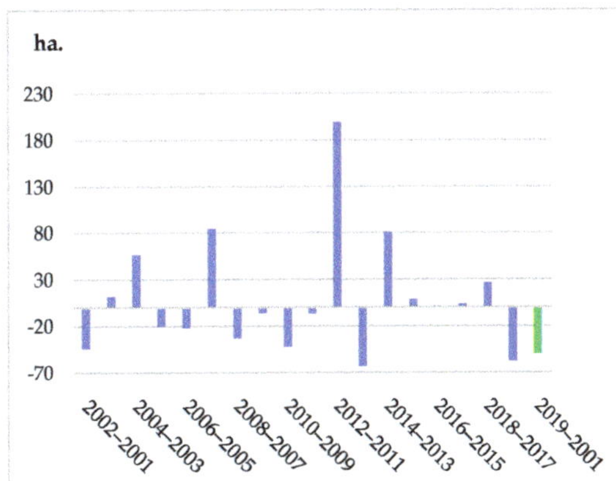

Bulgaria

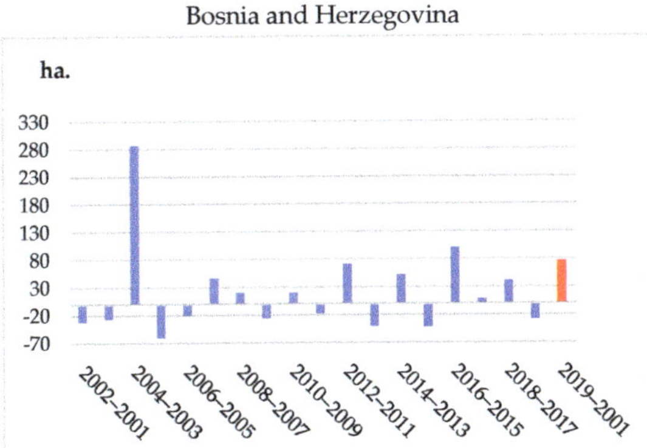

Croatia

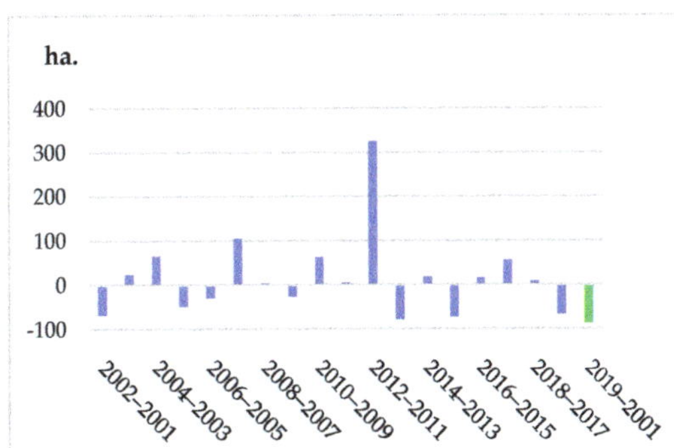

Egypt

Figure A2. *Cont.*

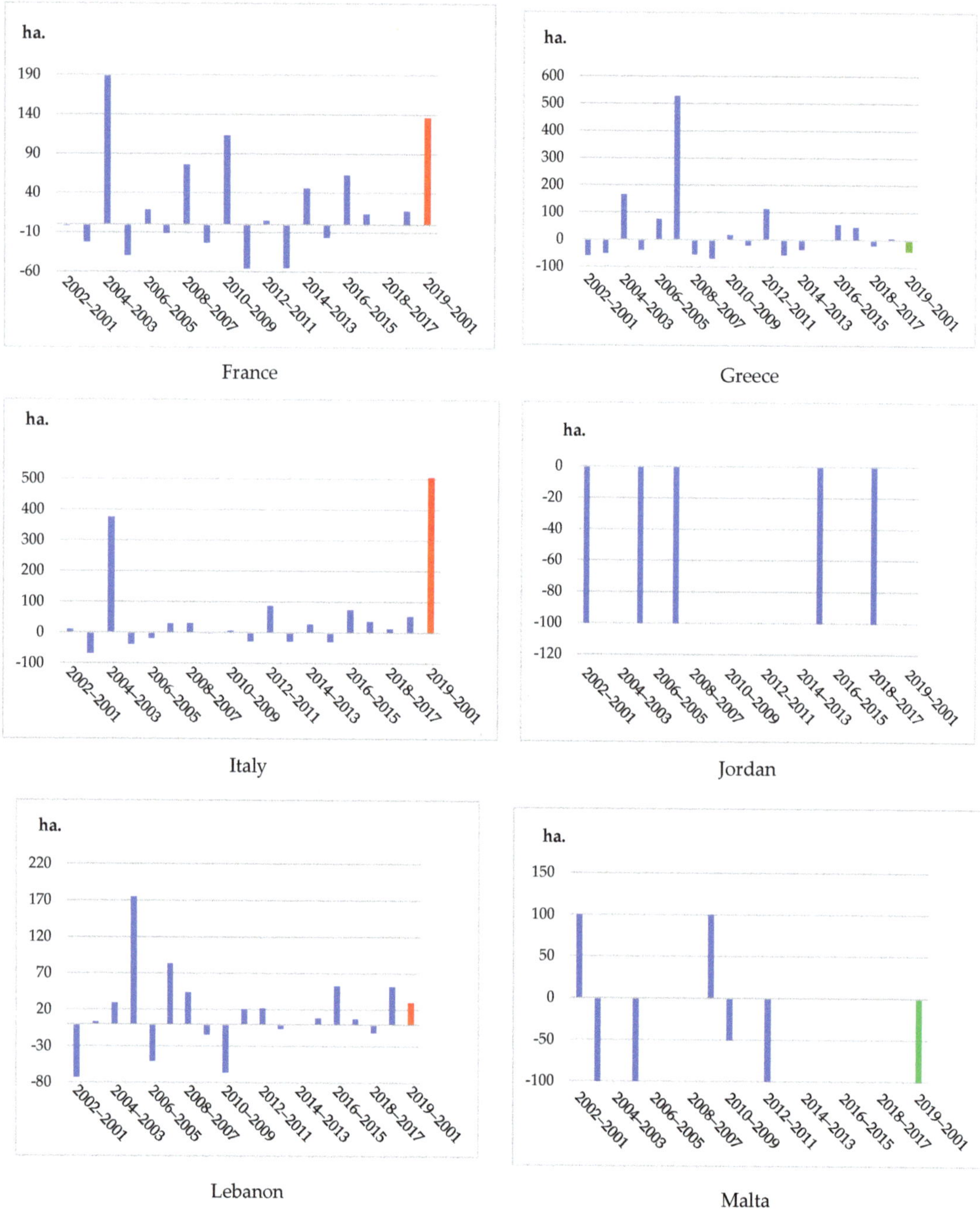

Figure A2. *Cont.*

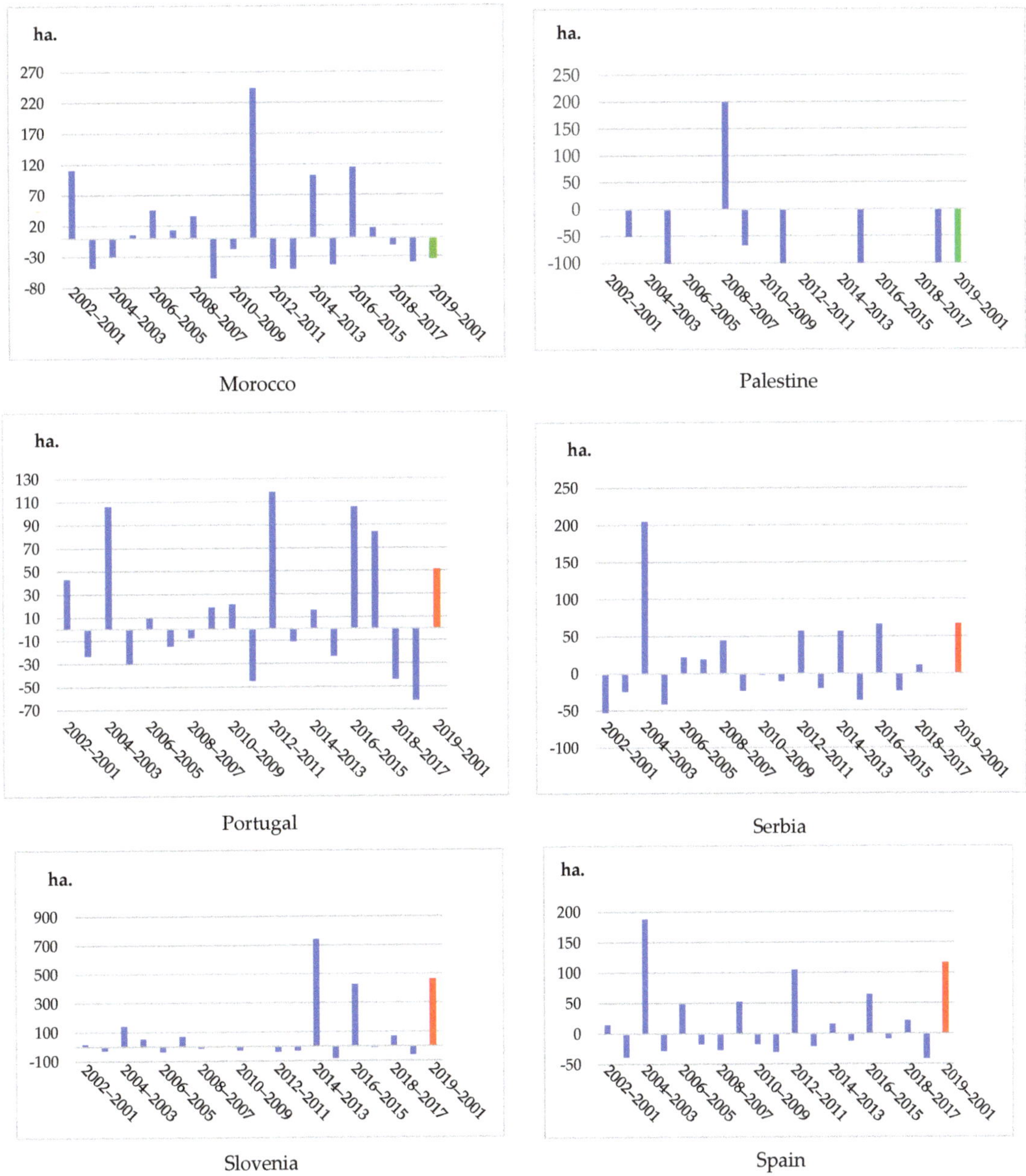

Figure A2. *Cont.*

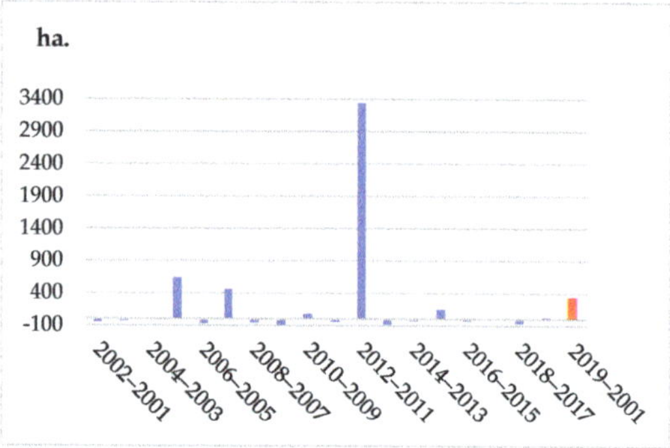

Syria

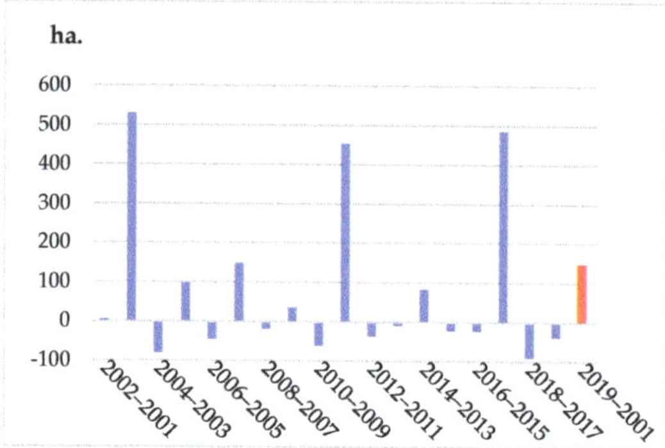

Tunisia

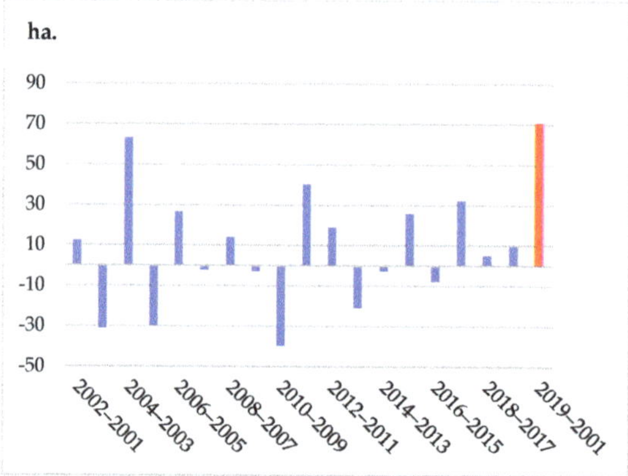

Turkey

Figure A2. The annual evolution of the TCLR for Mediterranean countries (2001–2019): the blue bars present the annual evolution of TCLR; the green and red bars present the evolution trend of TCLR for the entire period.

References

1. Vásquez-Grandón, A.; Donoso, P.J.; Gerding, V. Forest Degradation: When Is a Forest Degraded? *Forests* **2018**, *9*, 726. [CrossRef]
2. Lindenmayer, D.B. Landscape change and the science of biodiversity conservation in tropical forests: A view from the temperate world. *Biol. Conserv.* **2010**, *143*, 2405–2411. [CrossRef]
3. Otero, X.L.; Mendez, A.; Nobrega, G.N.; Ferreira, T.O.; Santiso-Taboada, M.J.; Melendez, W.; Macias, F. High fragility of the soil organic C pools in mangrove forests. *Mar. Pollut. Bull.* **2017**, *119*, 460–464. [CrossRef] [PubMed]
4. Margono, B.A.; Turubanova, S.; Zhuravleva, I.; Potapov, P.; Tyukavina, A.; Baccini, A.; Goetz, C.; Hansen, M.C. Mapping and monitoring deforestation and forest degradation in Sumatra (Indonesia) using Landsat time series data sets from 1990 to 2010. *Environ. Res. Lett.* **2012**, *7*, 034010. [CrossRef]
5. Simion, A.G.; Nedelcu, I.D.; Grecu, A.; Popa, M.C.; Peptenatu, D. Analysis of the Dynamics of Forest Areas in the Territorial Administrative Units of The Apuseni Mountains, Romania. In Proceedings of the Conference on Public Recreation and Landscape Protection-with Sense Hand in Hand? Krtiny, Czech Republic, 13–15 May 2019; Fialova, J., Ed.; Mendel University: Brno, Czech Republic, 2019; pp. 29–34.
6. FAO and UNEP. The State of the World's Forests 2020. Forests, Biodiversity and People. Rome. 2020. Available online: http://www.fao.org/documents/card/en/c/ca8642en (accessed on 26 June 2020).
7. Kumar, P.; Krishna, A.P.; Rasmussen, T.M.; Pal, M.K. An Approach for Fraction of Vegetation Cover Estimation in Forest Above-Ground Biomass Assessment Using Sentinel-2 Images. In *Computer Vision and Image Processing. CVIP 2020. Communications in Computer and Information Science*; Singh, S.K., Roy, P., Raman, B., Nagabhushan, P., Eds.; Springer: Singapore, 2021; p. 1376. [CrossRef]
8. GFW. Global Forest Watch. World Resources Institute. 2014. Available online: www.globalforestwatch.org (accessed on 26 June 2020).

9. Food and Agriculture Organization of the United Nations-State of Mediterranean Forests 2018. Available online: http://www.fao.org/3/CA2081EN/ca2081en.PDF?fbclid=IwAR2XiKuQHU5SWkDCxb6-ALrhsmqFNb0CsQhyhcqi09TWckQKukRdBK5TOgw (accessed on 22 August 2021).
10. Fabbio, G.; Merlo, M.; Tosi, V. Silvicultural management in maintaining biodiversity and resistance of forests in Europe—The Mediterranean region. *J. Environ. Manag.* **2003**, *67*, 67–76. [CrossRef]
11. Katsanevakis, S.; Coll, M.; Piroddi, C.; Steenbeek, J.; Lasram, F.B.; Zenetos, A.; Cardoso, A.C. Invading the Mediterranean Sea: Biodiversity patterns shaped by human activities. *Front. Mar. Sci.* **2014**, *1*. [CrossRef]
12. Rundel, P.W.; Arroyo, M.T.K.; Cowling, R.M.; Keeley, J.E.; Lamont, B.B.; Vargas, P. Mediterranean Biomes: Evolution of Their Vegetation, Floras, and Climate. *Annu. Rev. Ecol. Evol. Syst.* **2016**, *47*, 383–407. [CrossRef]
13. Curtis, P.G.; Slay, C.M.; Harris, N.L.; Tyukavina, A.; Hansen, M.C. Classifying drivers of global forest loss. *Science* **2018**, *361*, 1108–1111. [CrossRef]
14. Ginesu, S.; Carboni, D.; Marin, M. Coastline modifications in Sardinia starting from archaeological data: A progress report. In Proceedings of the International Conference of Environment, Landscape, European Identity/Annual Scientific Meeting of the Faculty-of-Geography, Bucharest, Romania, 4–6 November 2011; Patru-Stupariu, I., Patroescu, M., Ioja, C.I., Rozylowicz, L., Eds.; Elsevier Science: Asterdam, The Netherlands, 2011; Volume 14, pp. 132–142.
15. Ginesu, S.; Carboni, D.; Marin, M. Erosion and use of the coast in the northern Sardinia (Italy). In Proceedings of the 25th International Conference on Environment at Crossroads-SMART Approaches for a Sustainable Future, Bucharest, Romania, 12–15 November 2015; Ioja, I.C., Comanescu, L., Dumitrache, L., Nedelea, A., Nita, M.R., Eds.; Elsevier Science: Asterdam, The Netherlands, 2015; Volume 32, pp. 230–243.
16. Khorchani, M.; Nadal-Romero, E.; Tague, C.; Lasanta, T.; Zabalza, J.; Lana-Renault, N.; Domínguez-Castro, F.; Choate, J. Effects of active and passive land use management after cropland abandonment on water and vegetation dynamics in the Central Spanish Pyrenees. *Sci. Total Environ.* **2020**, *717*, 137–160. [CrossRef]
17. Scarascia-Mugnozza, G.; Oswald, H.; Piussi, P.; Radoglou, K. Forests of the Mediterranean region: Gaps in knowledge and research needs. *For. Ecol. Manag.* **2000**, *132*, 97–109. [CrossRef]
18. Thompson, J.D. *Plant Evolution in the Mediterranean*; Oxford University Press: Oxford, UK, 2005.
19. Falcucci, A.; Maiorano, L.; Boitani, L. Changes in land-use/land-cover patterns in Italy and their implications for biodiversity conservation. *Landsc. Ecol.* **2007**, *22*, 617–631. [CrossRef]
20. Allan, J.R.; Venter, O.; Maxwell, S.; Bertzky, B.; Jones, K.; Shi, Y.C.; Watson, J.E.M. Recent increases in human pressure and forest loss threaten many Natural World Heritage Sites. *Biol. Conserv.* **2017**, *206*, 47–55. [CrossRef]
21. Abd-Elaziz, S.; Zeleňáková, M.; Mésároš, P.; Purcz, P.; Abd-Elhamid, H.F. Anthropogenic Activity Effects on Canals Morphology, Case Study: Nile Delta, Egypt. *Water* **2020**, *12*, 3184. [CrossRef]
22. Mostafa, S.M.; Wahed, O.; El-Nashar, W.Y.; El-Marsafawy, S.M.; Zeleňáková, M.; Abd-Elhamid, H.F. Potential Climate Change Impacts on Water Resources in Egypt. *Water* **2021**, *13*, 1715. [CrossRef]
23. Kaplan, J.O.; Krumhardt, K.M.; Zimmermann, N. The prehistoric and preindustrial deforestation of Europe. *Quat. Sci. Rev.* **2009**, *28*, 3016–3034. [CrossRef]
24. Raftoyannis, Y.; Nocentini, S.; Marchi, E.; Sainz, R.C.; Guemes, C.G.; Pilas, I.; Peric, S.; Paulo, J.A.; Moreira-Marcelino, A.C.; Costa-Ferreira, M.; et al. Perceptions of forest experts on climate change and fire management in European Mediterranean forests. *iForest-Biogeosciences For.* **2013**, *7*, 33–41. [CrossRef]
25. Prefac, Z.; Dumitru, S.; Chendeș, V.; Sîrodoev, I.; Cracu, G. Assessment of Landslide Susceptibility Using the Certainty Factor Model: Rascuta Catchment (Curvature Subcarpathians) Case Study. *Carpathian J. Earth Environ. Sci.* **2016**, *11*, 617–626.
26. Kaplan, J.O.; Krumhardt, K.M.; Gaillard, M.J.; Sugita, S.; Trondman, A.K.; Fyfe, R.; Marquer, L.; Mazier, F.; Nielsen, A.B. Constraining the Deforestation History of Europe: Evaluation of Historical Land Use Scenarios with Pollen-Based Land Cover Reconstructions. *Land* **2017**, *6*, 91. [CrossRef]
27. Rezapour, S.; Alipour, O. Degradation of Mollisols quality after deforestation and cultivation on a transect with Mediterranean condition. *Environ. Earth Sci.* **2017**, *76*, 22. [CrossRef]
28. Palahi, M.; Mavsar, R.; Gracia, C.; Birot, Y. Mediterranean forests under focus. *Int. For. Rev.* **2008**, *10*, 676–688. [CrossRef]
29. Reyer, C.P.O.; Bathgate, S.; Blennow, K.; Borges, J.G.; Bugmann, H.; Delzon, S.; Faias, S.P.; Garcia-Gonzalo, J.; Gardiner, B.; Gonzalez-Olabarria, J.R.; et al. Are forest disturbances amplifying or canceling out climate change-induced productivity changes in European forests? *Environ. Res. Lett.* **2017**, *12*, 3. [CrossRef] [PubMed]
30. Torun, P.; Altunel, A.O. Effects of environmental factors and forest management on landscape-scale forest storm damage in Turkey. *Ann. For. Sci.* **2020**, *77*, 2. [CrossRef]
31. Ogaya, R.; Peñuelas, J. Climate Change Effects in a Mediterranean Forest Following 21 Consecutive Years of Experimental Drought. *Forests* **2021**, *12*, 306. [CrossRef]
32. Munyua, H. ICTs and Small-Scale Agriculture in Africa: A Scoping Study. Final Report. International Development Research Centre (IDRC), Ottawa. 2007. Available online: https://idl-bnc-idrc.dspacedirect.org/bitstream/handle/10625/50998/IDL-50998.pdf?sequence=1 (accessed on 26 June 2020).
33. CBO-Congressional Budget Office, The Congress of the United States, Deforestation and Greenhouse Gases. Available online: https://www.cbo.gov/sites/default/files/112th-congress-2011-2012/reports/1-6-12-forest.pdf (accessed on 26 June 2020).

34. Peptenatu, D.; Sirodoev, I.; Pravalie, R. Quantification of the aridity process in South-Western Romania. *J. Environ. Health Eng.* **2013**, *11*. [CrossRef] [PubMed]
35. Pravalie, R.; Peptenatu, D.; Sirodoev, I. The Impact of Climate Change on the Dynamics of Agricultural Systems in South-Western Romania. *Carpathian J. Earth Environ. Sci.* **2013**, *8*, 175–186.
36. Comanescu, L.; Nedelea, A. Floods and public perception on their effect. Case Study: Tecuci Plain (Romania), year 2013. In Proceedings of the 25th International Conference on Environment at Crossroads-SMART Approaches for a Sustainable Future, Bucharest, Romania, 12–15 November 2015; Ioja, I.C., Comanescu, L., Dumitrache, L., Nedelea, A., Nita, M.R., Eds.; Elsevier Science: Asterdam, The Netherlands, 2015; Volume 32, pp. 190–199.
37. Diakakis, M.; Priskos, G. Public perception of flood risk in flash flood prone areas of Eastern Mediterranean: The case of Attica Region in Greece. *Int. J. Disaster Risk Reduct.* **2018**, *28*, 404–413. [CrossRef]
38. Kastridis, A.; Stathis, D. Evaluation of Hydrological and Hydraulic Models Applied in Typical Mediterranean Ungauged Watersheds Using Post-Flash-Flood Measurements. *Hydrology* **2020**, *7*, 12. [CrossRef]
39. Lionello, P.; Malanotte-Rizzoli, P.; Boscolo, R.; Alpert, P.; Artale, V.; Li, L.; Luterbacher, J.; May, W.; Trigo, R.; Tsimplis, M.; et al. The Mediterranean climate: An overview of the main characteristics and issues. In *Mediterranean Climate Variability*; Lionello, P., Malanotte-Rizzoli, P., Boscolo, R., Eds.; Elsevier: Amsterdam, The Netherlands, 2006. [CrossRef]
40. Ramdani, M.; Elkhiati, N.; Flower, R.J. Lakes of Africa: North of Sahara. In *Encyclopedia of Inland Waters*; Likens, G.E., Ed.; Elsevier: Amsterdam, The Netherlands, 2009; Volume 2, pp. 544–554.
41. Underwood, E.C.; Viers, J.H.; Klausmeyer, K.R.; Cox, R.L.; Shaw, M.R. Threats and biodiversity in the mediterranean biome. *Divers. Distrib.* **2009**, *15*, 188–197. [CrossRef]
42. Azuara, J.; Lebreton, V.; Peyron, O.; Mazier, F.; Combourieu-Nebout, N. The Holocene history of low altitude Mediterranean Fagus sylvatica forests in southern France. *J. Veg. Sci.* **2018**, *29*, 438–449. [CrossRef]
43. Bugalho, M.N.; Caldeira, M.C.; Pereira, J.S.; Aronson, J.; Pausas, J.G. Mediterranean cork oak savannas require human use to sustain biodiversity and ecosystem services. *Front. Ecol. Environ.* **2011**, *9*, 278–286. [CrossRef]
44. Ciobotaru, A.-M. *Impactul Defrisarilor Asupra Dinamicii Structurale a Economiilor Locale din Judetul Suceava*; Editura Transversal: Targoviste, Romania, 2020; p. 262.
45. Hansen, M.C.; Potapov, P.V.; Moore, R.; Hancher, M.; Turubanova, S.A.; Tyukavina, A.; Thau, D.; Stehman, S.V.; Goetz, S.J.; Loveland, T.R.; et al. High-Resolution Global Maps of 21st-Century Forest Cover Change. *Science* **2013**, *342*, 850–853. Available online: http://earthenginepartners.appspot.com/science-2013-global-forest (accessed on 21 June 2020). [CrossRef] [PubMed]
46. Moreira, F.; Viedma, O.; Arianoutsou, M.; Curt, T.; Koutsias, N.; Rigolot, E.; Barbati, A.; Corona, P.; Vaz, P.; Xanthopoulos, G.; et al. Landscape-wildfire interactions in southern Europe: Implications for landscape management. *J. Environ. Manag.* **2011**, *92*, 2389–2402. [CrossRef] [PubMed]
47. El Shaer, H.M. Land desertification and restoration in Middle East and North Africa (MENA) region. *Sci. Cold Arid Reg.* **2015**, *7*, 7–15. [CrossRef]
48. Iglesias, A.; Mougou, R.; Moneo, M.; Quiroga, S. Towards adaptation of agriculture to climate change in the Mediterranean. *Reg. Environ. Chang.* **2011**, *11*, 159–166. [CrossRef]
49. Fernández-García, V.; Marcos, E.; Huerta, S.; Calvo, L. Soil-vegetation relationships in Mediterranean forests after fire. *For. Ecosyst.* **2021**, *8*, 18. [CrossRef]
50. Cerri, C.E.P.; Cerri, C.C.; Maia, S.M.F.; Cherubin, M.R.; Feigl, B.J.; Lal, R. Reducing Amazon Deforestation through Agricultural Intensification in the Cerrado for Advancing Food Security and Mitigating Climate Change. *Sustainability* **2018**, *10*, 989. [CrossRef]
51. Parente, L.; Ferreira, L. Assessing the Spatial and Occupation Dynamics of the Brazilian Pasturelands Based on the Automated Classification of MODIS Images from 2000 to 2016. *Remote Sens.* **2018**, *10*, 606. [CrossRef]
52. Borrelli, P.; Modugno, S.; Panagos, P.; Marchetti, M.; Schütt, B.; Montanarella, L. Detection of harvested forest areas in Italy using Landsat imagery. *Appl. Geogr.* **2014**, *48*, 102–111. [CrossRef]
53. Geist, H.J.; Lambin, E.F. Proximate Causes and Underlying Driving Forces of Tropical Deforestation: Tropical forests are disappearing as the result of many pressures, both local and regional, acting in various combinations in different geographical locations. *Bioscience* **2002**, *52*, 143–150. [CrossRef]
54. Kouassi, J.-L.; Gyau, A.; Diby, L.; Bene, Y.; Kouamé, C. Assessing Land Use and Land Cover Change and Farmers' Perceptions of Deforestation and Land Degradation in South-West Côte d'Ivoire, West Africa. *Land* **2021**, *10*, 429. [CrossRef]

forests

MDPI

Article

Single-Track Bike Trails in the Moravian Karst as Part of Forest Recreation

Petr Hrůza *, Petr Pelikán and Lucie Olišarová

Department of Landscape Management, Faculty of Forestry and Wood Technology, Mendel University in Brno, 613 00 Brno, Czech Republic; petr.pelikan@mendelu.cz (P.P.); lucie.olisarova@gmail.com (L.O.)
* Correspondence: petr.hruza@mendelu.cz; Tel.: +420-545-134-085

Abstract: Recently, cycling has become a popular recreation activity, and mountain biking provides an experience that is sought by an increasing number of people. Bike trails constructed for mountain bikers in access areas lead mostly through the forest and provide not only an extraordinary riding experience but the opportunity to admire the surrounding nature. The reason for constructing such trails from a landowner's point of view is to help keep bikers' movements within a defined access area and to ensure adjacent areas are left free for other forest functions. It also helps distribute groups of visitors with other interests to other parts of the forest. This is what we call "controlled recreation". In this example, it means that if cyclists come to the locality to use the bike trails, they should ride only along the designated trails; however, they may leave these trails and ride on the surrounding land. This article studied the movements of bikers in an accessible area of the Moravian Karst and the regulation of their movements by controlled recreation. Attendance in the area was measured using automatic counters. These were placed at the entry points to the accessible area and just behind the routes where the trails branch off. The results showed that bikers mostly stayed on the formal routes and that the trails were effective, i.e., there was no uncontrolled movement of bikers into the surrounding forest stands. We also noted the time of day that cyclists were active. These results can be used to better plan work in the forest, for example, harvesting and logging. To further the suitable development of accessible areas of the forest, we also compared the usual size of trail areas in two other European countries and the increasing width of bike trails due to the transverse slope of the terrain.

Keywords: forest recreation; ecosystem services; forest management; visitors monitoring; single-track bike trails

Citation: Hrůza, P.; Pelikán, P.; Olišarová, L. Single-Track Bike Trails in the Moravian Karst as Part of Forest Recreation. *Forests* **2021**, *12*, 1601. https://doi.org/10.3390/f12111601

Academic Editors: Radu-Daniel Pintilii and Diego Varga

Received: 21 October 2021
Accepted: 16 November 2021
Published: 19 November 2021

1. Introduction

Mountain biking has become a popular leisure-time activity in many countries [1], in the countryside as well as in protected areas close to cities [2]. It encompasses many specialised disciplines including down-hill biking, tour and cross-country riding and competition styles such as free-riding and four-cross [3,4].

The increasing demands of holidaymakers are an integral part of modern life. As living standards increase, the demand for adrenaline-producing hobbies increases. Forest management is used not only to protect its use for timber production but also to promote the recreation functions of forests and the welfare of visitors [5]. Most outdoor recreational activities in forests related to paths and trails [6].

Monitoring visitor attendance provides basic information about the number of visitors and their spatial distribution within the forest [7]. Tourism in large, protected areas has been described by Navrátil et al. [8].

The modern trend in forests is to construct bike trails for cyclists. Březina et al. [9] monitored visitor arrivals in forests. Their data show the potential for future cash flows from the city to forest areas. Their research offers a potential tool for investigating cash flows in the local economy and a method for determining the potential of the socio-economic functions of forest management as a local multiplier [10].

According to German research, the existence of such trails can be harmonious with the needs of foresters. Especially in heavily visited regions, attractive advertising on the trails and other paths contributes to the effectiveness of visitor management [11]. But it is very important to monitor the numbers of people in the forest [12], as visitors can affect aspects of forest management such as harvesting and logging.

Narrow one-way mountain bike trails, called single tracks, are designed and constructed according to special, long-proven methods, making them suitable for outdoor recreation. Ideally, they are set in the terrain sensitively, so that they do not interfere too much with the character of the landscape in the areas they go through [13].

The interest of tourists in forest localities is increasing. Forest stands, parks and other green areas are the most valuable for recreation [14,15]. People prefer forest and well-kept green areas [16] and water amenities, such as lakes and rivers [17], to city landscapes [18]. An important and relevant issue for contemporary tourism, sport and recreation planners is how to develop trails and mountain bike areas in a way that is in keeping with the demands of proficient mountain bike riders. In [19], the authors offer an overview of the affective experiences that come from mountain biking over a range of common ride obstacles and terrains.

As mountain biking is popular in many natural areas, it can remain controversial, at least in part, due to the divergent views about its environmental impacts [20], one of the most visible of which is the modification of the landscape's structure and quality [21]. In Austria, official trails do not necessarily meet the needs of mountain bikers, who often ride on unofficial trails or along paths where biking is not allowed. This behaviour can result in conflicts with other countryside users, landowners, hunters and conservationists [4]. Mountain biking is seen as an important activity for growing participation in outdoor recreation. Increasingly, places are marketed as suitable for mountain biking and have supporting legal rights of access in place. However, in a study of the Cairngorms National Park in Scotland, as in many places, it was found that mountain bikers were not always made welcome by managers and other visitors [22].

In Slovenia, access to forest skidding tracks and signposted mountain paths, which are greatly preferred by mountain bikers, is generally not legal. There is also a lack of mountain biking management and infrastructure at the national level. The increased interest in mountain biking on trails in natural areas necessitates a systematic approach to management. An important challenge for such management in natural areas is conflicts with other user groups, particularly hikers [23].

Interviews carried out in the forest of Allschwil near Basel, Switzerland, revealed that, in some areas, there was conflict among different user groups, particularly between dog owners, cyclists and groups of hikers or joggers. Knowledge of the habits and preferences of forest visitors allows for the planning of measures that separate different forest user groups and prevent them from entering areas with a high conservation value [24].

Today, forest owners have to cope with an increased demand for recreation. This brings many obligations, which are enshrined in law. One task is to identify potential users and quantify their movements across forest property. From the perspective of the Forest Act No. 289/1995 Coll. [25], Section 20 of the Forest Act, point j) states: "It is forbidden to ride bicycles, horses, skis or sleighs except on forest roads and marked trails in the forest stands". This implies that off-road bikers can only ride on forest roads if they are not on open terrain and specially constructed trails for cyclists, which include trails.

Hrůza [26,27] deals with the legal aspects of purpose-built forestry communications and their position in the law, which includes, among other things, bike trails. Currently, the number of conflicts among forest users is increasing due to the new and growing societal demands for forest recreation in addition to the traditional forest function of wood production. Outdoor sports and forest education programmes are adding to the demands on forest use. Other authors [28] mention conflict between recreational users, e.g., between bikers and hikers. These conflicts are expected to become more acute in the future, which poses new challenges to both forest policymakers and forest managers. Forest owners' and managers' efforts should be to manage and disperse the visitors from various interest groups within the forest environment so that each visitor can engage in their activity without a negative impact on the surrounding area, other visitors or commercial forestry activities. Reference [29] reports on how government managers of protected areas in South Western Australia engaged with the mountain biking community. This included the development of a user compatibility matrix that facilitated park management decision making to reduce negative social and environmental impacts while, at the same time, providing for a range of recreational opportunities in the protected area. A multi-step, methodological triangulation conflict model from US recreation management was applied and tested in the Black Forest Nature Park [30]. The results from two groups, hikers and mountain bikers, were analysed in depth. The main potential conflicts were due to the fact of infrastructure and differing values. These were influenced by various visitor characteristics such as resource attachment, experiences, style of activity, expectations and motives.

One of the reasons forest owners construct bike trails on their land is to have control over the movements of cyclists through forest stands.

- The aims of the present study were to determine whether bikers' movements can be directed by constructing single-track bike trails and to determine the quantity of visitors who left the trails to set out across the wider surroundings of forest ecosystem.
- To achieve this, we needed to determine the number of visitors and where they entered or left the area.
- At the same time, the movement and the spatial distribution of cyclists within the area of interest, according to the time of day and working and non-working days, were investigated, as this is important for forest management such as harvesting and logging.
- Single tracks are usually described as a type of mountain-biking trail approximately the width of a bike. This narrow design makes them easy to construct in the forest. But the width of a trail increases with an increasing transverse terrain slope; thus, we wanted to understand how a single-track width changed according to an increasing transverse terrain slope.
- We compared the usual area of single-track bike trails in two other Europe countries to provide forest owners with an idea of how large these access areas for controlled recreation should be.

2. Materials and Methods

The Moravian Karst Single Track Centre was constructed in 2015 across 547.74 ha at the Training Forest Enterprise Masaryk Forest Křtiny (TFE), which is an organisational part of the Mendel University in Brno (MENDELU), with the aim of directing the activities of mountain bike visitors to specialised narrow trails. The TFE is located near Brno in the South Moravian region and manages all together a forest area of 10,492 ha. The representation and distribution of woody plants reflect the vegetation gradation and habitat conditions. Deciduous tree species (33% dominated by beech) cover 62% of the forest area, and 38% are covered by conifers (19% spruce). Sustainable forest management aims at management measures ensuring the fulfilment of forest ecosystem services and other functions in changing environmental and social conditions and optimal use of the production potential of forests, with particular regard to the needs of the university (teaching and research) and the public (recreation). The TFE closely collaborates with the Faculty of Forestry and Wood Technology MENDELU to offer experimental forest stands and other establishments for educational and research purposes.

Our aim was to determine whether bikers in the centre moved outside the site to use other forest roads or other parts of the forest or whether they stayed mostly in the trail area. For forest management purposes, we also investigated the quantitative movements of visitors in the Moravian Karst Single Track Centre and their time distribution to provide forest managers with information on land use at various times.

The four trails consist of single tracks that go through the forest stand and, partially, on forest roads in the northeast TFE part. The starting point for these trails is located near Camp Olšovec in Jedovnice, where there is a sport–recreation centre for mountain bikers (boarding place) with technical facilities. The single-track sections, which go directly through the forest stand, are referred to in the study as Track 0 (770 m), Track 1 (7250 m), Track 2 (3730 m) and Track 3 (2010 m) (Figure 1). The single tracks measure 13.8 km in total length and are connected by forest roads. The total length of all trails together with forest roads is 21.4 km.

Figure 1. Location of the study area the Moravian Karst Single Track Centre in the Czech Republic.

The hypothetical assumption was that cyclists would only use the marked trail routes prepared for them to join a ride and would not leave the single-track area of 547.74 ha to go into the surrounding forest ecosystem. The single-track bike trails constructed in one part of the forest would be sufficiently attractive that they would not leave them, and the surrounding forest stands would be undisturbed by bikers. If this was not confirmed, the question would be: how many visitors left the area for other parts of the forest?

First, the method of monitoring cyclists in the selected area was decided, and the assumption of their direction of movement was made.

In cooperation with the monitoring coordinator of cyclists from the Partnership Foundation, two types of counters were chosen and locations suitable for placing them were selected. Their placement was agreed with the administrator of the Moravian Karst Single Track Centre, and the necessary facilities were leased.

Monitoring was performed for one month in the high season for cycling activity (15 June–15 July). The presence of visitors was monitored with respect to the time of day and whether the day was a working or non-working day.

Four locations were selected (i.e., U Kempu, Trojúhelníky, U Proklestu and Na Mokřadní) to install the first type of counter, Eco-counters, to cover the cyclist entry points to the trail area; their movements were captured further on the routes in the area (Figure 2). Thus, sensors were set-up to scan all incoming and outgoing visitors to the trail area.

Counters were placed at narrow places so that cyclists could not ride past them side by side. They were mounted on trees next to the road or on a nearby pole if a tree was unavailable. The counters can measure reliably over four meters. The devices were locked and marked with a contact number to prevent them from being accidentally damaged by people. All counter locations were photographically documented (Figure 3).

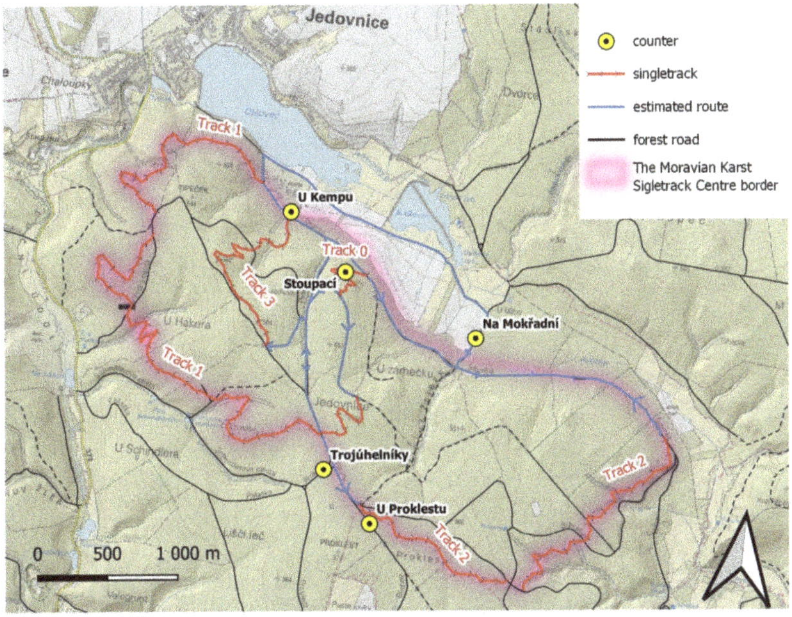

Figure 2. Locations of counters for data measuring in the Moravian Karst Single Track Centre.

Figure 3. Examples of counter installation.

An Eco-counter is a device for counting all users of a route (cyclists, pedestrians and cars). It has two sensors and can determine the direction of its target's motion. The counters were manually calibrated to identify individual groups of visitors and a calibration coefficient was set to ensure the maximum validity of the collected data and minimise errors, i.e., manual calibration counts were performed to determine the numbers of individual users as cyclists, pedestrians or cars. An error can arise in wider places if cyclists ride close together. The counter then records them as one person. The calibration prevents this deficiency and, thus, ensures the maximum accuracy of the measured data. At each site, data were collected for eight hours per day. Data were collected continuously throughout the measurement period of six days. Subsequently, the counters were removed from each site, the data were downloaded and, using the calibration coefficient, calculated to the actual values of the individual participants and the direction of their travel. Due to the fact that this study focused on the movement of cyclists in the given area, the results subsequently analyse only the movements of cyclists.

We placed one of the second type of counter, a TRAFx Mountain Bike Counter, at Stoupací on Track 0 to discover whether cyclists used the ascending single-track bike trail to access downhill single tracks or whether they also used the surrounding forest road.

This counter responds to metal wheel parts, so it is able to detect bikes with a carbon frame. The width of the recording field was up to 1 m from the counter, so when installed in the middle of the track, it can cover a track up to 2 m wide. It was installed under the surface of the trail, hiding it from human eyes, minimising the risk of damage or vandalism.

Both types of counters consisted of a sensor that recorded the passage of a cyclist and were connected to the data unit, which stored the measured data and supplied power to the counter.

During the measurement phase, there was one month of continuous inspection and manual data collection. The Eco-Visio online application (https://www.eco-counter.com/applications; accessed on 20 October 2021) was used to download, analyse and implement monitoring-related data.

As single tracks are often represented as narrow paths following the shape of the terrain, with minimum influence on their surroundings, we focused on the real width of constructed single tracks in the Moravian Karst Centre. The width of single tracks is generally stated to be up to 1 m, which only relates to one bike track width of 0.8 m, but the total width varies depending on the single track's inclination and the transverse slope of the terrain. It is necessary to take into account the cut-and-fill slopes and take as the width the entire width of the single-track formation. This greatly changes the total area take-up as evidenced by the results of our investigation.

One of the aims of this research was to determine a suitable and sustainable size for a single-track area for mountain biking, allowing for other visitors, timber production and other forest functions, as there is now a great demand for this type of sports activity in the forest environment.

For comparison, we chose 10 such centres abroad. Five of these were in the UK, where this issue is historically captured systemically and on the basis of a national strategy. Trails were chosen in Wales, which has been written about in the sense of "here it all began, 25 years ago we saw the start of the trail centre revolution, and it was here in Wales. Coed y Brenin led the way in setting up waymarked trails specifically for mountain biking" (https://www.mbwales.com/2016/07/18/trails-centres-began/; accessed on 20 October 2021). We can describe these trails as closed circles, often far from large towns and habitations. The location of the trails in forest stands was not a condition, but most were located in forest. To select suitable single-track areas, a Welsh government website, Mountain Bike Wales (https://www.mbwales.com; accessed on 20 October 2021), was used. The site promotes mountain biking and provides information about each site, including maps. Five centres were selected: Abercarn, Afan, Brechfa, Cw Merfyn and Ganllwyd.

The other five trail areas examined were in Denmark, on slightly hilly terrain in suburban forests, so their location and the shape of terrain were more comparable with the Moravian Karst Single Track Centre. The Danish facilities were located close to areas of habitation; being within forest stands was a condition of their selection. The Mountain Bike Project website (https://www.mtbproject.com; accessed on 20 October 2021) was used to select five suitable areas: Djævlesporet, Egebjerg, Kongshøj, Silkeborg and Vodskov.

We did not compare the size of the areas, but trail density was established for the total length of all trails and the efficiency established for the trail distribution. Specially, the efficiency provided us with information about single-track land use and based on this value, it is possible to decide whether it is appropriate to further expand the single-track bike centre or to have single tracks more concentrated if needed according to the demand.

The total area of the biking amenity was taken as that enclosed by the trails themselves and forest roads that were used as connecting lines between the trails and forest roads leading to the start of the trails from the car park.

The total length of the trails was calculated as the length of all trails, including forest roads, so that the trails calculated formed a connected whole. None of the trails or their parts were counted twice, although some sections of forest roads could be used as part of several trail routes.

The density of trails within a facility was established as the total length of the trails divided by the area of the facility and expressed in m/ha.

The density H is given by the Equation (1):

$$H = \frac{D}{S} \ \left[\mathrm{m \cdot ha^{-1}}\right] \tag{1}$$

where: D is the total length of the trail network (m), S is the accessed area (ha).

As the parameters do not provide any information on the trail network's distribution, a parameter for trail efficiency was added.

The efficiency U is given as the proportion between the average geometric (shortest) distance from a regular geometric 10 ha square network (grid) and the theoretical distance, the calculation of which was based on the ideal regular distribution of the trails in the area according to Equation (2).

$$U = \frac{D_t}{\overline{D_g}} \cdot 100 \ [\%] \tag{2}$$

where D_t is the theoretical distance (m) calculated as the average distance due to the optimal distribution of trails (Figure 4) in the accessed area and depending on the trail density H according to Equation (3).

$$D_t = \frac{10{,}000}{4H} \ [\mathrm{m}] \tag{3}$$

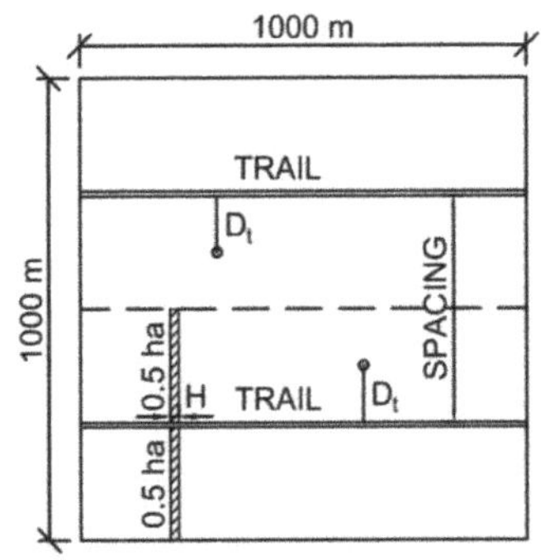

Figure 4. Theoretical distance, D_t, due to the optimal trail distribution in the forest access area and trail density, H 20 m/ha.

In addition, the geometric distance, D_g, represents the direct distance from the centre point of a regular geometric 10 ha square grid to the trail. Its average value (Equation (4)) depends on the trail distribution and is generally higher than the theoretical distance.

$$\overline{D_g} = \frac{D_{g1} + D_{g2} + \ldots D_{gn}}{n} \ [\mathrm{m}] \tag{4}$$

To analyse the trail areas, maps on the OpenStreetMap WMS server were used in the QGIS application environment. The geometric distances from grid points were determined using QGIS software and the vector analysis tool *"distance to nearest hub"*. The same analyses were performed for the Moravian Karst Single Track Centre for comparison.

3. Results

The comparison of the monitored visitor entry points into the Moravian Karst Single Track Centre showed (Table 1) large differences in total attendance in the monitored period. The site at U Kempu, located on a forest road at the entrance to the Training Forest Enterprise Křtiny, MENDELU by the Jedovnice Olšovec Lake, dominated attendance figures with 21,559 visits. Another monitored site near Olšovec at Na Mokřadní recorded a significantly lower value of 4172 visitors. But the fewest visitors were recorded on the opposite side at Trojúhelníky (2440) and U Proklestu (1070), which is the side from which people leave the Moravian Karst Single Track Centre to visit other parts of Forest Training Entrprice Křtiny.

Table 1. Overall monitoring results of visitors on selected forest roads in the Training Forest Enterprise Křtiny MENDELU during the period 15 June–15 July (calibrated data).

Locality	Total Attendance			Average Daily Attendance		
	Total Records	Working Days	Non-Working Days	Daily Average	Working Days	Non-Working Days
U Kempu	21,559	7107	14,452	695	374	1204
Stoupací	8522	2566	5615	264	135	468
Na Mokřadní	4172	1757	2415	135	92	201
Trojúhelníky	2440	1443	1034	79	64	102
U Proklestu	1070	539	531	35	28	44

Focusing only on cyclists (e.g., data cleaned of vehicles and pedestrians using calibration counting), we counted 9714 cyclists at U Kempu travelling towards Jedovnice; going in the opposite direction, the figure was 9599 cyclists. On the Jedovnice side at Na Mokřadní, it was 1807 cyclists in and 1207 out. On the opposite side of the Centre, the counters at Trojúhelníky and U Proklestu recorded only 1133 cyclists riding into the area and 927 riding out (Table 2).

Table 2. Number of cyclists according to each counter.

	U Kempu	Stoupací	Trojúhelníky	U Proklestu	Na Mokřadní
Cyclists					
In *	9714	8522	938	195	1807
Out **	9599	- ***	672	255	1127

* Direction into the area. ** Direction out the area. *** Counter on the single-track Stoupací-bikers counted in one direction.

From these numbers, we can conclude that 92% of cyclists stayed at the Moravian Karst Single Track Centre and the construction of single-track trails in response to the need for controlled recreation fulfilled its purpose.

At most locations, Saturday was the busiest day. The exception was at U Proklestu, where Monday (65) and Sunday (162) were the busiest days. This shows that this route was favoured mostly by trekking cyclists, who mainly used the forest roads at the Centre.

The average daily visit rate at the sites corresponds to the total visit rate. At all localities, the number of visitors on non-working days was less than 2/3 of total attendance.

The maps (Figures 5 and 6) and measurements show that bikers used the marked trail route. The main purpose for them was to enjoy these trails. The existing three single tracks go through the forest stand and, partially, on forest roads, so bikers do not need to use other forest roads.

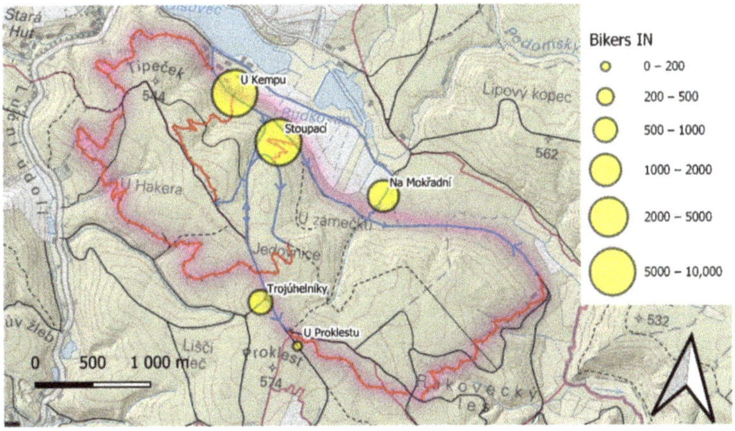

Figure 5. Number of cyclists in the direction into the area.

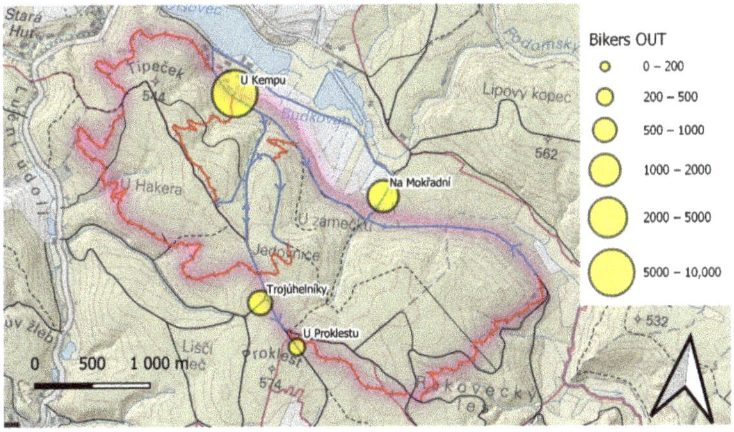

Figure 6. Number of cyclists in the direction out of the area.

From a total of 12,654 incoming cyclists, 8522 used single-track 0 instead of the adjacent forest road. This gives us interesting information that bikers use single tracks rather than ascending forest roads to reach the top of the downhill single tracks near U Proklestu.

The single-track area was mostly visited during daylight hours, which is consistent with the guidelines for visitors from trail managers. The peak hours are given in Table 3.

Table 3. Visitor layout by peak hours.

Locality	Peak Hours	In General
U Kempu	11:00–12:00; 16:00–17:00	10:00–18:00
Stoupací	10:00–12:00	9:00–17:00
Trojúhelníky	10:00–11:00; 14:00–16:00	10:00–18:00
U Proklestu	10:00–12:00; 15:00–18:00	10:00–18:00
Na Mokřadní	14:00–16:00; 17:00–18:00	10:00–19:00

The real width of most constructed single tracks, taking cut-and-fill slopes into account, fell in the width category 1–2 m, with 2–3 m the next most common as seen in Table 4.

Table 4. Trail width parameters.

Width (m)		Ratio of Total Single-Track Length (%)			
	Limit Transverse Terrain Slope (%)	Trail 0	Trail 1	Trail 2	Trail 3
<1	<16	0.0	0.2	35.4	0.0
1–2	16–42	76.9	70.2	21.8	61.2
2–3	42–50	20.7	22.7	28.3	23.1
3–4	50–53	1.7	5.3	11.1	11.5
>4	>53	0.7	1.6	3.3	4.1
Total length (%)		100.0	100.0	100.0	100.0

The 770 m Track 0 represents a part that is common to all of the trails and is used to access downhill single-tracks 1, 2 and 3. The section has a long rise that makes up 83% of the trail length with variable width including cut-and-fill slopes of between 1.0 and 3.0 m. The mean total width of the road formation is 1.8 m, with 77% of the trail length being wider than 1.0 m (Table 4).

Track 1 has the character of a ridge trail with a balanced ascent and descent ratio going over the north-western slopes of a peak. The total length is 7250 m, and the width ranges from 0.7 to 9.2 m with a mean of 1.8 m. For 435 m of the trail, the width is greater than 3.0 m (Table 4).

Track 2 has a total length of 3730 m and resembles the main downhill part of a single track. The trail is 1.0–6.2 m wide, with a mean of 2.4 m. Sixty-six percent of the trail length is wider than 2.0 m, and for a length of 535 m, the trail is wider than 3.0 m (Table 4).

Track 3 first runs along the ridge forest road, diverging from it after approximately 600 m. The total section length is 2010 m, and the trail width ranges between 0.8 and 6.4 m (Table 4) with a mean of 2.1 m. For 315 m, the width of the trail is greater than 3.0 m.

Based on our measurements, we can conclude that the width of all single tracks over their entire length is greater than 1.0 m. On average, 90% of the trail lengths are 1.0–3.0 m wide (Table 4). Almost 1.4 km of the trail is wider than 3.0 m (11% of the total length of the single track located in the forest stand).

The characteristic parameters of biking areas in Wales and Denmark are compared in Table 5 with those of the Moravian Karst Single Track Centre with its total area of 547.74 ha, total trail length of 21,365 m, density of 39 m/ha and efficiency of 40%.

Table 5. Characteristics of the single-track areas compared.

Country	Trail	Area (ha)	Length (m)	Density (m/ha)	Efficiency (%)
Wales	Abercarn	403.10	26,820	67	38
Wales	Afan	1661.42	100,467	60	37
Wales	Brechfa	594.34	44,352	75	29
Wales	Cw Merfyn	1381.55	35,062	25	29
Wales	Gannlwyd	1154.51	60,824	53	48
Denmark	Djaevlesporet	157.81	13,349	85	39
Denmark	Egebjerg	42.43	10,440	246	47
Denmark	Kongshoj	49.30	8255	167	47
Denmark	Silkeborg	201.38	13,806	69	32
Denmark	Vodskov	108.67	12,254	113	47
Czechia	Moravia Karst	547.74	21,365	39	40

4. Discussion

In previous years, data from the counter located on Track 0 were compared. According to Olišarová et al. [12], the most frequented months of 2017 and 2018 were July and August. A comparison of the study routes in the area showed a large difference in total attendance in the monitored period. In July 2017, a total of 9558 cyclists rode past the counters, and the following year (2018) showed 9170 entries. The data from this year in the first half of July show a total of 5056 passages.

There was a wide range of holidaymakers in the area. The highest attendance was recorded at U Kempu, near the camp gate. Nearly 90% of visitors recorded there were cyclists. On the contrary, the smallest number of visitors was counted at U Proklestu, where the number of cyclists was less than 50%. Almost two-thirds of visitors came to the area on a non-working day, while during the week only 1/3 were in the area.

There are a large number of researchers and university employees at the locality, many of whom come by car. The most frequent passages were recorded by the counter at Trojúhelníky, probably because this road connects to utility road number 373 leading out from the area to another part of the Training Forest Enterprise Křtiny MENDELU.

According to Olišarová et al. [12], some visitors travelled one route several times, others have tactically chosen their way according to their abilities and skills, as each route differs in the difficulty of the terrain and elevation.

The project has proven that visitors stay in the single-track area, and very few go on to adjacent locations.

It is necessary to realise that urban forests are often popular sites for recreational activities such as hiking, biking and motorised recreation. This can result in the formation of extensive trail networks, fragmenting vegetation into patches separated by modified edge effects and, ultimately, contributing to the degradation of the ecosystem, and management should seek to minimise the creation of informal trails by hardening popular routes, instigating stakeholder collaboration and centralising visitor flow [31]. This statement can be supported by our findings that the regulation of bikers' movements by controlled recreation fulfilled its objective and works. [31] adds, the forest lost to informal biking and hiking trails reached an area of 5%. The trail distribution in a single-track bike trail centre, evaluated by trail efficiency, can help to make a decision regarding new single-track bike trail design in a bike centre or its further development if needed. Informal trails generally had worse surface conditions and were poorly designed and located. Per site, formal and informal trails resulted in similar loss of forest strata, with wider trails resulting in greater loss of forest [32]. Choosing the right corridor for a single-track lay-out according to transverse slope terrain can minimise the width of single-trail formations. In [33], the authors present a study where they analysed the spatial overlap and social conflicts between mountain bikers and runners. This can help managers and decision makers design proper infrastructure for outdoor activities. Strategic management errors can be avoided by knowing user preferences and by offering improved conditions that meet the expectations and needs of different user groups. One of the reasons why we constructed single-track bike trails at TFE MENDELU was the same visible experience regarding conflict among groups of forest visitors, and we can state by this research that offering improved conditions for a leisure outdoor activity at a locality will keep visitors onside.

5. Conclusions

The presence of single tracks attracts mountain bikers who are willing to make a longer journey to pay their visit. Their goal is to enjoy the adrenaline-rush bike ride that single tracks unquestionably provide. Therefore, they do not need to venture into a surrounding area of the forest ecosystem and, thus, do not interfere with forest management or other interests in the rest of the forest. There are various groups of forest visitors with different reasons for moving deeper into the forest ecosystem such as pedestrians, runners, families with children and people walking dogs. All these recreational functions are provided alongside forest management activities focusing mainly on timber production.

The construction of bike trails with single-track parts makes it possible to manage the movements of bikers on a specific trail (i.e., single tracks). As seen from this study, cyclists usually follow the marked routes and do not affect the surrounding forest.

The assumption of the narrow design of a single trail up to 1 m in width applied only in the case of a transverse slope of the terrain of up to approximately 16%. However, to increase the attractiveness, the trails are traced through a morphologically hilly area. Only 35% of the Track 3 length meets the 1 m width assumption due to the fact of its location on the plateau. All other trails are actually much wider. With an increasing transverse slope, the overall width increases rapidly. The most common width, in the range of 1–2 m, is situated in transverse terrain slopes of 16–42%. Approximately 20–30% of the trail lengths are traced in transverse slopes of 42–50%, resulting in an overall width of 2–3 m.

When comparing the land use of bike centres, as we can see, the Danish bike trail centres are smaller than those in Wales, which are located at remote sites, and the Moravian Karst Single Track Centre, but the design and density of their layout means the area is better used. When an extension of the bike centre is required, the efficiency parameter can help forest owners decide whether to extend trails inside or outside of the centre. When the efficiency parameter is less than 50%, construction inside should be recommended. At the same time, trails which divert bikers away from forest roads and avoid crossings of single tracks with forest roads should be given preference. The aim should be to achieve a uniform distribution of trails so that the forest ecosystem in the related parts of a single-track centre is not impaired.

It is clear from the collected data that the presence of individual routes is beneficial for the regulation of cyclists on forest paths seeking excitement; at the same time, there is no conflict with other visitors, who are expecting different experiences from visiting the forest.

The practical benefits of cycling monitoring for forest managers can be summarised as:

- Use for the controlled movement of bikers in the forest area.
- Planning measures to prevent/limit forestry management collision with forest visitors.
- Planning for conservation measures.
- Maintenance priority planning for both single tracks and used forest roads.
- Proposed solutions and recommendations in terms of land development.
- Reporting traffic data to the media for positive public relations.

Author Contributions: P.H. initiated the investigation, drafted and carried out the methodology and analysis and finalized the manuscript; P.P. carried out the analysis and visualized pictures L.O. carried out the terrain investigation and drafted the manuscript. All authors have read and agreed to the published version of the manuscript.

Funding: The research was funded by Mendel University in Brno (grant number: LDF_VP_2018029).

Institutional Review Board Statement: Ethical review and approval were waived for this study, due to anonymity of this study and no harm or collection of ethical data on humans.

Informed Consent Statement: Not applicable.

Data Availability Statement: The data are available and stored by the correspondent author.

Conflicts of Interest: The founding sponsor, Mendel University in Brno, had no role in the collection, analyses or interpretation of data; in the writing of the manuscript; in the decision to publish the results.

References

1. Fialová, J.; Březina, D.; Žižlavská, N.; Michal, J.; Machar, I. Assessment of visitor Preferences and Attendance to Singletrails in the Moravian Karst for the Sustainable Development Proposals. *Sustainability* **2019**, *11*, 3560. [CrossRef]
2. Pickering, C.M.; Rossi, S. Mountain biking in peri-urban parks: Social factors influencing perceptions of conflicts in three popular National Parks in Australia. *J. Outdoor Recreat. Tour.* **2016**, *15*, 71–83. [CrossRef]
3. Quinn, M.; Chernoff, G. *Mountain Biking: A review of the Ecological Effects*; Parks Canada, National Office, University of Calgary: Calgary, AB, Canada, 2010; p. 42.

4. Koemle, D.B.A.; Morawetz, U.B. Improving mountain bike trails in Austria: An assessment of trail preferences and benefits from trail features using choice experiments. *J. Outdoor Recreat. Tour.* **2016**, *15*, 55–65. [CrossRef]
5. Wilkes-Allemann, J.; Putz, M. Governance of forest recreation: Analysing the case of mountain bike trails. *Schweiz. Z. Forstwes.* **2017**, *168*, 321–328. [CrossRef]
6. Cordell., H.K.; Lewis, B.; McDonald, B. Long-term outdoor recreation participation trends. In Proceedings of the Fourth International Outdoor Recreation Tuorism Trends Symposium and the 1995 National Recreation Resource Planning Conference, Saint Paul, MN, USA, 14–17 May 1995; pp. 35–38.
7. Zahradník, D.; Banas, M.; Jirásková, E. Back and front of visitor Monitoring-Examples of good and less successful visitor monitoring in the Czech Protected areas. *J. Landsc. Manag.* **2012**, *3*, 14–19.
8. Navrátil, J.; Knotek, J.; Pícha, K.; Fialová, J. *Gentle Tourism in Large Protected Areas of Nature: South Bohemia and Šumava*; Publishing House JIH: České Budějovice, Czech Republic, 2014; 143p.
9. Březina, D.; Michal, J.; Fialová, J. Monitoring of visitor arrivals at the area of Městské lesy Hradec Králové a. s. using field counting devices. In Proceedings of the Public Recreation and Landscape Protection—With Sense Hand in Hand, Křtiny, Czech Republic, 13–15 May 2019; Mendel University in Brno: Brno, Czech Republic, 2019; pp. 250–253.
10. Březina, D.; Fialová, J.; Michal, J.; Slováčková, H. The methodology of quantification the influence of Hradec Králové city forests plc. On the regional economics. In *SilvaNet—WoodNet 2018: Proceedings Abstracts of Student Scientific Conference*; Mendel University in Brno: Brno, Czech Republic, 2018; pp. 71–72.
11. Kluge, T. Flow-, Singletrail und das Grundrecht auf Naturgenuss. In Proceedings of the Conference Dialogforum Wald: Wald Freizeit, Erholung, Gesundheit, Berlin, Germany, 20–21 April 2016.
12. Olišarová, L.; Kala, L.; Hrůza, P. Moving of visitors single trail Moravian Karst on the forestry roads in Křtiny. In Proceedings of the Public Recreation and Landscape Protection—With Sense Hand in Hand, Křtiny, Czech Republic, 13–15 May 2019; Mendel University in Brno: Brno, Czech Republic, 2019; pp. 254–257.
13. IMBA. Available online: https://www.imba.com/explore-imba/resource-hub (accessed on 2 April 2021).
14. Lynn, N.A.; Brown, D. Effect of recreational use impact on hiking experiences in natural areas. *Landsc. Urban Plan.* **2003**, *64*, 77–87. [CrossRef]
15. Majnoonian, H. *Discussion of Parks and Green Spaces and Promenades*; Department of Parks and Green Spaces of Tehran City-Iran Press: Tehran, Iran, 1995; 251p.
16. Laval, D.; Umar, M.A.N.; Balogun, A.L. A Geographic Information System and Multi-Criteria Decision Analysis in Proposing New Recreational Park Sites in Universiti Teknologi Malaysia. *Mod. Appl. Sci.* **2011**, *5*, 39–55. [CrossRef]
17. Šlezingr, M.; Uhmannová, H. The Revitalization of Rivers—Bankside Trees and Shrubs. In Proceedings of the Public Recreation and Landscape Protection—With Man Hand in Hand? Křtiny, Czech Republic, 13–15 May 2019; Mendel University in Brno: Brno, Czech Republic, 2019.
18. Gül, A.; Gezer, A. Modeling proposal for the selection of urban forest location and its evaluation using Isparta city example. In Proceedings of the First National Urban Forestry Congress in Turkey, Ankara, Turkey, 9–11 April 2004; pp. 365–382.
19. Hagen, S.; Boyes, M. Affective ride experiences on mountain bike terrain. *J. Outdoor Recreat. Tour.* **2016**, *15*, 89–98. [CrossRef]
20. Pickering, C.; Ansong, M.; Wallace, E. Experimental assessment of weed seed attaching to a mountain bike and horse under dry conditions. *J. Outdoor Recreat. Tour.* **2016**, *15*, 66–70. [CrossRef]
21. Schneider, J.; Fialová, J.; Vyskot, I. *Landscape Recreology I*; Mendel University in Brno: Brno, Czech Republic, 2008; 140p.
22. Brown, K.M. The role of belonging and affective economies in managing outdoor recreation: Mountain biking and the disengagement tipping point. *J. Outdoor Recreat. Tour.* **2016**, *15*, 34–46. [CrossRef]
23. Zajc, P.; Berzelak, N. Riding styles and characteristics of rides among Slovenian mountain bikers and management challenges. *J. Outdoor Recreat. Tour.* **2016**, *15*, 10–19. [CrossRef]
24. Rusterholz, H.P.; Baur, B. Characterization and preferences of the visitors in three areas of a recreation forest: Results of a survey in the forest of Allschwil. *Schweizerische Zeitschrift Forstwesen.* **2003**, *154*, 397–404.
25. The Parliament of the Czech Republic. *Forest Act and Amendments to Certain Acts (Forest Act)*; Act No. 289/1995 Coll; on Forests and on Amendments to Certain Acts (Forest Act); The Parliament of the Czech Republic: Prag, Czech Republic, 1995.
26. Hrůza, P. *Making the Forest Accessible and Its Complex Concept*; Mendel University in Brno: Brno, Czech Republic, 2014; 123p.
27. Hrůza, P. Technologies for the reinforcement of forest timber hauling roads and their recreational suitability. In Proceedings of the Public Recreation and Landscape Protection—With Man Hand in Hand, Křtiny, Czech Republic, 1–3 May 2016; Mendel University in Brno: Brno, Czech Republic, 2016; pp. 238–241.
28. Wilkes-Allemann, J.; Putz, M.; Hirschi, C.; Fischer, C. Conflict situations and response strategies in urban forests in Switzerland. *Scand. J. For. Res.* **2015**, *30*, 204–216. [CrossRef]
29. Newsome, D.; Stender, K.; Annear, R.; Smith, A. Park management response to mountain bike trail demand in South Western Australia. *J. Outdoor Recreat. Tour.* **2016**, *15*, 26–34. [CrossRef]
30. Mann, C.; Absher, J.D. Recreation conflict potential and management implications in the northern/central Black Forest Nature Park. *J. Environ. Plan. Manag.* **2008**, *513*, 363–380. [CrossRef]
31. Ballantyne, M.; Gudes, O.; Pickering, C.M. Recreational trails are an important cause of fragmentation in endangered urban forests: A case-study from Australia. *Landsc. Urban Plan.* **2014**, *130*, 112–124. [CrossRef]

32. Ballantyne, M.; Pickering, C.M. Differences in the impacts of formal and informal recreational trails on urban forest loss and tree structure. *J. Environ. Manag.* **2015**, *159*, 94–105. [CrossRef] [PubMed]
33. Santos, T.; Mendes, R.N.; Vasco, A. Recreational activities in urban parks: Spatial interactions among users. *J. Outdoor Recreat. Tour. Res. Plan. Manag.* **2016**, *15*, 1–9. [CrossRef]

forests MDPI

Article

Exploring the Outdoor Recreational Behavior and New Environmental Paradigm among Urban Forest Visitors in Korea, Taiwan and Indonesia

Doyeon Kim [1], Ricky Avenzora [2] and Ju-hyoung Lee [3,*]

1 Korea Arboreta and Gardens Institute, Sejong-si 30129, Korea; todoyeon@koagi.or.kr
2 Department of Forest Resources Conservation and Ecotourism, Faculty of Forestry, Bogor Agricultural University, IPB Dramaga Campus, Bogor 16680, Indonesia; rickyav@apps.ipb.ac.id
3 Department of Forest Resources, Yeungnam University, 280 Daehak-ro, Gyeongsan 38541, Korea
* Correspondence: jhlee9@yu.ac.kr; Tel.: +82-53-810-2927

Abstract: This is international comparative research on the perception of local residents toward the natural environment in South Korea, Taiwan, and Indonesia. Through the New Ecological Paradigm (NEP) investigation, perceptions of natural environmental conservation and utilization of 664 urban forest visitors were analyzed, and the relationship between recreational behavior, NEP scores, and demographic characteristics was investigated. The three countries, with different histories, cultures, and economic development, showed statistically significant differences in all items. In terms of the NEP response score, Taiwan showed the most positive results with an average of 4.08. Frequent visits by the elderly and family were common significant factors of high NEP score for all survey locations. In the confirmatory factor analysis of latent variables for NEP, 'limits to growth' were significant in South Korea while 'ecological crisis' was more significant in Taiwan and Indonesia. Forest experience frequency was a common factor affecting NEP, indicating that frequent forest visits during leisure time are a major factor in improving the ecological paradigm.

Keywords: NEP; nature experience; Asia; outdoor recreation; urban forest

Citation: Kim, D.; Avenzora, R.; Lee, J.-h. Exploring the Outdoor Recreational Behavior and New Environmental Paradigm among Urban Forest Visitors in Korea, Taiwan and Indonesia. *Forests* **2021**, *12*, 1651. https://doi.org/10.3390/f12121651

Academic Editors: Radu-Daniel Pintilii and Diego Varga

Received: 27 October 2021
Accepted: 25 November 2021
Published: 28 November 2021

Publisher's Note: MDPI stays neutral with regard to jurisdictional claims in published maps and institutional affiliations.

1. Introduction

A number of studies on human psychological and physical health recovery effects in forests have reported that experiences and recreational activities in natural environments have a positive influence on humans [1–6]. Exposure to and recreational experiences in natural environments are directly associated with people's perception in which the correlation between the perception of natural environment and recreational behaviors, such as forest experience frequency, visit duration, and activities, has been empirically studied in previous research [7–10]. In an international comparative study on 741 individuals, experiences in nature were reported to improve environment-friendly behavior and the perception of experiences in nature [9]. Furthermore, a study which suggested the measure for environment-friendly perception also reported that the level of environmental awareness has a valid correlation with behaviors in natural environments such as forest experience frequency [7].

In summary, positive perception of nature experiences and natural environment are closely related to environment-friendly behaviors including outdoor recreational activities [10–18].

The dominant social paradigm (DSP), which shows rapid economic growth and high resource requirements after the industrial revolution, anthropocentrically evaluates the ecological environment, and regards nature as a commodity to be developed [19–22]. The development of modern civilization and cities, based on the growth of scientific techniques, led to the blind faith that technical progress could unconditionally help human society. It was also criticized as a technofix (technological fix) [23–25].

Catton and Dunlap (1978) explained people's perception of the natural environment with the 'Environmental Worldview' concept that people's attitude toward the natural environment is changing from a DSP to a 'New Ecological Paradigm (NEP)' [26]. Furthermore, Dunlap and Van Liere (1978) claimed that the previous economic-centered and development-oriented conquer paradigm regarding the natural environment transitioned to an environment-centered NEP [27].

The transition from a DSP to an NEP, starting from the late 20th century, shows a shift from the idea of human superiority to the recognition of the value of nature itself and from acknowledging the inevitability of environmental pollution for economic profit to emphasize the importance of maintaining the health of ecosystems [28–31]. The fundamental difference between these two paradigms is whether humans are viewed separately from or as part of nature [32]. Comparative research on the perception of the natural environment in industrialized developed countries and developing countries has established that countries with a low DSP are highly interested in the environment and have relatively high motivation to resolve environmental issues, while countries with high DSP have lower personal interest in the natural environment due to economic and technical progress [19,20].

After World War II, South Korea achieved the highest rapid economic development among underdeveloped countries [33]. A distinctive phenomenon of Korean society in its rapid economic development can be explained by turbo-capitalism theory [34–36], which is characterized by dichotomous thinking, an unconditional preference for speed, nonreflective judgment, and the sacrifice of the few for the profit of the many. Such a goal-directed rapid development paradigm of Korean society led to insufficient leisure time, excessive labor time, and extremely insufficient time to experience nature [29]. These characteristics were considered severance with nature and hatred toward natural ecology [37–41].

To investigate the correlation between the perception of countries with different GDPs and levels of industrial development toward the natural environment, technical progress and the level of capitalist development were based on the conventional view of nature. Overcoming material poverty is the top priority of developing countries worldwide; therefore, the focus is usually on industrial techniques for economic development, despite environmental pollution, rather than techniques to protect the natural environment [42–44]. A study observed a U-shaped relationship between income level and environmental perception, which can be explained by the difference in economic levels and is similar to the hypothesis of the Environmental Kuznets Curve (EKC) [45]. Developing countries in Asia are implementing growth-oriented development policies; the resulting rapid industrialization and urbanization have increased the severity of environmental problems, urban expansion, and pollution [46–48]. However, it has become an important reference point for establishing the natural environment protection policy of Asian countries that have experienced rapid industrial and economic development. Approximately 76% (414 million hectare) of Asian forests are located in the southeast region, and the decrease in the forest area of this region accounts for 25% of the global total decrease [49]. Since Asia had the highest net gain of forest area in 2010–2020, by 2.4 million hectares per year [50], the perceived status of Asian people in southeast area toward natural environmental protection policies is drawing more attention. Milbrath (1989) emphasized the transition of perception toward the NEP, which positively impacts the natural environment and formation of nature-friendly values, from the previous human-centered DSP for sustainable development [51]. Because such a paradigm shift results in daily environmentally friendly practices, it has been suggested as the ultimate solution for sustainable development [42–59].

Meanwhile, previous international comparative research on this topic has revealed the relationship between the difference in the natural environment, historical culture, level of economic development, and perception of the natural environment. Compared to Brazil and Mexico, the most dichotomous view of the human exception paradigm was observed in the United States [60]. In a study targeting university students in Mexico, Nicaragua, Peru, Spain, and the United States, it was found that environmentally friendly perceptions are reflected in daily practical behaviors [61]. Another similar study result, by Vicente-Molina et al. (2013), surveyed three different economic levels, including pro-environmental behavior and environmental knowledge, of university students in the United States, Mexico, Spain, and Brazil [62]. In a large-scale international comparative study targeting university students from 14 countries in Europe and South America, including the United States, a correlation between positive perception toward the natural environment and nature-friendly attitude was found [63].

For sustainable technical development and technology transfer of the Asian Forest Cooperation Organization (AFoCo) and Official Development Assistance (ODA), the extent of the natural environmental perception of residents in the corresponding society needs to be proactively considered. However, research on natural environmental perception in Asia is insufficient and has only provided simple comparisons until today [64,65]. Consequently, an augmentation strategy for nature-friendly environmental perception based on the level of perception in different Asian regions is required to decrease deforestation and positive responses to climate change. Even though the same Asian countries, the perception of nature and forest conservation might differ by NEP categories and countries, and there is a particular influence factor, which is the hypothesis of this study. Therefore, this study aimed to compare and analyze the characteristic differences of natural environmental perception and investigate popular ecological paradigm facets through a field survey study on the perception and natural recreational behavior in different Asian countries.

2. Materials and Methods

2.1. Survey Procedure

South Korea, Taiwan, and Indonesia, where the forest area of national land comprises >50% of forests, and urban forests located close to Seoul in South Korea, Taipei in Taiwan, Jakarta in Indonesia were selected as the survey locations (Table 1 and Figure 1). The survey was conducted by researchers each country in the morning and afternoon on weekends when there was a high traffic of visitors; the trained researchers were supported from Yeungnam University in Korea, Dong Hwa University in Taiwan, and Bogor Agricultural University in Indonesia.

Table 1. Description of the surveyed urban forest areas and survey schedule.

Survey Location	Areas	Survey Date	Area (m^2)	Dominant Vegetation
Seoul forest	Seoul, Korea	2016 June 2017 June 2018 August	480,994	*Cercidiphyllum japonicum, Pinus parviflora, Pinus densiflora*
Bukhan mountain forest			76,922,000	*Forsythia saxatilis*
Da'an urban forest	Taipei, Taiwan	2016 August 2017 May	258,940	*Magnolia kobus, Gardenia jasminoides*
Taman Menteng forest	Jakarta, Indonesia	2016 September 2018 August	24,546	*Pithecellobium dulce, Ficus lyrata Warb., Bauhinia blakeana*
Taman Suropati forest			16,328	*Swietenia mahagoni, Terminalia catappa*
Hutan Kota Srengseng city forest			150,000	*Agathis Dammara, Aleurites moluccanus*

A structured questionnaire covering general demographic characteristics, visiting behavior (frequency, length of stay, transportation means, companions, and visit motivation), and NEP scale was developed for the survey (Table 2). Only those who voluntarily consented to participate took the survey.

Figure 1. The survey locations.

Table 2. Interview questionnaire.

Categories	Question Subjects
Characteristics	Age, gender, education, marital status
Visiting behavior	Frequency, length of stay, motivation, companions
New Ecological Paradigm (NEP)	
Limits to growth	We are approaching the limit of the number of people Earth can support. Earth has plenty of natural resources if we just learn how to develop them. Earth is like a spaceship with very limited room and resources.
Anti-anthropocentrism	Humans have the right to modify the natural environment to suit their needs. Plants and animals have as much right as humans to exist. Humans were meant to dominate the rest of nature.
Balance of nature	When humans interfere with nature it often produces disastrous consequences. The balance of nature is strong enough to cope with the impacts of modern industrial nations. The balance of nature is very delicate and easily upset.
Anti-exceptionalism	Human ingenuity will ensure that we do not make Earth unliveable. Despite our special abilities, humans are still subject to the laws of nature. Humans will eventually learn enough about how nature works to be able to control it.
Eco-crisis	Humans are severely abusing the environment. The so-called 'ecological crisis' facing humankind has been greatly exaggerated. If things continue in their present course, we will soon experience a major ecological catastrophe.

NEP questions were translated into Korean, Chinese, and Indonesian, and survey results collected from a total of 664 respondents were analyzed. The sample comprised 169 people (84 males and 75 females) from South Korea, 217 (73 males and 144 females) from Taiwan, and 278 (156 males and 12 females) from Indonesia.

The survey was conducted between 2016 and 2018. The data acquisition took approximately three years due to limiting conditions such as the general survey procedure of data acquisition, analysis, errors found, and resurvey; natural weather conditions, including dry and rainy seasons; religious and cultural characteristics such as Ramadan; and difficulties such as communication and time differences in performing international research.

To select the survey locations, we consulted research teams from Yeungnam University in Korea, Dong Hwa University in Taiwan, and Bogor Agricultural University in Indonesia. The selection criteria included a total size of >1 ha, located at the center of the capital with high accessibility and diverse vegetation. The field survey was conducted on Saturdays and Sundays.

2.2. Characteristics of Six Surveyed Urban Forests in Seoul, Taipei and Jakarta

Seoul Forest, Seoul; this is a large-scale urban forest located on the Han River at the center of Seoul, Korea. After transforming a golf course and horse-riding course into nature-friendly civil recreational areas in 2005, this place attracted an average of 0.25 million visitors per day (7–8 million visitors per year) [66]. Ecological experience education programs are provided in the five themes of cultural art, ecological forest, field study, wetland ecology, and waterside areas. There were 415,795 trees of 95 different species in this forest (Table 1). Bukhan Mountain, Seoul; located in the north of downtown Seoul, this functions as a wild natural environmental downtown forest with high accessibility. This location contains more than 1300 species of flora and fauna. There are also many historical and cultural sites, including the Bukhansanseong Fortress with over 2000 years of history and over 100 Buddhist temples [67]. The ecological education center of the Ministry of Environment is located here, providing ecological education to citizens. It is a representative forest recreational area with 5.5 million annual visitors and 71.8 km of walking and hiking trails.

Da'an urban forest, Taipei; this forest is located to the south of downtown Taipei and known as 'Taipei's lung'; approximately 10,000 to 30,000 people visit daily. Natural recreational experiences, such as flower exhibitions, are provided, as well as lakes, playgrounds, and walking trails. The forest consists of 60 different forest tree types [68,69].

Taman Menteng, Jakarta; located at the center of Jakarta, this was a soccer stadium transformed into an urban forest. Here, opportunities for the promotion of health and nature experiences are provided through several sports facilities, greenhouses, lawn plazas, and bicycle and horse-riding trails, and approximately 2200 people visit this urban forest every day (0.8 million visitors per year) [70]. Taman Suropati, Jakarta; in 1920, the Burgemeester Bisschopplein park was made public. It now functions as a recreational leisure space as Taman Suropati in Jakarta [71]. Hutan Kota Srengseng, Jakarta; this forest was created in 1995 and is located in the west of Jakarta. There are various inhabits and high tree density with 2570 trees per ha [72].

2.3. Survey Instrument and Statistical Analysis

The NEP scale test, revised by Dunlap et al. (2000), was used to investigate the level of natural environmental perception [11]. A total of 15 questions were classified under five categories: 'limits to growth', 'anti-anthropocentrism', 'balance of nature', 'anti-exceptionalism', and 'ecological crisis' [11]. A five-point scale was used as the response method. The even-numbered questions were composed for reverse scoring, securing an improved balance compared with the existing NEP test. Further, visitor demographic information—that is, gender, age, educational level, and marital status, along with information on the type of visit, such as the motivation to visit, frequency of visit, type of company, and length of visit—were investigated (Table 2).

The collected data were subjected to variance analysis using the statistic software SPSS 25 (Statistical Package for the Social Sciences ver.25 by IBM SPSS Statistics) to determine the difference between countries and NEP categories, and AMOS 25 (Analysis of Moment Structure ver.25 by IBM SPSS AMOS) was used to confirm the validity of the CFA (Confirmatory Factor Analysis) model. The reliability verification of the NEP scale for the investigation of natural environmental perception was performed. The internal reliability of each survey location was verified using Cronbach's alpha, and the Cronbach's alphas were 0.798, 0.811, and 0.760 for Seoul, Taipei, and Jakarta, respectively, suggesting its suitability for this research.

3. Results

3.1. Visiting Behavior Comparison and NEP Scores in Three Countries

The demographic and behavioral characteristics of the respondents are shown in Table 3. In terms of visiting behavior, families visiting together were the most frequent in South Korea (36.5%) and Taiwan (53%), whereas individual visits were the most frequent in Indonesia (49.0%). The motivations for visits were outdoor activities (46.8%), such as jogging and walking in South Korea, and experiences in nature such as rest and recharge, relaxation, and breathing fresh air in Taiwan (50.0%) and Indonesia (70.3%). The visiting frequency is closely related to the nature experience volume; thus, it is an important factor in natural environmental perception [41]. The most common frequency of visits was once per month, once per week, and 3–5 times per year in South Korea (42.9%), Taiwan (47.8%), and Indonesia (56.7%), respectively. This indicates that the nature experience volume of the Taiwanese respondents was the highest and that of the Indonesian respondents was the lowest. In terms of visit time, short visits lasting less than 1 h were the most frequent in South Korea (42.1%) and Taiwan (40.1%), and visits lasting less than 2 h were the most frequent in Indonesia (46.7%).

Based on the visiting frequency and time, it was concluded that people visit more frequently and stay for a shorter time in South Korea and Taiwan, while in Indonesia, the frequency of nature experience is low, but visitors stay for a longer time.

Different results in terms of the perception of the natural environment in three countries with different histories, cultures, and natural environments were expected (Table 4).

Table 3. Characteristics and visiting behavior of respondents in three countries (%).

Variables			Korea (n = 169)	Taiwan (n = 217)	Indonesia (n = 278)
Characteristics	Gender	Male	58.7	33.5	56.0
		Female	41.3	66.8	44.0
	Age	20s	9.5	39.7	6.0
		30s	21.4	26.8	39.7
		40s	30.2	17.2	31.7
		50s	29.4	9.5	15.7
		60s and over	9.5	6.9	7.0
	Education	High school or below	11.1	20.3	50.0
		College or above	88.9	79.7	50.0
	Marital status	Single	25.4	54.3	9.3
		Married or living together	70.6	44.8	83.0
		Separated or divorced	1.6	0.9	4.7
		Widowed	2.4	0.9	3.0
Behavior	Companions	Alone	11.9	12.1	49.0
		With family	36.5	53.0	36.0
		With neighbor, relative	15.9	1.7	-
		With friend, colleague	35.7	33.2	15.0
	Motivation	Outdoor activity	46.8	31.5	10.3
		Nature experience	33.3	50.0	70.3
		For the children	9.5	6.9	19.3
		Others	10.3	11.6	-
	Frequency	>1/week	24.6	47.8	14.7
		1/month	42.9	38.8	11.3
		3–5/a	20.6	7.8	56.7
		<1/year	11.9	5.6	17.3
	Visit duration	<1 h	42.1	40.1	33.3
		>2 h	37.3	38.4	46.7
		<4 h	20.6	21.6	20.0

Table 4. Comparison of the average and variance analysis for NEP scores in three countries.

Categories	Korea	Taiwan	Indonesia	*F*-Value
Limits to growth	3.51	4.00	3.34	112.727 ***
Anti-anthropocentrism	3.79	4.20	2.21	990.216 ***
Balance of Nature	3.66	4.06	3.65	43.120 ***
Anti-exceptionalism	3.11	3.98	3.01	257.784 ***
Eco-crisis	3.87	4.16	3.33	175.015 ***
F-value	112.238 ***	8.951 ***	80.912 ***	
Average	3.59	4.08	3.11	505.686 ***

Notes: *** $p < 0.001$.

3.2. Analysis of the NEP in Korea

The NEP results according to the demographic and behavioral characteristics of South Korea are shown in Table 5. Thus far, several studies have examined the relationship between the level of education and natural environmental perception [73–76]; in this study, the positive perception of highly educated respondents was verified in four NEP categories ($p < 0.01$). No statistically significant response with respect to gender, age, or marital status was found. The difference based on the type of company was verified in two NEP categories, and respondents visiting with family (3.73) showed the most positive perception of the natural environment in the total average NEP score ($p < 0.05$).

Table 5. Variance analysis for NEP scores with characteristics and behavior in Korea.

	Category	Limits to Growth	Anti-Anthropocentrism	Balance of Nature	Anti-Exceptionalism	Eco-Crisis	Average
Education	High school	3.19	3.29	3.33	2.74	3.50	3.21
	College	3.54	3.85	3.70	3.15	3.92	3.63
	t-value	−2.854 **	−2.798 **	−1.836	−3.375 **	−2.345 *	−4.059 **
	(*p*)	(0.008)	(0.006)	(0.069)	(0.002)	(0.021)	(0.001)
Companions	Alone	3.67	3.87	3.58	2.73	3.82	3.53
	Family	3.64	3.90	3.76	3.26	4.10	3.73
	Neighbor	3.02	3.47	3.32	2.87	3.62	3.26
	Friend	3.51	3.84	3.68	3.23	3.78	3.61
	F-value	3.245 *	1.423	1.790	2.380	2.817 *	2.924 *
	(*p*)	(0.014)	(0.230)	(0.135)	(0.055)	(0.028)	(0.024)
Motivation	Activity	3.51	3.70	3.59	2.99	3.93	3.54
	Nature	3.49	3.87	3.70	3.18	3.64	3.58
	Children	3.56	3.67	3.89	2.97	4.06	3.63
	Others	3.49	4.03	3.62	3.51	4.18	3.76
	F-value	0.028	1.036	0.642	2.205	3.483 *	0.616
	(*p*)	(0.994)	(0.379)	(0.589)	(0.091)	(0.018)	(0.606)
Visit frequency	1/week <	3.77	4.20	3.85	3.31	4.02	3.83
	1/month	3.51	3.78	3.77	3.06	4.04	3.63
	3–5/year	3.32	3.55	3.47	2.97	3.65	3.39
	1/year >	3.27	3.38	3.18	3.09	3.33	3.25
	F-value	2.763 *	6.601 *	4.415 **	1.198	7.310 ***	5.941 **
	(*p*)	(0.045)	(0.000)	(0.006)	(0.313)	(0.000)	(0.001)

Notes: * $p < 0.05$, ** $p < 0.01$, *** $p < 0.001$; High school: high school or below; College: college or above; Activity: outdoor activity, nature: nature experience.

In the ecological crisis category of the motivation to visit, other (4.18) and time spent with children (4.06) were the highest, and no significant difference was found in the other four NEP categories. In terms of visiting frequency, the statistical significance was verified in four NEP categories, excluding anti-exceptionalism. Respondents who visited once per week gave the most positive response (3.83, $p < 0.01$), which corresponded to previous research on the frequency of experiences in nature and perceptions of the natural environment [41,77–81].

Thus, based on the investigation results, level of education, type of company, motivation to visit, and frequency of visits were found to be statistically significant factors affecting the level of natural environmental perception in South Korea.

3.3. Analysis of the NEP in Taiwan

The NEP results, depending on the demographic and behavioral characteristics of the survey participants in Taiwan, are presented in Table 6. There was no significant difference in NEP scores depending on age, motivation to visit, or length of stay. The NEP scores with age and level of education were statistically significant only in the eco-crisis category, and females (4.21) and university students (4.20) showed a higher NEP scores than males (4.05) and high school students (3.98), respectively ($p < 0.05$).

Married (4.17) and single (4.05) respondents had a positive natural environmental perception, presenting obvious contrast with bereaved respondents (3.37) ($p < 0.05$). Marital status generally affects human quality of life [82–84]. Based on the analysis results of Taiwanese respondents, it was concluded that marital status is significantly positively correlated with the perception of the natural environment.

A significant difference was found in the limits to the growth category of the company type. Visiting alone was the most common (4.13), followed by visiting friends (4.00), family (3.99), and neighbors and groups (3.33) ($p < 0.05$). Partial correlation with the type of company was identified, and individual visitors were found to sensitively recognize the limitations of the growth of mankind in the global environment.

Table 6. Variance analysis for NEP scores with characteristics and behavior in Taiwan.

	Category	Limits to Growth	Anti-Anthropocentrism	Balance of Nature	Anti-Exceptionalism	Eco-Crisis	Average
Gender	Male	4.03	4.15	4.00	3.91	4.05	4.03
	Female	3.98	4.23	4.09	4.01	4.21	4.10
	t-value	0.689	−1.269	−1.151	−1.622	−2.083 *	−1.445
	(*p*)	(0.491)	(0.206)	(0.251)	(0.106)	(0.038)	(0.150)
Education	High school	3.94	4.09	4.01	3.96	3.98	4.00
	College	4.01	4.23	4.07	3.98	4.20	4.10
	t-value	−0.739	−1.865	−0.677	−0.266	−2.543 *	−1.679
	(*p*)	(0.463)	(0.063)	(0.499)	(0.790)	(0.012)	(0.095)
Marital status	Single	3.98	4.20	4.01	3.95	4.14	4.05
	Married	4.11	4.11	4.17	4.17	4.28	4.17
	Divorced	3.33	4.17	4.17	3.83	4.17	3.93
	Widowed	3.00	3.83	3.33	3.50	3.17	3.37
	F-value	3.508 **	0.443	1.889	1.274	1.920	2.598 *
	(*p*)	(0.008)	(0.777)	(0.113)	(0.281)	(0.108)	(0.037)
Companions	Alone	4.13	4.21	4.05	4.11	4.20	4.14
	Family	3.99	4.19	4.12	4.00	4.18	4.09
	Neighbor	3.33	4.08	3.83	3.83	3.83	3.78
	Friend	4.00	4.22	3.98	3.91	4.12	4.05
	F-value	3.208 *	0.190	1.466	1.683	0.713	1.391
	(*p*)	(0.024)	(0.903)	(0.225)	(0.172)	(0.545)	(0.246)
Frequency	1/week <	4.15	4.26	4.30	4.26	4.41	4.27
	1/month	3.87	4.13	3.96	3.87	4.01	3.97
	3–5/year	4.02	4.28	4.09	4.15	4.23	4.16
	1/year >	4.08	4.03	4.18	4.01	4.23	4.11
	F-value	3.611 *	2.259	2.639	4.634 **	4.275 **	4.848 **
	(*p*)	(0.014)	(0.082)	(0.050)	(0.004)	(0.006)	(0.003)

Notes: * $p < 0.05$, ** $p < 0.01$; High school: high school or below; College: college or above; Marial status: married or living together, divorced, or separated.

In terms of frequency of visits, the limits to growth ($p < 0.05$), anti-exceptionalism ($p < 0.01$), and eco-crisis categories ($p < 0.01$) were all statistically significant, and a more positive perception appeared with a higher frequency of visits. Similar to the case of South Korea, the frequency of visits was deduced to be a major influential factor for the NEP (Table 4) in Taiwan, which is in line with previous studies on the frequency of visits and preference for the natural environment [41,77–81]. However, those who visited once a month had a lower perception in all categories compared to groups with a higher frequency of visits, showing different patterns compared to extant research results.

Thus, the statistical significance between NEP scores and factors such as gender, level of education, marital status, type of company, and frequency of visits in Taiwan was not verified completely, but by categories partly.

3.4. Analysis of the NEP in Indonesia

The NEP results according to the demographic and behavioral characteristics of Indonesia are presented in Table 7. According to the age-group comparison, the total average natural environmental perception of the respondents in their 50s (3.15) and over 60s (3.15) was the most positive ($p < 0.05$), and those in their 50s in the balance of nature category (3.85, $p < 0.05$), in their 20s and 60s in the anti-exceptionalism category (3.19, $p < 0.05$), and in their 40s in the eco-crisis category (3.49, $p < 0.05$) displayed a high NEP scores. These results are in good agreement with previous research results, which reported more positive perceptions of the natural environment and nature-based leisure with increasing age [85–90].

In the analysis of different levels of education, high school graduate respondents had a more positive natural environmental perception in the balance of nature (3.71, $p < 0.05$) and anti-exceptionalism (3.06, $p < 0.05$) categories compared to the university graduate respondents, in contrast with the results from South Korea and Taiwan. The positive groups for the NEP categories differed in terms of marital status, displaying varied results compared to South Korea and Taiwan; hence, it was difficult to determine a pattern for interpretation.

Table 7. Variance analysis for NEP scores by characteristics and behavior in Indonesia.

	Category	Limits to Growth	Anti-Anthropocentrism	Balance of Nature	Anti-Exceptionalism	Eco-Crisis	Average
Age	20s	3.11	2.31	3.70	3.19	3.19	3.10
	30s	3.34	2.14	3.63	2.96	3.24	3.06
	40s	3.35	2.26	3.56	3.04	3.49	3.14
	50s	3.35	2.30	3.85	2.91	3.31	3.15
	60s over	3.41	2.13	3.67	3.19	3.33	3.15
	F-value	1.576	1.532	3.334 *	2.613 *	5.813 ***	2.824 *
	(*p*)	(0.181)	(0.193)	(0.011)	(0.036)	(0.000)	(0.025)
Education	High school	3.32	2.20	3.71	3.06	3.30	3.12
	College	3.36	2.22	3.59	2.95	3.37	3.10
	t-value	−0.743	−0.191	2.398 *	2.220 *	−1.467	1.047
	(*p*)	(0.458)	(0.849)	(0.017)	(0.027)	(0.143)	(0.296)
Marital status	Single	3.38	2.08	3.73	3.11	3.21	3.09
	Married	3.30	2.29	3.65	3.08	3.35	3.15
	Divorced	3.72	2.28	3.11	2.67	3.33	3.02
	Widowed	2.96	2.07	3.74	2.85	3.44	3.01
	F-value	4.015 **	1.241	4.680 ***	2.379 *	0.853	1.504
	(*p*)	(0.002)	(0.290)	(0.000)	(0.039)	(0.513)	(0.189)
Companions	Alone	3.33	2.20	3.58	2.96	3.32	3.08
	Family	3.31	2.24	3.71	3.03	3.40	3.14
	Friend	3.45	2.16	3.74	3.11	3.21	3.13
	F-value	2.102	0.526	3.414 *	2.382	3.025	3.172 *
	(*p*)	(0.124)	(0.591)	(0.034)	(0.094)	(0.050)	(0.043)
Motivation	Activity	3.18	2.24	3.56	3.12	3.34	3.09
	Nature	3.37	2.21	3.69	2.96	3.31	3.11
	Children	3.29	2.20	3.55	3.14	3.41	3.12
	F-value	3.480 *	0.067	2.989	5.009 **	1.410	0.195
	(*p*)	(0.032)	(0.935)	(0.052)	(0.007)	(0.246)	(0.823)
Frequency	>1/week	3.26	2.21	3.77	3.14	3.27	3.13
	1/month	3.23	2.28	3.62	3.18	3.40	3.14
	3–5/a	3.37	2.18	3.67	2.99	3.35	3.11
	<1/a	3.37	2.25	3.51	2.85	3.29	3.05
	F-value	1.874	0.527	2.639	5.463 **	0.810	1.740
	(*p*)	(0.134)	(0.664)	(0.050)	(0.001)	(0.489)	(0.159)
Visit duration	in 1 h	3.35	2.20	3.74	3.00	3.30	3.12
	in 2 h	3.34	2.23	3.62	3.01	3.32	3.10
	over 4 h	3.32	2.17	3.58	3.00	3.42	3.10
	F-value	0.065	0.381	3.047 *	0.029	1.554	0.250
	(*p*)	(0.937)	(0.684)	(0.049)	(0.971)	(0.213)	(0.779)

Notes: * $p < 0.05$, ** $p < 0.01$, *** $p < 0.001$; High school: high school or below; College: college or above, marriage status: married or living together, divorced, or separated, activity: outdoor activity, nature: nature experience.

In terms of comparison with different types of companies, those who visited with family (3.14, $p < 0.05$) and friends (3.13, $p < 0.05$) had a more positive perception of the natural environment than those who made 'individual' visits, and these results were also obtained for the balance of nature category. In terms of motivation to visit, the group visiting for a nature experience showed a positive response in the limits to growth ($p < 0.05$) and the balance of nature ($p = 0.052$) categories, and the group visiting to spend time with their children (3.14, $p < 0.01$) showed a high value in the anti-exceptionalism category. Although there was no statistical significance ($p = 0.503$), these results are similar to the NEP results in Taiwan, where nature experience (4.18) and children (4.09) were the main motivations for visiting.

Similar to South Korea and Taiwan, which displayed high NEP scores with an increasing frequency of visits, statistical significance was verified in the anti-exceptionalism ($p < 0.01$) and balance of nature ($p = 0.050$) categories. In terms of the length of visit, only the balance of nature category was statistically significant ($p < 0.05$), and the NEP score of respondents who stayed for less than 1 h (3.74) was high.

Thus, the factors affecting the level of natural environmental perception of the participants from Indonesia were age, level of education, marital status, type of company, motivation to visit, frequency of visit, and length of visit.

3.5. Confirmatory Factor Analysis of Differences in Three Countries

To verify the validity of the NEP measurement variables, a CFA was individually conducted for South Korea, Taiwan, and Indonesia (Table 8). The final model consisted of a total of 15 observed variables, three from each of the five latent variables, that is, limits to growth, anti-anthropocentrism, balance of nature, anti-exceptionalism, and eco-crisis, according to the NEP question structure. To verify the discriminant validity, construct reliability (CR) and average variance extracted (AVE) were calculated by the formula of Hair et al. [91]. Table 9 summarizes the suitability of the models for each country.

$$CR = \frac{(\sum standardized\ factor)^2}{(\sum standardized\ factor)^2 + (\sum Error\ variance)}$$

$$AVE = \frac{\sum standardized\ factor^2}{Number\ of\ items}$$

Table 8. CFA model regression coefficient and model reliability for NEP scores of survey locations (Korea/Taiwan/Indonesia).

Category		NEP Number	Korea	Taiwan	Indonesia
Limits to growth	ß-coeff.	1	0.820	0.569	0.601
		6	0.829	0.480	0.773
		11	0.387	0.699	0.430
		CR	0.895	0.796	0.815
		AVE	0.757	0.671	0.609
Anti-anthropo-centrism	ß-coeff.	2	0.670	0.673	0.605
		7	0.638	0.660	0.445
		12	0.692	0.587	0.588
		CR	0.864	0.850	0.767
		AVE	0.680	0.655	0.527
Balance of nature	ß-coeff.	3	0.563	0.821	0.737
		8	0.687	0.589	0.581
		13	0.402	0.641	0.453
		CR	0.821	0.873	0.811
		AVE	0.616	0.701	0.598
Anti-exceptionalism	ß-coeff.	4	0.626	0.632	0.655
		9	0.678	0.465	0.518
		14	0.511	0.735	0.795
		CR	0.834	0.823	0.852
		AVE	0.630	0.615	0.665
Eco-crisis	ß-coeff.	5	0.758	0.848	0.442
		10	0.476	0.640	0.749
		15	0.658	0.798	0.735
		C.R	0.834	0.920	0.824
		AVE	0.635	0.795	0.621

ß-coeff.: Standardized coefficient, CR: Construct reliability, AVE: Average variance extracted.

Table 9. CFA model suitability for NEP scores of survey locations (Korea/Taiwan/Indonesia).

	Chi-Square	d.f.	IFI Dlta^2	TLI rho^2	CFI	RMSEA
Korea	119.002 **	80	0.937	0.913	0.934	0.054
Taiwan	103.735 *	80	0.970	0.960	0.969	0.037
Indonesia	136.922 ***	80	0.915	0.880	0.911	0.052

Notes: * $p < 0.05$, ** $p < 0.01$, *** $p < 0.001$.

The root mean square error of approximation (RMSEA), which is the interpretation criterion for suitability, was <0.6, and thus satisfactory for all three models [92]. The IFI (Incremental Fit Index), TLI (Tucker–Lewis index), and CFI (Comparative Fit Index) were almost >0.9 in all cases, except for Indonesia's TLI at 0.880, confirming good suitability [93].

Based on the standardized lambda analysis results for each location, the limits to growth of South Korea and the eco-crisis of Taiwan and Indonesia were relatively high (Table 8). South Korea showed loaded values of 0.829 to 0.387, and the factor loaded values for Questions 1 and 6 in limits to growth and Question 5 in eco-crisis were high. Taiwan showed loaded values of 0.848 to 0.465, and the factor loaded values for Question 3 in balance of nature, Question 14 in anti-exceptionalism, and Questions 5 and 15 in eco-crisis were high. Indonesia displayed loaded values of 0.795 to 0.430 and high factor loading values for Question 6 in limits to growth, Question 3 in balance of nature, Question 14 in anti-exceptionalism, and Questions 10 and 15 in eco-crisis. None of the survey locations showed a major explanation for anti-anthropocentrism.

It is difficult to deduce a model that is applicable to three different Asian countries, and in our models, the factor with a loaded value <0.5, appeared in all survey locations. Nevertheless, the models presented an AVE value >0.5, and CR >0.7, satisfying the internal consistency and convergent validity (Table 9).

4. Discussion

4.1. International Comparative Study in Asian Countries

Based on the research results on the relationship between nature recreational behavioral patterns of residents, New Environmental Paradigm, and the demographic characteristics, the NEP scores of South Korea, Taiwan, and Indonesia showed a statistically significant difference in all categories. Previous studies identified gender, age, educational level, and economic power as the factors affecting perception toward the natural environment [94,95], and the current study also obtained the identical results. Although they are all Asian communities, different results were obtained because of their different histories, cultures, and natural environments [60,61].

The difference in NEP response scores can be explained by the difference in economic level, and similar to the hypothesis of Environmental Kuznets Curve [45], the relationship between economic level and environmental perception was observed. Indonesia is economically developing, and citizens' perceptions of the natural environment are not positive when compared with South Korea and Taiwan. However, Taiwan, which achieved rapid economic growth earlier than South Korea, showed the highest NEP scores. 'Limits to growth' in Korea and 'eco-crisis' in Taiwan and Indonesia were analyzed with a high standardized coefficient in the CFA model.

The level of education, type of company, and frequency of visits were the most common factors affecting NEP in all three countries. A higher NEP score was observed with a higher frequency of visits and when respondents were accompanied by family and friends. The South Korean and Taiwanese participants presented a higher perception of environmental conservation with a higher level of education (see Table 3), similar to the results of previous studies [73–76], while the opposite results were observed in Indonesia.

4.2. Critical Elements: Having Opportunity for Nature Experience Influencing NEP

Many previous studies have thus far reported that a higher frequency of visits to a green area leads to an increase in nature experience, positive natural environmental perception, and nature-friendly behavior [41,77–81,96]. This is similar to Wade and Swanston's (2013) claim that experience influences perception, which refers to the correlation between the frequency, or forest experience, and the perception of the natural environment [97].

The frequency of visits in Indonesia, where leisure time was insufficient due to economic development, and its NEP score was the lowest. This is in line with a previous study that revealed the relationship between the difference in forest experience opportunities by GDP and perception of the natural environment [98].

The GDP per capita based on purchasing power parity was highest in Taiwan at $55,078, followed by Korea at $43,124 and Indonesia at $12,074 [99], which correspond to the order of the three countries for the frequency of visits to urban forests as well as the level of NEP. It can be interpreted as a correlation between the economic level of ordinary citizens, securing opportunities to experience forests (frequency of visits), and perception of the natural environment (NEP).

In South Korea, during the rapid industrial development phase in the 70s, people had excessive working hours and insufficient leisure time for any nature experience [33,41]. The increase in the frequency of forest experience is closely related to the improvement of natural environmental perception and is also in accordance with the augmentation of leisure time and household income through national economic development.

As demonstrated by Bentler and Speckart (1979), frequent visits to green areas would inspire the perception of the natural environment, which is expected to induce nature-friendly behavior [100]. Therefore, a policy that can increase opportunities to experience nature is necessary.

4.3. Study Limitations

There are two major limitations in this study that could be addressed by future research. For this international comparative study, different samples in the six locations in three countries. To compare countries, sampling locations, customs, climate, history, and culture should consider how representative the samples are of the respective populations because these may have influenced the respondents' mindset. However, it was impossible to completely match the survey period and the additional survey by location because it did not sufficiently consider each location's culture, religious event period, and seasonal weather.

However, this study presents a meaningful international comparative study of Asia before the COVID-19 outbreak nevertheless, since the paradigm has changed significantly in the field of natural recreation before and after COVID-19 [101–103].

In addition, these survey locations demonstrate significant differences, preventing the development of a single CFA model to explain all three Asian countries. A follow-up study may consider applying various and detailed indicators of each country's social, economic, leisure, and tourism fields in addition to the factors used in the current study.

5. Conclusions

The purpose of this study was to compare and analyze the differences of environmental paradigms through a field survey on the perception and recreational behavior of visitors in six different urban forests in three other Asian countries. Overall, this study found that an increase in the frequency of forest experiences was closely related to improving positive environmental perception, and this is also in accordance with the increase in leisure time through national economic development. Understanding natural environmental perceptions and socio-economic conditions in South Korea, Taiwan, and Indonesia will lead to easy accessibility and a high frequency of visits to secure sufficient nature experience capacity. Developing an explanatory model that fits the current status of each country is another further research.

Author Contributions: Conceptualization, J.-h.L.; methodology, J.-h.L.; validation, J.-h.L. and R.A.; formal analysis, J.-h.L.; investigation, D.K. and R.A.; resources, D.K. and R.A.; data curation, J.-h.L.; writing—original draft preparation, D.K. and J.-h.L.; writing—review and editing, J.-h.L.; visualization, D.K., R.A. and J.-h.L.; supervision, J.-h.L.; project administration, J.-h.L.; funding acquisition, J.-h.L. All authors have read and agreed to the published version of the manuscript.

Funding: This research received no external funding.

Institutional Review Board Statement: Not applicable.

Informed Consent Statement: Not applicable.

Data Availability Statement: Not applicable.

Acknowledgments: The author would like to thank Chieh-Lu Li and Hung-Hua Chen, National Dong Hwa University, Taiwan, for obtaining the data.

Conflicts of Interest: The authors declare no conflict of interest.

References

1. Shin, W.S.; Yeoun, P.S.; Yoo, R.W.; Shin, C.S. Forest experience and psychological health benefits: The state of the art and future prospect in Korea. *Environ. Health Prev. Med.* **2010**, *15*, 38. [CrossRef]
2. Hartig, T.; Mitchell, R.; de Vries, S.; Frumkin, H. Nature and health. *Annu. Rev. Public Health* **2014**, *35*, 207–228. [CrossRef]
3. Shin, C.S.; Pyoung, S.Y.; Jo, M.N.; Kim, J.Y. Effects of forest healing activity on women's menopausal symptoms and mental health recovery. *J. Korean Soc. People Plants Environ.* **2015**, *18*, 319–325. [CrossRef]
4. Dolling, A.; Nilsson, H.; Lundell, Y. Stress recovery in forest or handicraft environments–An intervention study. *Urban For. Urban Green.* **2017**, *27*, 162–172. [CrossRef]
5. Park, S.; Kim, S.; Kim, G.; Choi, Y.; Kim, E.; Paek, D. Evidence-Based Status of Forest Healing Program in South Korea. *Int. J. Environ. Res. Public Health* **2021**, *18*, 10368. [CrossRef]
6. Shosha, M. Forest Bathing Therapy: The Healing Power of Nature. *Int J. Psychiatr. Res.* **2021**, *4*, 1–2.
7. Nisbet, E.K.; Zelenski, J.M.; Murphy, S.A. The nature relatedness scale: Linking individuals' connection with nature to environmental concern and behavior. *Environ. Behav.* **2009**, *41*, 715–740. [CrossRef]
8. Thapa, B. The mediation effect of outdoor recreation participation on environmental attitude-behavior correspondence. *J. Environ. Educ.* **2010**, *41*, 133–150. [CrossRef]
9. Colléony, A.; White, R.; Shwartz, A. The influence of spending time outside on experience of nature and environmental attitudes. *Landsc. Urban Plan.* **2019**, *187*, 96–104. [CrossRef]
10. Rosa, C.D.; Collado, S. Experiences in nature and environmental attitudes and behaviors: Setting the ground for future research. *Front. Psychol.* **2019**, *10*, 763. [CrossRef]
11. Dunlap, R.E.; Van Liere, K.D.; Mertig, A.G.; Jones, R.E. New trends in measuring environmental attitudes: Measuring endorsement of the new ecological paradigm: A revised NEP scale. *J. Soc. Issues* **2000**, *56*, 425–442. [CrossRef]
12. Bjerke, T.; Thrane, C.; Kleiven, J. Outdoor recreation interests and environmental attitudes in Norway. *Manag. Leis.* **2006**, *11*, 116–128. [CrossRef]
13. Bamberg, S.; Möser, G. Twenty years after Hines, Hungerford, and Tomera: A new meta-analysis of psycho-social determinants of pro-environmental behaviour. *J. Environ. Psychol.* **2007**, *27*, 14–25. [CrossRef]
14. Larson, L.R.; Whiting, J.W.; Green, G.T. Exploring the influence of outdoor recreation participation on pro-environmental behaviour in a demographically diverse population. *Local Environ.* **2011**, *16*, 67–86. [CrossRef]
15. Markle, G.L. Pro-environmental behavior: Does it matter how it's measured? Development and validation of the pro-environmental behavior scale (PEBS). *Hum. Ecol.* **2013**, *41*, 905–914. [CrossRef]
16. Larson, L.R.; Stedman, R.C.; Cooper, C.B.; Decker, D.J. Understanding the multi-dimensional structure of pro-environmental behavior. *J. Environ. Psychol.* **2015**, *43*, 112–124. [CrossRef]
17. Pensini, P.; Horn, E.; Caltabiano, N.J. An exploration of the relationships between adults' childhood and current nature exposure and their mental well-being. *Child. Youth Environ.* **2016**, *26*, 125–147. [CrossRef]
18. Otto, S.; Pensini, P. Nature-based environmental education of children: Environmental knowledge and connectedness to nature, together, are related to ecological behaviour. *Glob. Environ. Chang.* **2017**, *47*, 88–94. [CrossRef]
19. Kilbourne, W.E.; Beckmann, S.C.; Thelen, E. The role of the dominant social paradigm in environmental attitudes: A multinational examination. *J. Bus. Res.* **2002**, *55*, 193–204. [CrossRef]
20. Kilbourne, W.E.; Carlson, L. The dominant social paradigm, consumption, and environmental attitudes: Can macromarketing education help? *J. Macromarketing* **2008**, *28*, 106–121. [CrossRef]
21. Polonsky, M.; Kilbourne, W.; Vocino, A. Relationship between the dominant social paradigm, materialism, and environmental behaviours in four Asian economies. *Eur. J. Mark.* **2014**, *48*, 522–551. [CrossRef]
22. Thyroff, A.E.; Kilbourne, W.E. Understanding pro-environmental intentions through growth, competitiveness, and concern. *Australas. Mark. J.* **2017**, *25*, 97–105. [CrossRef]
23. Ehrenfeld, D. *The Arrogance of Humanism*; Oxford University Press: New York, NY, USA, 1978; p. 304.
24. Bansal, P.; Kilbourne, W.E. The ecologically sustainable retailer. *J. Retail. Consum. Serv.* **2001**, *8*, 139–146. [CrossRef]
25. Huesemann, M.; Huesemann, J. *Techno-Fix: Why Technology Won't Save Us or the Environment*; New Society Publishers: Gabriola, BC, Canada, 2011; p. 464.
26. Catton, W.R., Jr.; Dunlap, R.E. Environmental sociology: A new paradigm. *Am. Sociol.* **1978**, *13*, 41–49.
27. Dunlap, R.E.; Van Liere, K.D. The new environmental paradigm. *J. Environ. Educ.* **1978**, *9*, 10–19. [CrossRef]
28. Izadpanahi, P.; Tucker, R. NEP (Children@ School): An instrument for measuring environmental attitudes in middle childhood. *Aust. J. Environ. Educ.* **2018**, *34*, 61–79. [CrossRef]
29. Xiao, C.; Dunlap, R.E.; Hong, D. Ecological worldview as the central component of environmental concern: Clarifying the role of the NEP. *Soc. Nat. Resour.* **2019**, *32*, 53–72. [CrossRef]
30. Ntanos, S.; Kyriakopoulos, G.; Skordoulis, M.; Chalikias, M.; Arabatzis, G. An application of the new environmental paradigm (NEP) scale in a Greek context. *Energies* **2019**, *12*, 239. [CrossRef]

31. Kemper, J.A.; Ballantine, P.W.; Hall, C.M. Sustainability worldviews of marketing academics: A segmentation analysis and implications for professional development. *J. Clean. Prod.* **2020**, *271*, 122568. [CrossRef]
32. La Trobe, H.L.; Acott, T.G. A modified NEP/DSP environmental attitudes scale. *J. Environ. Educ.* **2000**, *32*, 12–20. [CrossRef]
33. Engelhard, K. *South Korea: From Developing to Industrial Status*; Waxmann: Munich, Germany, 2004; p. 399. (In German)
34. Martin, H.P.; Schumann, H. *The Global Trap: Globalization and the Assault on Democracy and Prosperity*; Zed Books: London, UK, 1997; p. 280.
35. Hae-Joang, C.H. 'You are entrapped in an imaginary well': The formation of subjectivity within compressed development—A feminist critique of modernity and Korean culture. *Inter-Asia Cult. Stud.* **2000**, *1*, 49–69. [CrossRef]
36. Bates, R. Administering the global trap: The role of educational leaders. *Educ. Manag. Adm.* **2002**, *30*, 139–156. [CrossRef]
37. McKnight, D.M. Overcoming ecophobia: Fostering environmental empathy through narrative in childre's science literature. *Front. Ecol. Environ.* **2010**, *8*, e10–e15. [CrossRef]
38. Gray, T. Overcoming ecophobia. *Wildl. Aust.* **2013**, *50*, 42–44. [CrossRef]
39. Sobel, D. *Beyond Ecophobia: Reclaiming the Heart in Nature Education*; The Orion Society: Great Barrington, MA, USA, 2013; p. 61.
40. Estok, S.C. Tracking ecophobia: The utility of empirical and systems studies for ecocriticism. *Comp. Lit.* **2015**, *67*, 29–36. [CrossRef]
41. Lee, J.H.; Lee, S.J. Nature experience influences nature aversion: Comparison of South Korea and Germany. *Soc. Behav. Personal.* **2018**, *46*, 161–176. [CrossRef]
42. Panayotou, T. Conservation of biodiversity and economic development: The concept of transferable development rights. *Environ. Resour. Econ.* **1994**, *4*, 91–110. [CrossRef]
43. Haggard, S.; Kaufman, R.R. *Development, Democracy, and Welfare States: Latin America, East Asia, and Eastern Europe*; Princeton University Press: Princeton, NJ, USA, 2008; p. 496.
44. Luo, L.; Tang, Q.; Lan, Y.C. Comparison of propensity for carbon disclosure between developing and developed countries. *Account. Res. J.* **2013**, *26*, 6–34. [CrossRef]
45. Cole, M.A.; Rayner, A.J.; Bates, J.M. The environmental Kuznets curve: An empirical analysis. *Environ. Dev. Econ.* **1997**, *2*, 401–416. [CrossRef]
46. Zhang, Z.; Abuduwaili, J.; Jiang, F. Determination of occurrence characteristics of heavy metals in soil and water environments in Tianshan Mountains, Central Asia. *Anal. Lett.* **2013**, *46*, 2122–2131. [CrossRef]
47. Zhao, H.; Guo, S.; Zhao, H. Characterizing the influences of economic development, energy consumption, urbanization, industrialization, and vehicles amount on PM2.5 concentrations of China. *Sustainability* **2018**, *10*, 2574. [CrossRef]
48. Liang, W.; Yang, M. Urbanization, economic growth, and environmental pollution: Evidence from China. *Sustain. Comput. Inform. Syst.* **2019**, *21*, 1–9. [CrossRef]
49. FAO. *Global Forest Resources Assessments 2000*; FAO: Rome, Italy, 2001; p. 479.
50. FAO. *Global Forest Resources Assessments 2020*; FAO: Rome, Italy, 2020; p. 184. [CrossRef]
51. Milbrath, L.W. *Envisioning a Sustainable Society: Learning our Way Out*; SUNY Press: New York, NY, USA, 1989; p. 403.
52. Bonfield, E.H. Attitude, social influence, personal norm, and intention interactions as related to brand purchase behavior. *J. Mark. Res.* **1974**, *11*, 379–389. [CrossRef]
53. Ajzen, I.; Fishbein, M. Attitude–behavior relations: A theoretical analysis and review of empirical research. *Psychol. Bull.* **1977**, *84*, 888–918. [CrossRef]
54. Park, U.A. The Married Women's Pro-Environmental Consumer Behavior. Ph.D. Thesis, Seoul National University, Seoul, Korea, 1997. (In Korean with English Abstract).
55. Kollmuss, A.; Agyeman, J. Mind the gap: Why do people act environmentally and what are the barriers to pro-environmental behavior? *Environ. Educ. Res.* **2002**, *8*, 239–260. [CrossRef]
56. Chen, X.; Peterson, M.N.; Hull, V.; Lu, C.; Lee, G.D.; Hong, D.; Liu, J. Effects of attitudinal and sociodemographic factors on pro-environmental behaviour in urban China. *Environ. Conserv.* **2011**, *38*, 45–52. [CrossRef]
57. Barr, S. Sustainable lifestyles. In *International Encyclopedia of the Social & Behavioral Sciences*, 2nd ed.; Wright, J.D., Ed.; Elsevier Science: Boston, MA, USA, 2015; pp. 828–833. [CrossRef]
58. Ones, D.S.; Wiernik, B.M.; Dilchert, S.; Klein, R. Pro-environmental behavior. In *International Encyclopedia of the Social & Behavioral Sciences*, 2nd ed.; Wright, J.D., Ed.; Elsevier Science: Boston, MA, USA, 2015; pp. 82–88. [CrossRef]
59. Lange, F.; Dewitte, S. Measuring pro-environmental behavior: Review and recommendations. *J. Environ. Psychol.* **2019**, *63*, 92–100. [CrossRef]
60. Bechtel, R.B.; Verdugo, V.C.; de Queiroz Pinheiro, J. Environmental belief systems: United States, Brazil, and Mexico. *J. Cross-Cult. Psychol.* **1999**, *30*, 122–128. [CrossRef]
61. Schultz, P.W.; Zelezny, L.C. Values and proenvironmental behavior: A five-country survey. *J. Cross-Cult. Psychol.* **1998**, *29*, 540–558. [CrossRef]
62. Vicente-Molina, M.A.; Fernández-Sáinz, A.; Izagirre-Olaizola, J. Environmental knowledge and other variables affecting pro-environmental behaviour: Comparison of university students from emerging and advanced countries. *J. Clean. Prod.* **2013**, *61*, 130–138. [CrossRef]
63. Schultz, P.W.; Zelezny, L.C. Values as predictors of environmental attitudes: Evidence for consistency across 14 countries. *J. Environ. Psychol.* **1999**, *19*, 255–265. [CrossRef]
64. Fontaine, R.; Richardson, S. Cross-cultural research in Malaysia. Cross Cultural Management. *Int. J.* **2003**, *10*, 75–89. [CrossRef]

65. Lee, J.H.; Kim, B.S. Understanding the perspectives on forests among migrants in Korea: Immigrants from China, Vietnam, and Mongolia. *For. Sci. Technol.* **2014**, *10*, 9–13. [CrossRef]
66. Sim, J. The park governance in the changing process of Seoul Forest Park management system. Ph.D. Thesis, Seoul National University, Seoul, Korea, 2018. (In Korean with English Abstract).
67. Korea National Park Research Institute. *Bukhan-San National Park White Paper*; Korea National Park Research Institute: Seoul, Korea, 2019; p. 125. (In Korean)
68. Jim, C.Y.; Chen, W.Y. Pattern and divergence of tree communities in Taipei's main urban green spaces. *Landsc. Urban Plan.* **2008**, *84*, 312–323. [CrossRef]
69. Hsu, H.P. Providing an attractive environment for people to engage in health activities: Serving with landscape. *Int. J. Concept. Struct. Smart Appl.* **2016**, *4*, 38–54. [CrossRef]
70. Nasrullah, N.; Gunawan, A. Perceptions and preferences of user to the thermal comfort in Menteng Park and Honda Tebet Park. *J. Lanskap. Indones.* **2017**, *9*, 24–35. [CrossRef]
71. Oktavia, R.C.; Siregar, H.; Sunarminto, T.; Hermawan, R. Analysis of recreational carrying capacity of urban parks and urban forests in DKI Jakarta Province. *Int. J. Sci. Basic Appl. Res.* **2019**, *46*, 38–56.
72. Kurniastuti, A.E. Pengelolaan Hutan Kota di Jakarta (Studi Kasus Hutan Kota Srengseng di Jakarta Barat). *J. Pembang. Wil. Kota* **2013**, *9*, 439–450. (In Indonesian) [CrossRef]
73. Packer, J.; Ballantyne, R.; Hughes, K. Chinese and Australian tourists' attitudes to nature, animals, and environmental issues: Implications for the design of nature-based tourism experiences. *Tour. Manag.* **2014**, *44*, 101–107. [CrossRef]
74. Masud, M.M.; Kari, F.B. Community attitudes towards environmental conservation behaviour: An empirical investigation within MPAs, Malaysia. *Mar. Policy* **2015**, *52*, 138–144. [CrossRef]
75. Mutanga, C.N.; Vengesayi, S.; Muboko, N.; Gandiwa, E. Towards harmonious conservation relationships: A framework for understanding protected area staff-local community relationships in developing countries. *J. Nat. Conserv.* **2015**, *25*, 8–16. [CrossRef]
76. Chen, B.X.; Qiu, Z.M. Community attitudes toward ecotourism development and environmental conservation in nature reserve: A case of Fujian Wuyishan National Nature Reserve, China. *J. Mt. Sci.* **2017**, *14*, 1405–1418. [CrossRef]
77. Newhouse, N. Implications of attitude and behavior research for environmental conservation. *J. Environ. Educ.* **1990**, *22*, 26–32. [CrossRef]
78. Wolsko, C.; Lindberg, K. Experiencing connection with nature: The matrix of psychological well-being, mindfulness, and outdoor recreation. *Ecopsychology* **2013**, *5*, 80–91. [CrossRef]
79. Lee, J.H.; Lee, D.J. Nature experience, recreation activity and health benefits of visitors in mountain and urban forests in Vienna, Zurich, and Freiburg. *J. Mt. Sci.* **2015**, *12*, 1551–1561. [CrossRef]
80. Soga, M.; Gaston, K.J. Extinction of experience: The loss of human–nature interactions. *Front. Ecol. Environ.* **2016**, *14*, 94–101. [CrossRef]
81. Lee, D.; Lee, J.H. A structural relationship between place attachment and intention to conserve landscapes—a case study of Harz National Park in Germany. *J. Mt. Sci.* **2017**, *14*, 998–1007. [CrossRef]
82. Hodge, C.J.; Duerden, M.D.; Layland, E.K.; Lacanienta, A.; Goates, M.C.; Niu, X.M. The association between family leisure and family quality of life: A meta-analysis of data from parents and adolescents. *J. Fam. Theory Rev.* **2017**, *9*, 328–346. [CrossRef]
83. Roberson, P.N.; Norona, J.C.; Lenger, K.A.; Olmstead, S.B. How do relationship stability and quality affect well-being? Romantic relationship trajectories, depressive symptoms, and life satisfaction across 30 years. *J. Child Fam. Stud.* **2018**, *27*, 2171–2184. [CrossRef]
84. Georgiadou, E.; Schmitt, G.M.; Erim, Y. Does the separation from marital partners of Syrian refugees with a residence permit in Germany have an impact on their quality of life? *J. Psychosom. Res.* **2020**, *130*, 109936. [CrossRef]
85. Hung, K.; Crompton, J.L. Benefits and constraints associated with the use of an urban park reported by a sample of elderly in Hong Kong. *Leis. Stud.* **2006**, *25*, 291–311. [CrossRef]
86. Robin, M.; Matheau-Police, A.; Couty, C. Development of a scale of perceived environmental annoyances in urban settings. *J. Environ. Psychol.* **2007**, *27*, 55–68. [CrossRef]
87. Liechty, T.; Yarnal, C.; Kerstetter, D. 'I want to do everything!': Leisure innovation among retirement-age women. *Leis. Stud.* **2012**, *31*, 389–408. [CrossRef]
88. Dzhambov, A.M.; Dimitrova, D.D. Elderly visitors of an urban park, health anxiety, and individual awareness of nature experiences. *Urban For. Urban Green.* **2014**, *13*, 806–813. [CrossRef]
89. Cox, H.G. *Later Life: The Realities of Aging*; Routledge: Milton Park, UK, 2015; p. 432.
90. Arnberger, A.; Allex, B.; Eder, R.; Ebenberger, M.; Wanka, A.; Kolland, F.; Wallner, P.; Hutter, H.P. Elderly residents' uses of and preferences for urban green spaces during heat periods. *Urban For. Urban Green.* **2017**, *21*, 102–115. [CrossRef]
91. Hair, J.F.; Black, W.C.; Babin, B.J.; Anderson, R.E.; Tatham, R.L. *Structural equation modeling: An introduction. In. Multivariate Data Analysis*, 6th ed.; Hair, J.F., Ed.; Pearson Prentice Hall: Hoboken, NJ, USA, 2006; pp. 752–753.
92. Hu, L.T.; Bentler, P.M. Cutoff criteria for fit indexes in covariance structure analysis: Conventional criteria versus new alternatives. *Struct. Equ. Modeling* **1999**, *6*, 1–55. [CrossRef]
93. Bentler, P.M. Comparative fit indexes in structural models. *Psychol. Bull.* **1990**, *107*, 238–246. [CrossRef]

94. Van Liere, K.D.; Dunlap, R.E. The social bases of environmental concern: A review of hypotheses, explanations and empirical evidence. *Public Opin. Q.* **1980**, *44*, 181–197. [CrossRef]
95. Schahn, J.; Holzer, E. Studies of individual environmental concern the role of knowledge, gender, and background variables. *Environ. Behav.* **1990**, *22*, 767–786. [CrossRef]
96. Luo, Y.; Deng, J. The New Environmental Paradigm and nature-based tourism motivation. *J. Travel Res.* **2008**, *46*, 392–402. [CrossRef]
97. Wade, N.J.; Swanston, M. *Visual Perception: An Introduction*; Psychology Press: London, UK, 2013; p. 321.
98. Zandersen, M.; Tol, R.S. A meta-analysis of forest recreation values in Europe. *J. For. Econ.* **2009**, *15*, 109–130. [CrossRef]
99. Worldbank. Available online: https://data.worldbank.org/indicator/NY.GDP.PCAP.PP.CD (accessed on 24 October 2021).
100. Bentler, P.M.; Speckart, G. Models of attitude-behavior relations. *Psychol. Rev.* **1979**, *86*, 452. [CrossRef]
101. Morse, J.W.; Gladkikh, T.M.; Hackenburg, D.M.; Gould, R.K. COVID-19 and human-nature relationships: Vermonters' activities in nature and associated nonmaterial values during the pandemic. *PLoS ONE* **2020**, *15*, e0243697. [CrossRef] [PubMed]
102. Venter, Z.S.; Barton, D.N.; Gundersen, V.; Figari, H.; Nowell, M. Urban nature in a time of crisis: Recreational use of green space increases during the COVID-19 outbreak in Oslo, Norway. *Environ. Res. Lett.* **2020**, *15*, 104075. [CrossRef]
103. Landry, C.E.; Bergstrom, J.; Salazar, J.; Turner, D. How Has the COVID-19 Pandemic Affected Outdoor Recreation in the US? A Revealed Preference Approach. *Appl. Econ. Perspect. Policy* **2021**, *43*, 443–457. [CrossRef]

forests

MDPI

Article

Differences in Airborne Particulate Matter Concentration in Urban Green Spaces with Different Spatial Structures in Xi'an, China

Bo Jiang [1], Chang Sun [1], Sen Mu [1], Zixin Zhao [2], Yingyuan Chen [1], Yiwei Lin [1], Ling Qiu [1,*,†] and Tian Gao [1,*,†]

1 College of Landscape Architecture and Art, Northwest A&F University, No.3 Taicheng Road, Xianyang 712100, China; ASL6006@nwafu.edu.cn (B.J.); 2019059347@nwafu.edu.cn (C.S.); musen@nwafu.edu.cn (S.M.); cyydy@nwafu.edu.cn (Y.C.); linyiwei@nwafu.edu.cn (Y.L.)

2 Accademia di Belle Arti di MILANO "Brera", Via Brera, 28, 20121 Milano, Italy; jennifer951129@163.com

* Correspondence: qiu.ling@nwsuaf.edu.cn (L.Q.); tian.gao@nwsuaf.edu.cn (T.G.); Tel.: +86-29-87080269 (L.Q.); +86-29-87082997 (T.G.)

† These two authors jointly supervised this work.

Abstract: With the acceleration of urbanization and industrialization, air pollution is becoming one of the most serious problems in cities. Urban green spaces, as "green infrastructure", are an important part of urban ecosystems for air purification. Therefore, 10 typical green spaces of urban parks in the city of Xi'an, China, were selected as study areas according to vegetation structure and species composition. Considering meteorological factors and time changes, the effects of the selected green spaces with different vegetation structures of different heights on the reduction in airborne particulate matter concentration were explored. The results showed that the following: (1) Temperature, relative humidity, wind speed, and air pressure had significant correlation with the concentration of airborne particulate matter at the different heights, and the correlations were the same at 1.5 m and 5 m. (2) After heating in winter, the concentration of airborne particulate matter with different particle sizes increased significantly. The concentration of airborne particulate matter showed different trends throughout the day, and the small particles (PM_1 and $PM_{2.5}$) had a trend of "lower in the morning and evening, and higher at noon", while the large particles (PM_{10} and TSP) gradually decreased over time. (3) In the selected green spaces with different vegetation structure types, the concentration of airborne particulate matter below the canopy (1.5 m) was generally higher than that in the middle of the canopy (5 m), but the effects of reducing the concentration of airborne particulate matter were consistent at the different heights. (4) The adsorption capacity of PM_1 and $PM_{2.5}$ concentration was strong in the partially closed broad-leaved one-layered forest (PBO), and poor in the partially closed broad-leaved multi-layered forest (PBM). Partially closed broad-leaved multi-layered forest (PBM) and partially closed coniferous and broad-leaved mixed multi-layered forest (PMM) also had strong dust-retention effect on PM_{10} and TSP, while closed broad-leaved one-layered forest (CBO) had a poor dust-retention effect. The results showed that the reduction effects of urban green spaces with different spatial structures on air particles were different, and were restricted by various environmental factors, which could provide a theoretical basis for the optimization of urban green space structure and the improvement of urban air quality.

Keywords: green space; airborne particulate matter; meteorological parameters; height

Citation: Jiang, B.; Sun, C.; Mu, S.; Zhao, Z.; Chen, Y.; Lin, Y.; Qiu, L.; Gao, T. Differences in Airborne Particulate Matter Concentration in Urban Green Spaces with Different Spatial Structures in Xi'an, China. *Forests* **2022**, *13*, 14. https://doi.org/10.3390/f13010014

Academic Editors: Radu-Daniel Pintilii and Diego Varga

Received: 1 December 2021
Accepted: 20 December 2021
Published: 22 December 2021

1. Introduction

With rapid economic development and urbanization, ecological problems are becoming increasingly prominent, and air pollution has become one of the most serious problems to be faced by the whole world—especially by largely urbanized and densely populated countries, such as China [1]. According to the state of China's Ecological Environment Bulletin 2019, 157 cities were not up to the desired standards of ambient air quality, accounting for 46.6% of the total number of cities in China, with the overall situation not being optimistic. Particulate matter with small aerodynamic diameter is the primary pollutant, with an uneven surface and strong adsorption ability; it can make a variety of harmful substances in the air attach to its surface—such as polycyclic aromatic hydrocarbons (PAHs), bacteria, and viruses—and enter the human body through the respiratory tract, directly endangering human health, and resulting in an increase in death rates from lung cancer [2]. It has been confirmed that air pollution poses a serious threat to human quality of life, and decreases the life expectancy of inhabitants who live in urban areas [3].

Depending on its aerodynamic diameter, airborne particulate matter can be classified into total suspended particulate matter (TSP; diameter $\leq$ 100 μm), inhalable particulate matter (PM_{10}; diameter $\leq$ 10 μm), fine particles ($PM_{2.5}$; diameter $\leq$ 2.5 μm), and ultrafine particles (PM_1; diameter $\leq$ 1 μm). PM originates from both natural sources—such as wildfires and dust storms [4]—and anthropogenic activities, such as mining and the burning of fossil fuels [5,6] The burning of fossil fuels for heating in winter releases waste, which provides direct or indirect conditions for the growth of airborne particulate matter—especially in the northern part of China. Many practical strategies have been proposed by the public to reduce particulate pollution in cities. Environmental laws have been passed to lower the amounts of toxic emissions from factories, modify energy resource structures, and limit vehicle numbers [7].

However, the control of air pollution is not an overnight action, but a long and arduous process. Urban green spaces, as "green infrastructure", are an important part of urban ecosystems, which play a significant role in improving the ecological environment, beautifying and optimizing the living environment. Several studies have shown that urban areas with high green coverage helped to reduce the concentration of airborne particulate matter, and often had a negative correlation with the latter [8,9]. It was found that the size, shape, and microstructure of plant leaves in green spaces—such as leaf surface roughness, waxy layer, and leaf hair—had significant effects on the capture of airborne particulate matter [10]. Therefore, using green spaces to purify dust in the air is a practical and effective measure, and the dust-retention ability of plants has become an important index of plant selection in urban green spaces [11].

The dust-catching ability of different plants in urban green spaces varies greatly. Some studies have found that the dust-catching ability of different plant communities is as follows: arbors > shrubs > herbs [12,13]. Tall trees mainly retard and filter the particulate matter and drifting particulate matter from the outside world, while shrubs and grasses can effectively intercept the particulate matter from the ground [14]. Przybysz et al. found that urban meadows accumulate PM from the ambient air more effectively than tradition lawns [15]. Terzaghi et al. found that an arbor–shrub–grass planting model can better reduce the concentration of airborne particulate matter [16]. However, McDonald et al. showed that more particulate matter will be retained in the compound structure of green spaces with a combination of trees, shrubs, and grass [17]. Therefore, these inconsistencies require further attention in order to expose the underlying factors. According to the study of Gao et al., vegetation structure plays a significant role in regulating air pollution, and the leaf dust retention of even the same species of plants was different in the urban green spaces with different vegetation structures (e.g., enclosed green spaces had lower dust retention than open green spaces) [18]. Moreover, Selmi et al. claimed that the dust retention in the "low" position of the same plant leaves was significantly higher than that in the "high" and "middle" positions [19]; in other words, different vertical heights of plants in the green spaces also showed great differences in their ability to reduce the concentration of airborne particulate matter. Therefore, it is necessary to systematical study the ability of green spaces with different spatial structures—including vegetation heights—in order to reduce the concentration of airborne particulate matter.

In addition, meteorological environmental parameters in urban green spaces also affect the concentration of particulate matter in different sections of vegetation. Wind speed, temperature, relative humidity, and air pressure affect the diffusion and settlement of particles with different sizes [20,21]. Wind speed plays an important role in horizontal transmission and dilution diffusion [22]. Changes in temperature affect convection in the vertical direction of the atmosphere which, in turn, affects the concentration of airborne particulate matter [23]. Changes in relative humidity also increase the concentration of fine particles [24]. When the air pressure change is obvious, the atmosphere is in an unstable state, facilitating the vertical diffusion of pollutants [25]. Given the above, the influence of meteorological factors dust retention should also be comprehensively taken into account.

Therefore, combined with the meteorological parameters, the concentration changes of particles with different sizes in urban green spaces with different vegetation structures and types at different heights should be explored in order to provide the optimal urban green space planning for future air quality improvement. The main objectives of this study are to investigate the following:

- The factors of meteorological parameters, monitoring time, vegetation structure, and vegetation height influencing the concentration of airborne particulate matter;
- The differences in the concentration distribution of airborne particulate matter in urban green spaces with different vegetation structures at different heights.

2. Materials and Methods

2.1. Study Area

Xi'an, the capital of Shaanxi Province, is located in the central part of the Guanzhong Plain, between 107.40°–109.49° E and 33.42°–34.45° N, bordering the Wei River in the north and the Qinling Mountains in the south of China; it is one of the important birthplaces of the Chinese civilization and nation, and the starting point of the Silk Road. Xi'an belongs to the semi-humid continental monsoon climate of the warm temperate zone, with distinct four seasons: spring is variable, summer is hot and rainy, autumn is cool and rainy, and winter is dry and cold with little rain or snow [26].

2.2. Classification and Selection of Urban Green Spaces

Through Google satellite image interpretation and field investigation, combined with the characteristics of urban green spaces in the city of Xi'an (Figure 1), the green spaces were first divided according to their spatial vegetation structure and species composition. By using the LAI-2200 Plant Canopy Analyzer and a fisheye camera (Figure 2), the canopy cover ratios of trees and shrubs were first divided into open green spaces (<10% canopy cover of trees/shrubs), partially open green spaces (10–40% canopy cover of trees/shrubs), partially closed green spaces (40–70% canopy cover of trees/shrubs), and closed green spaces (>70% canopy cover of trees/shrubs). According to the species composition, open green spaces and partially open green spaces were then divided into two subtypes—lawn, and grassland—while partially closed green spaces and closed green spaces were divided into three subtypes: coniferous forest, broad-leaved forest, and coniferous and broad-leaved mixed forest. Partially closed green spaces and closed green spaces were further subdivided into two types—one-layered forest, and multi-layered forest—forming a set of unified standard urban green space classification systems (Table 1) [27].

According to the actual situation of the city of Xi'an, 10 typical and abundant sample types of sites were selected, including open lawn (O), partially open green space (PO), partially closed broad-leaved one-layered forest (PBO), partially closed broad-leaved multi-layered forest (PBM), partially closed coniferous and broad-leaved mixed multi-layered forest (PMM), closed broad-leaved one-layered forest (CBO), closed broad-leaved multi-layered forest (CBM), closed coniferous one-layered forest (CCO), closed coniferous and broad-leaved mixed one-layered forest (CMO), and closed coniferous and broad-leaved mixed multi-layered forest (CMM). Two or three sample plots of each type were selected as duplicates, and two public squares with hard pavement were taken as the control groups (CK), which were quite open and covered by green space around the square. A total of 26 sample plots were finally selected (Figure 3).

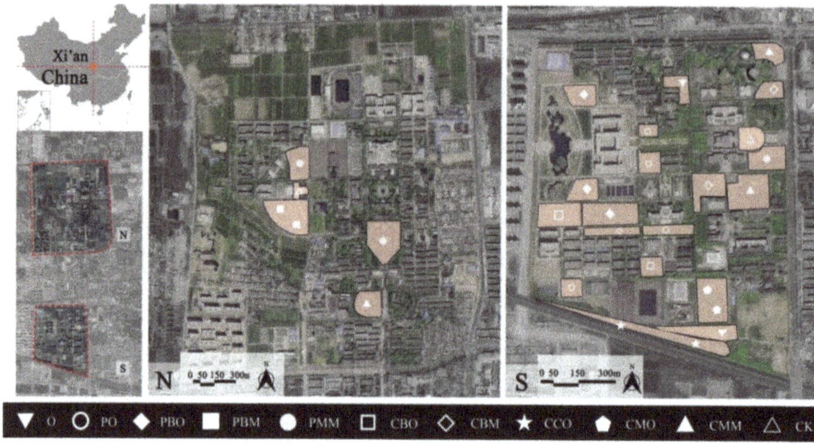

Figure 1. Map showing sampling locations and their repetition.

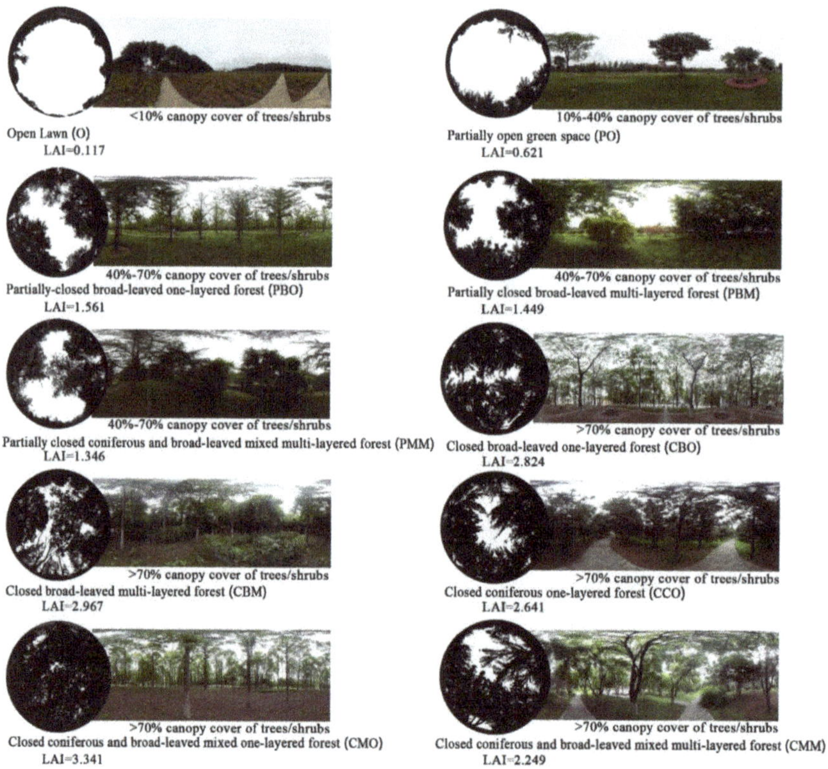

Figure 2. Fisheye camera view of urban green space classification.

Table 1. Classification with three levels shown for urban green spaces.

	Horizontal Structure	Species Composition	Vertical Structure
Urban green spaces	Open green spaces (<10% canopy cover of trees/shrubs)	Lawn mainly dominated by *Cynodon dactylon* Grass flowers mainly dominated by *Veronica persica*	–
	Partially open green spaces (10–40% canopy cover of trees/shrubs)	Lawn mainly dominated by *Arachis hypogaea* Grass flowers mainly dominated by *Oxalis corymbosa*	
	Partially closed green spaces (40–70% canopy cover of trees/shrubs)	Broad-leaved trees mainly dominated by *Melia azedarach* Coniferous trees mainly dominated by *Pinus tabuliformis* Mixed plants mainly dominated by *Ligustrum sinense* and *Cedrus deodara*	One-layered Multi-layered
	Closed green spaces (>70% canopy cover of trees/shrubs)	Broad-leaved trees mainly dominated by *Platanus orientalis* Coniferous trees mainly dominated by *Picea asperata* Mixed plants mainly dominated by *Koelreuteria paniculata* and *Platycladus orientalis*	One-layered Multi-layered

Figure 3. Panoramic views of 10 different vegetation structures (O: open lawn; PO: partially open green space; PBO: partially closed broad-leaved one-layered forest; PBM: partially closed broad-leaved multi-layered forest; PMM: partially closed coniferous and broad-leaved mixed multi-layered forest; CBO: closed broad-leaved one-layered forest; CBM: closed broad-leaved multi-layered forest; CCO: closed coniferous one-layered forest; CMO: closed coniferous and broad-leaved mixed one-layered forest; CMM: closed coniferous and broad-leaved mixed multi-layered forest; CK: the control groups).

2.3. Field Monitoring

In order to combine the influence of meteorological parameters on reducing the concentration of airborne particulate matter in urban green spaces, meteorological parameters (i.e., temperature, relative humidity, wind speed, and air pressure) were monitored in sunny, windy, or calm weather. The meteorological parameters were monitored through the use of handheld weather stations (Kestrel 5500), and the concentrations of airborne particulate matter (TSP, PM_{10}, $PM_{2.5}$, and PM_1) were measured with handheld particle counters (Aerocet 831) in the selected plots. In the grid pattern of each sample plot, four locations were uniformly selected as sampling points to represent the typical sample types. The monitors were set up at heights of 1.5 m and 5 m in each sample plot—1.5 m is the average height at which human respiration occurs, while 5 m is the average height at the middle of the plant canopy; thus, it is convenient to compare the effects of different heights of vegetation on reducing the concentration of airborne particulate matter. In addition, the surrounding environment of each sampling plot was consistent, without pollution sources, and at least 20 m away from the road and the water edge as a buffer, so as to avoid other factors affecting the experimental results. The data were collected during the three months from 1 October 2020 to 31 December 2020, which belonged to the pre-heating and heating periods. Two days were selected from the beginning and the end of each month, and three time periods per day were taken—from 8:00 to 10:00, 12:00 to 14:00, and 16:00 to 18:00. All sampling sites were monitored sequentially at the same time.

2.4. Statistical Analysis

In this study, Microsoft Office Excel 2010 software was used for all data recording and collection. Generalized linear analysis and correlation analysis were used in the statistical software package IBM SPSS Statistics 19 to model all variables, in order to explore whether each variable had a significant effect on airborne particulate matter, and to identify how each variable affected airborne particulate matter concentration. The acceptable significance level was $p < 0.05$.

3. Results

3.1. Effects of the Dominated Factors on Airborne Particulate Matter

In this study, it was found that the meteorological parameters, monitoring time, vegetation structure, and height had significant effects on the concentrations of PM_1 and $PM_{2.5}$ (Table 2). Some meteorological parameters had a certain influence on the concentrations of PM_{10} and TSP. Except for temperature, vegetation structure, and air pressure, the other factors had significant effects on PM_{10} and TSP concentrations.

Table 2. Variance analysis of factors affecting PM concentration.

Parameters	Df	PM_1		$PM_{2.5}$		PM_{10}		TSP	
		F-Value	*p*-Value	F-Value	*p*-Value	F-Value	*p*-Value	F-Value	*p*-Value
Temperature	1	2537.40	0.000	1979.89	0.000	2.88	0.090	38.45	0.000
Relative humidity	1	843.40	0.000	865.56	0.000	4.97	0.026	50.14	0.000
Wind velocity	1	247.26	0.000	224.79	0.000	9.12	0.003	5.73	0.017
Air pressure	1	7.42	0.006	4.79	0.029	1.43	0.023	2.16	0.142
Pre-heating and heating periods	1	3086.40	0.000	3245.85	0.000	8.56	0.003	4.81	0.028
Monitoring time	1	298.35	0.000	446.50	0.000	3.73	0.035	0.49	0.048
Vegetation structure	10	27.35	0.000	10.85	0.001	3.25	0.071	2.00	0.015
Height	1	74.24	0.000	10.97	0.001	23.96	0.000	29.34	0.000

Df: degree of freedom; F-Value: variance test volume; *p*-Value: significant test of regression equation.

3.2. Effects of Meteorological Parameters on Airborne Particulate Matter

The results showed that the meteorological parameters—including temperature, relative humidity, wind speed, and air pressure—had significant correlations with the concentration of airborne particulate matter at different vertical heights, and the correlations of meteorological parameters with the concentration of airborne particulate matter were consistent at the heights of 1.5 m and 5 m (Table 3).

Table 3. Correlation analysis of meteorological parameters and airborne particulate matter concentrations.

Height	Project	Spearman Correlation Test			
		Temperature (°C)	Humidity (%)	Wind Speed (m/s)	Air Pressure (mpa)
1.5 m	PM_1 (μg/m^3)	−0.087 **	0.175 **	−0.224 **	0.212 **
	$PM_{2.5}$ (μg/m^3)	−0.253 **	0.251 **	−0.188 **	0.311 **
	PM_{10} (μg/m^3)	−0.147 **	0.026	−0.051 *	−0.021
	TSP (μg/m^3)	−0.110 **	0.040	−0.040	−0.087 **
5 m	PM_1 (μg/m^3)	−0.177 **	0.168 **	−0.229 **	0.028
	$PM_{2.5}$ (μg/m^3)	−0.169 **	0.250 **	−0.193 **	0.098 **
	PM_{10} (μg/m^3)	−0.143 **	0.034	−0.024	−0.056 *
	TSP (μg/m^3)	−0.131 **	0.020	−0.038	−0.110 **

**: At level 0.01 (two-tailed), the correlation was significant; *: at level 0.05 (two-tailed), the correlation was significant.

During the October–December period of measurement, the temperature varied from −3.4 to 26.4 °C. The effect of temperature on the concentration of airborne particulate matter at different heights was consistent, showing a significant negative correlation—that is, the higher the temperature, the lower the concentration of airborne particulate matter (Figure 4a,b). Humidity had a significant positive correlation with the concentration of airborne particulate matter (Figure 4c,d). With the increase in relative humidity, PM_1 and $PM_{2.5}$ were more likely to condense, leading to an increase in the concentration of airborne particulate matter. In this study, the variation range of wind speed was 0–3.5 m/s. With the increase in wind speed, the diffusion of airborne particulate matter concentration was promoted, thus reducing the airborne particulate matter concentration (Figure 4e,f). The vertical atmospheric pressure ranged from 955.2 to 996.4 mPa. The change in air pressure caused different changes in the concentration of airborne particulate matter with different particle sizes (Figure 4g,h); it was positively correlated with the concentrations of fine particulate matter PM_1 and $PM_{2.5}$, but negatively correlated with the concentrations of particulate matter PM_{10} and TSP—that is, the higher the air pressure, the lower the concentrations of PM_{10} and TSP.

3.3. Effects of Diurnal Variation before and after the Heating Period on Airborne Particulate Matter

The generalized linear models of time—before and after heating—and the concentration of airborne particulate matter were established for statistical analysis, and meteorological parameters were taken into account. The results showed that time in both the pre-heating and heating periods had significant effects on the concentration of airborne particulate matter ($p < 0.01$). The concentration of airborne particulate matter increased significantly after heating, and the highest concentration reached 10 times that of the pre-heating period (Figure 5). The concentrations of airborne particles with different particle sizes were significantly different in different time periods. The smaller particles, such as PM_1 and $PM_{2.5}$, showed a trend of "lower in the morning and evening, and higher at noon"; however, PM_{10} and TSP showed a trend of gradual decrease (Figure 6).

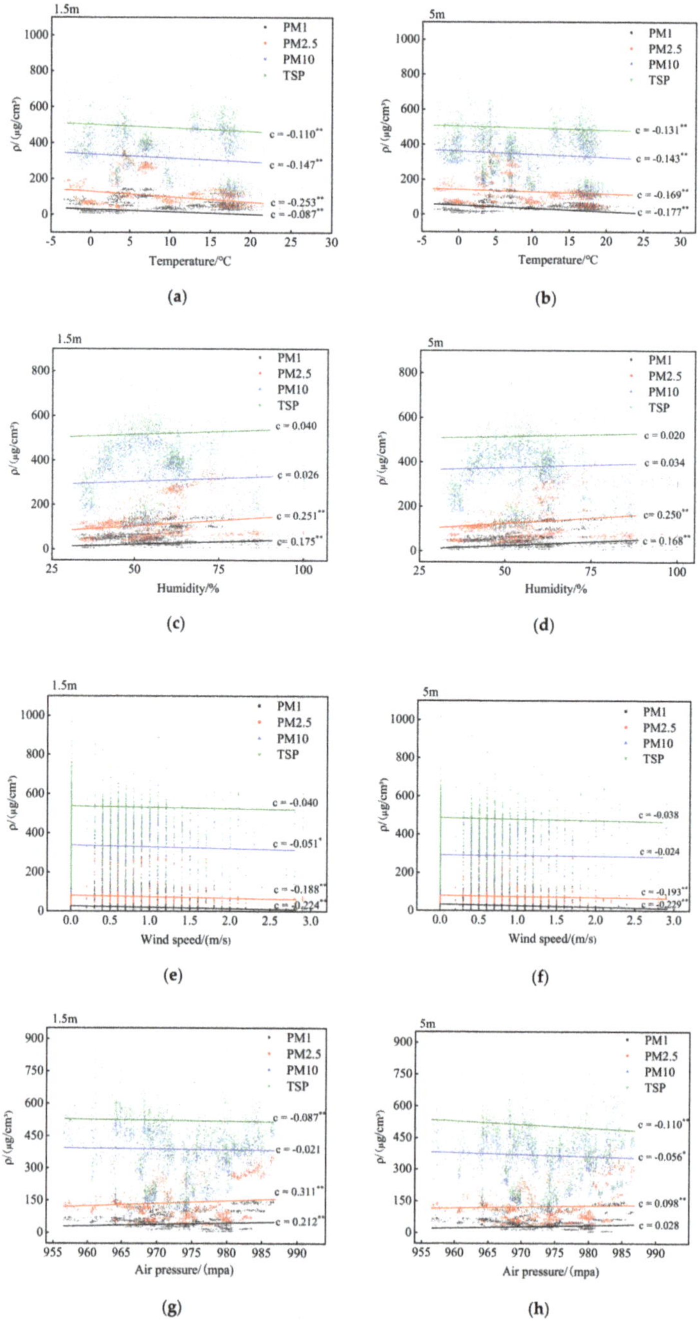

Figure 4. The relationship between meteorological factors and PM concentration in the different heights: (**a**,**b**) effect of temperature on the concentration of airborne particulate matter; (**c**,**d**) effect of humidity on the concentration of airborne particulate matter; (**e**,**f**) effect of wind speed on the concentration of airborne particulate matter; (**g**,**h**) effect of air pressure on the concentration of airborne particulate matter; "c" stands for correlation coefficient; **: at level 0.01 (two-tailed), the correlation was significant; *: at level 0.05 (two-tailed), the correlation was significant.

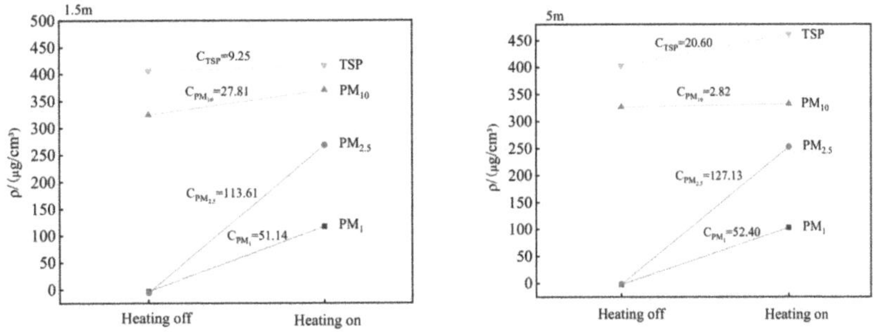

(a) (b)

Figure 5. Changes in airborne particulate matter concentration before and after heating at different heights. Note: C represents the generalized linear analysis coefficient of the change in airborne particulate matter concentration before and after heating. The absolute value of C represents the difference between the factor level and the population mean. (**a**) Changes in airborne particulate matter concentration before and after heating at 1.5m height; (**b**) Changes in airborne particulate matter concentration before and after heating at 5m height.

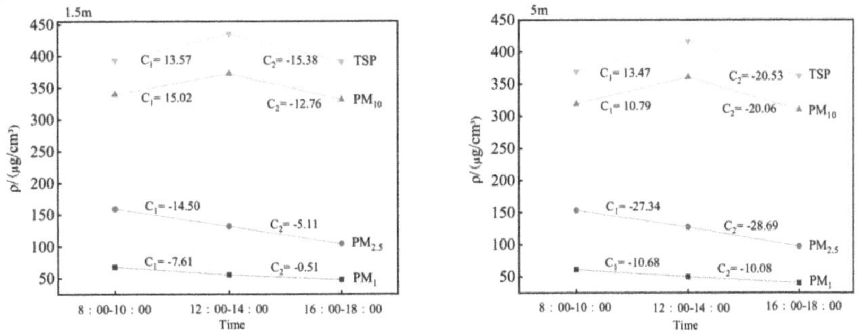

(a) (b)

Figure 6. Influence of different time periods on airborne particulate matter at different heights. Note: C1 represents the generalized linear analysis coefficient of the change in airborne particulate matter concentration in the morning and noon. C2 represents the generalized linear analysis coefficient of the change in airborne particulate matter concentration in the noon and evening. The absolute value of C represents the difference between the factor level and the population mean. (**a**) Influence of different time periods on airborne particulate matter at 1.5m height; (**b**) Influence of different time periods on airborne particulate matter at 5m height.

3.4. Effects of Different Vegetation Structures on the Concentration of Airborne Particulate Matter

The results showed that there were significant differences in height and airborne particulate matter concentration among the different vegetation structures (Figure 7), and the concentration of airborne particulate matter below the canopy (1.5 m) was much higher than that in the middle of the canopy (5 m). There was no significant difference in the concentration of airborne particulate matter between the control groups (CK) and the different vegetation structure types.

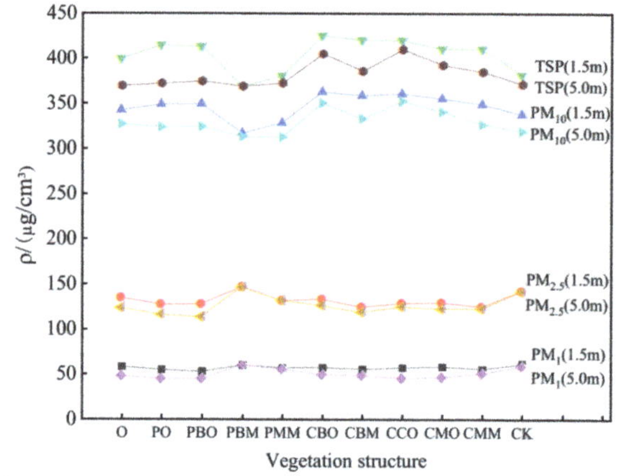

Figure 7. Distribution of mean changes in airborne particulate matter by vertical height.

In the sample plots with different vegetation structures, the concentrations of airborne particulate matter with different particle sizes at different heights were different (Figure 7). However, the concentration was almost the same in the partially closed broad-leaved multi-layered forest (PBM). The concentration of PM_1 in the closed coniferous one-layered forest (CCO) had the largest difference, reaching 11.60 μg/cm^3. In other vegetation structure plots, the differences in PM_1 concentration at different heights were small. The difference in $PM_{2.5}$ concentration was mainly concentrated in the open (O) and partially open (PO) green spaces, but in the partially closed broad-leaved one-layered forest (PBO), the concentration at 1.5 m height was much higher than that at 5 m height, and the difference value could reach 14.27 μg/cm^3. In addition to partially closed broad-leaved multi-layered forest (PBM), the largest difference in PM_{10} concentration was found in other multi-layered structural forests, such as PMM, CBM, and CMM. Except for the PBM, partially closed coniferous and broad-leaved mixed multi-layered forest (PMM) and closed coniferous one-layered forest (CCO), the difference in TSP concentration in other plots was more than 10 μg/cm^3.

Moreover, there were differences in the reduction effect of green spaces with different spatial vegetation structures on the concentration of airborne particulate matter with different particle sizes. At the height of 1.5 m, the reduction effects of the 10 different vegetation structures on PM_1 and $PM_{2.5}$ concentrations were similar, but the reduction effects on PM_{10} and TSP concentrations were significantly different; among them, the partially closed broad-leaved one-layered forest (PBO) and the partially open green spaces (PO) had the most significant negative effects on PM_1 and $PM_{2.5}$, as well as the strongest adsorption capacity. The closed broad-leaved one-layered forest (CBO) had a positive effect on PM_{10} and TSP, and the concentrations of PM_{10} and TSP were the highest in this type of green space. In the partially closed broad-leaved multi-layered forest (PBM), the concentrations of PM_{10} and TSP were the lowest, and the dust-retention ability was strong (Figure 8a).

At the height of 5 m, there were significant differences in PM_1 reduction among the 10 green spaces with different vegetation structures (Figure 8b). Except for the partially closed broad-leaved multi-layered forest (PBM) and the partially closed coniferous and broad-leaved mixed multi-layered forest (PMM), PM_1 concentrations were negatively affected by other vegetation structures. In particular, the partially closed broad-leaved one-layered forest (PBO) had the best dust-retention effect for PM_1 and $PM_{2.5}$. In addition, the partially closed broad-leaved multi-layered forest (PBM), partially closed coniferous and broad-leaved mixed multi-layered forest (PMM), and closed broad-leaved one-layered forest (CBO) had positive impacts on $PM_{2.5}$ concentration—that is, $PM_{2.5}$ concentration was higher in these types of green spaces. In the closed broad-leaved one-layered forest

(CBO), closed coniferous one-layered forest (CCO,) and closed coniferous and broad-leaved mixed one-layered forest (CMO) there were significant differences in PM_{10} concentration reduction, while the other seven types of green spaces had moderate and similar reduction effects on PM_{10} concentration. In the closed coniferous one-layered forest (CCO) and the closed broad-leaved one-layered forest (CBO), TSP concentration reached the maximum value, and dust-retention ability was weak; however, the partially closed broad-leaved multi-layered forest (PBM) had a strong reduction effect on PM_{10} and TSP.

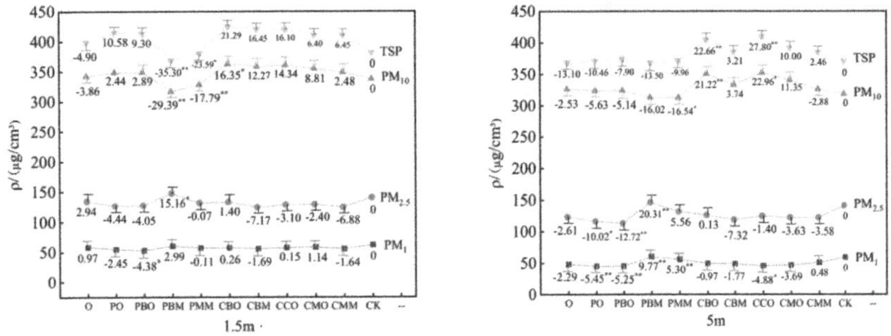

(a) **(b)**

Figure 8. Comparison of airborne particulate matter concentrations in vegetation structures at different vertical heights. Note: The values in the figure represent the correlation coefficients of generalized linear analysis between green spaces with different vegetation structures and concentrations of airborne particulate matter with different particle sizes—that is, the distance between different factor levels and the population mean. **: At level 0.01 (two-tailed), the correlation was significant; *: At level 0.05 (two-tailed), the correlation was significant. (**a**) Comparison of airborne particulate matter concentrations in the different vegetation structures at 1.5m height; (**b**) Comparison of airborne particulate matter concentrations in the different vegetation structures at 5m height.

In general, the reduction effects of different vegetation structures on the concentrations of particulate matter with different particle sizes were significantly different, but the reduction effect was consistent at different heights. Regardless of whether the height was 1.5 m or 5 m, the partially closed broad-leaved one-layered forest (PBO) had the strongest dust-retention ability for small PM_1 and $PM_{2.5}$ particles, followed by partially open green spaces (PO). The concentrations of PM_1 and $PM_{2.5}$ were the highest in the partially closed broad-leaved multi-layered forest (PBM). For the large PM_{10} and TSP particles, the partially closed broad-leaved multi-layered forest (PBM) and the partially closed coniferous and broad-leaved mixed multi-layered forest (PMM) had strong adsorption capacity and a good dust-retention effect, while the closed broad-leaved one-layered forest (CBO) structure had a weak reduction effect.

4. Discussion

The results show that the concentrations of airborne particulate matter with different particle sizes in urban green spaces were all affected by meteorological parameters, monitoring time, vegetation structure, and height, indicating that the control of air pollution by urban green spaces is indeed a complex process, due to the synthetic action of these factors. Therefore, more factors and their synergies should be taken into consideration for improving air quality via urban green spaces in the future.

4.1. The Influence of Meteorological Parameters on the Concentration of Airborne Particulate Matter

Our results showed that temperature, relative humidity, wind speed, and air pressure had significant differences in their effects on the concentration of airborne particulate matter.

With the increase in temperature, air convection in the vertical direction was more frequent. Such gas circulation exchange accelerated the transport of airborne particulate matter, which was conducive to reducing the concentration of airborne particulate matter [28]. Airborne particulate matter concentration and relative air humidity showed good consistency. When the relative humidity of the air increased, the concentration of airborne particulate matter also increased. The increase in humidity made the moisture in the air increase, the particles become moist, and the weight of the particles increase, thus reducing the diffusion of the particles, so that the particles gathered to a certain extent, leading to an increase in the concentration of the particles in the air [29]. When the relative humidity increased to a certain extent, the wet sedimentation increased, and then the concentration of airborne particulate matter decreased. In addition, the increase in wettability and relative humidity can trigger certain biological particle emission mechanisms, such as active wet ejection of fungal spores or hygroscopic expansion of pollen fragmentation, thereby increasing the concentration of airborne particulate matter around the plants [30]. The influence of wind speed on airborne particulate matter varied with the surrounding environment. In this study, the conditions of sunny, breezes, or no wind were selected for monitoring (the mean wind speed was 0.31 m/s), and it was still found that wind speed had a significant negative correlation with the concentration of airborne particulate matter. The greater the wind speed, the more conducive to the spread of pollutants, and the less the $PM_{2.5}$ concentration. Small wind or no wind, an obvious inversion layer limited vertical movement of the low-altitude atmosphere [31]. In airborne particulate matter stuck at low heights or close to the ground, fog/haze weather appeared frequently. The airborne particulate matter was difficult to spread and not conducive to dilution to the periphery, which facilitated the formation of local accumulated airborne particulate matter, subjecting the air quality to a high concentration of pollution [32]. The reason for this was that low-speed wind cannot carry away water vapor; thus, the increase in humidity was conducive to the formation of haze [33]. When the air pressure was lower, the concentrations of PM_1 and $PM_{2.5}$ decreased accordingly. After the change in air pressure, the particles converged to the middle in the horizontal direction, and moved upward in the vertical direction. At the same time, the wind speed in the horizontal direction was low, and the horizontal diffusion conditions of particles were unfavorable [34].

4.2. The Influence of Time on the Concentration of Airborne Particulate Matter

It can be found that time has a significant effect on the concentration of airborne particulate matter. Airborne particulate matter concentration in the heating period was much higher than that in the pre-heating period. The causes of this phenomenon were diverse. In autumn and winter, waste from burning fossil fuels in northern China contributed directly to the increase in the concentration of particulate matter in the air [35]. At the same time, the decrease in temperature and precipitation was not conducive to the settling of airborne particulate matter [36]. In the winter heating period, broad-leaved trees shed a large number of leaves, and only some evergreen broad-leaved trees and conifers act as dust traps; thus, this reduced the adsorption capacity of particles, resulting in a higher concentration of airborne particulate matter. PM is a kind of aerosol substance, and this kind of airborne particulate matter has a certain gravity effect. The daily variation of airborne particulate matter with small particle size reached the maximum value between 12:00 and 14:00 at noon, which may be related to human activities, and the flow of people was large at noon [37]. The convective exchange and vertical diffusion of the atmosphere were strengthened, and the thickness of the mixing layer increased [38]. At night, after sedimentation, the larger diameter particles accumulated continuously near the ground. At the same time, solar radiation could also be a factor in the diurnal variation of airborne particulate matter concentration on sunny days [39]. With the emergence of solar radiation during the day, the ground temperature rose and formed a warm air mass near the ground. The warm air mass near the ground rose with the particles, so the concentration of particles near the ground began to increase at noon, and reached its minimum in the evening with the influence of temperature and light.

4.3. The Influence of Vegetation Structure on the Concentration of Airborne Particulate Matter at Different Heights

Among the 10 types of green spaces in this study, the distributions of airborne particulate matter in different vegetation structures were different. Our results showed that the concentration of airborne particulate matter in the lower part of the canopy was much higher than that in the middle part of the canopy, which is consistent with the findings of Chan [40]. Due to the different underlying surface of the sample sites, with the increase in height, the atmospheric humidity first decreased, and the concentration of airborne particulate matter also decreased [41]. In autumn and winter, when the temperature dropped, fine particles had a strong suspension capacity in the atmosphere, and were also affected by various factors at different heights, such as leaf adsorption and branch blocking [42]. Coarse particulate matter was larger in size, and had obvious sedimentation in the atmosphere. The concentration near the ground was higher than that at the height of the plant canopy. With the passage of time, after the heating began, the branches of trees would also block the diffusion of particles. At the same time, when the temperature dropped, the particles kept colliding and undergoing friction, condensation, settlement, etc., and their concentration gradually increased [43].

Green spaces with different special structures had different effects on the reduction in the concentration of airborne particulate matter with various particle sizes. The larger the particle size, the stronger the reduction ability of the green spaces. The reason for this may be that plants could use their special micromorphological structures to play a certain retention role [44]. The quantity ratio and volume ratio of PM_{10} retained on leaf surfaces are often much higher than those of $PM_{2.5}$ [45], indicating that plants have a stronger effect on reducing airborne particulate matter with larger particles. The density of particulate matter was higher than that of the single-layer structure in the green spaces, which may be related to the higher plant density in the green spaces, along with the poor ventilation conditions in the forest, which was not conducive to the transport and diffusion of airborne particulate matter [46]. Sehmel et al. found that when the dust-containing air flow passed through the tree crown, some dust with larger particles was blocked by the branches/leaves and fell, and could be returned to the air due to the action of wind and other external forces, while the other part remained on the surface of the branches and leaves [47]. The amount of PM gathered on leaves depends on the quantity, size, and morphology of the leaves, and can also be increased by the presence of epicuticular waxes, in which PM can become stuck or immersed [48]. In the closed multi-layered structural forest, the vertical structure of the vegetation was complex and there were many plant species, which to some extent hinders the airborne particulate matter from settling to the ground [49]. The particles adsorbed on the surfaces of plant leaves were only temporarily retained, and were prone to bouncing back and then being suspended in the atmosphere, thus increasing the concentration of airborne particulate matter [50]. Urban flowering meadows are more structurally and botanically diverse than lawns. Influenced by natural ecosystems, urban flowering meadows are mowed less frequently, thus reducing the emission of particulate matter into the air [15]. However, in the open green spaces dominated by lawns, the settling effect of airborne particulate matter in the air was less hindered, and the airborne particulate matter directly settled on the ground due to the effect of gravity [51]. In addition, mosses showed a higher capability of trapping atmospheric particulate matter than certain trees [52]. Therefore, different vegetation structures had different distributions of airborne particulate matter at different heights of vegetation.

4.4. Limitations and Future Study

The shortcomings of this study are that there was a set period of monitoring the concentration of particulate matter over the spatial structure, and the setting of the height gradient was limited. The next step should be to increase the monitoring time and vegetation height for further study, in order to provide an optimal urban green space planning scheme for future reduction in the concentration of airborne particulate matter.

5. Conclusions

Taking the urban green spaces with different spatial structures in Xi'an as the study area, and considering meteorological parameters, this study quantitatively compared the effects of monitoring time, spatial structure, and vegetation height on the concentration of airborne particulate matter in the urban green spaces. The results showed that the following: (1) There were significant correlations between meteorological parameters and particulate concentrations. The concentrations of different sizes of particulate matter in northern China during the heating season are generally higher than those in pre-heating season. In the evening, the concentration of airborne particulate matter was low. At noon, the concentrations of PM_1 and $PM_{2.5}$ reached their maximum. The concentrations of PM_{10} and TSP reached their maximum in the morning. It is not recommended to go out during the morning and noon; rather, one should travel less and stay indoors (Figure 9). (2) The effects of different vegetation structures on reducing the concentration of airborne particulate matter with different particle sizes were significantly different, but the effect was consistent at different heights. Partially closed green spaces had strong adsorption capacity for particles with different sizes, which played a certain reduction role. Moreover, the concentration of particulate matter below the canopy was generally higher than that in the middle of the canopy. In order to improve the air quality in the future, the partially closed space enclosure model can be given priority in the planning and design of urban green spaces. For the adsorption of fine particulate matter (PM_1 and $PM_{2.5}$), it is recommended to plant the partially closed broad-leaved one-layered forest (PBO). For PM_{10} and TSP, the partially closed broad-leaved multi-layered forest (PBM) and the partially closed coniferous and broad-leaved mixed multi-layered forest (PMM) are recommended. At the same time, the terrain of urban forests can be modified to raise the ground for planting in order to provide potential fresh air during human recreation (Figure 10). The results of this study identified the dust-retention effects of urban green spaces with different spatial structures, which can provide parameterization information for air-improvement-oriented planning and design of urban green spaces in future.

Figure 9. Schematic diagrams of time variation of different particle size concentrations.

Figure 10. Schematic diagram of the influence of vegetation structure on airborne particulate matter concentration.

Author Contributions: Conceptualization, L.Q. and T.G.; methodology, L.Q. and T.G.; validation, B.J., L.Q. and T.G.; formal analysis, B.J., L.Q. and T.G.; investigation, B.J., C.S., S.M., Y.C., Z.Z., Y.L., L.Q. and T.G.; resources, L.Q. and T.G.; data curation, B.J., C.S., S.M., Z.Z., Y.C., Y.L., L.Q. and T.G.; writing—original draft preparation, B.J.; writing—review and editing, B.J., L.Q. and T.G.; visualization, B.J., L.Q. and T.G.; supervision, L.Q. and T.G.; project administration, L.Q. and T.G.; funding acquisition, L.Q. and T.G. All authors have read and agreed to the published version of the manuscript.

Funding: This research was funded by the National Natural Science Foundation of China (grant number: 31971722), the Science and Technology Innovation Program of Shaanxi Academy of Forestry (grant number: SXLK2021-0216), the Key Program of Science and Technology Innovation of the Shaanxi Academy of Forestry (grant number: SXLK2021-02-0X), the Key Research and Development Program of Xianyang (grant number: 2021ZDYF-SF-0022), and the Scientific Research Cooperation Agreement Project of the Xianyang Forestry Bureau (grant number: 20211221000007).

Institutional Review Board Statement: The study was conducted according to the guidelines of the Declaration of Helsinki, and approved by the Institutional Review Board of College of Landscape Architecture and Arts, Northwest A&F University.

Informed Consent Statement: Informed consent was obtained from all subjects involved in the study.

Data Availability Statement: The data presented in this study are available on request from the corresponding author. The data are not publicly available due to policy of the institute.

Acknowledgments: We are grateful to Katie Oswalt of Mississippi State University for helping to revise this manuscript. We also thank the 60 volunteers who helped us with the experiment.

Conflicts of Interest: The authors declare no conflict of interest.

References

1. Xing, Y.; Brimblecombe, P. Urban park layout and exposure to traffic-derived air pollutants. *Landscape Urban Plan* **2020**, *194*, 103682. [CrossRef]
2. Hu, M.G.; Jia, L.; Wang, J.F.; Pan, Y.P. Spatial and temporal characteristics of particulate matter in Beijing, China using the Empirical Mode Decomposition method. *Sci. Total Environ.* **2013**, *458–460*, 70–80. [CrossRef] [PubMed]
3. He, C.; Qiu, K.Y.; Alahmad, A.; Pott, R. Particulate matter capturing capacity of roadside evergreen vegetation during the winter season. *Urban For. Urban Green.* **2020**, *48*, 126510. [CrossRef]
4. Beckett, K.P.; Freer-Smith, P.H.; Taylor, G. Urban woodlands: Their role in reducing the effects of particulate pollution. *Environ. Pollut.* **1998**, *99*, 347–360. [CrossRef]
5. Li, X.; Wu, J.R.; Elser, M.; Feng, T.; Cao, J.J.; El-Haddad, I.; Huang, R.; Tie, X.X. Contributions of residential coal combustion to the air quality in Beijing–Tianjin–Hebei (BTH), China: A case study. *Atmos. Chem. Phys.* **2018**, *18*, 10675–10691. [CrossRef]
6. Sawidis, T.; Breuste, J.; Mitrovic, M. Trees as bioindicator of heavy metal pollution in three European cities. *Environ. Pollut.* **2011**, *159*, 3560–3570. [CrossRef]

7. Weber, S.; Litschke, T. Variation of particle concentrations and environmental noise on the urban neighbourhood scale. *Atmos. Environ.* **2008**, *42*, 7179–7183. [CrossRef]
8. Nowak, D.J.; Crane, D.E.; Stevens, J.C. Air pollution removal by urban trees and shrubs in the United States. *Urban For. Urban Green.* **2006**, *4*, 115–123. [CrossRef]
9. Chen, S.J. Effects of plants on atmospheric particulate matter based on simulation experiments. *Beijing For. Univ.* **2015**, 3–10.
10. Weerakkody, U.; Dover, J.W.; Mitchell, P. The impact of rainfall in remobilising particulate matter accumulated on leaves of four evergreen species grown on a green screen and a living wall. *Urban For. Urban Green.* **2018**, *35*, 21–31. [CrossRef]
11. Guo, W.; Shen, T.Y.J.; Zheng, S.Q.; Wang, W.; Liu, C.F. Progress in the research on the mechanism and law of dust retention in urban green space. *Ecol. Environ.* **2010**, *19*, 1465–1470.
12. Lovett, G.M.; Lindberg, S.E. Concentration and deposition of particles and vapors in a vertical profile through a forest canopy. *Atmos. Environ.* **1992**, *26*, 1469–1476. [CrossRef]
13. Li, H.E.; Wang, Z.Y.; Tan, J.D.; Hu, X.C.; Lu, Y.D.; Li, P.S. Effects of Main Urban Landscape Plants on Dust Blocking in Fosha. *Ecol. Sci.* **2006**, *25*, 395–399.
14. Su, J.X.; Jin, S.J.; Yan, J.G.; Zhang, Q.D.; Gao, R.R.; Lu, Y.M. Study on the Dust Catching Property of the Campus Plants in Shanxi Normal University. *J. Shanxi Norm. Univ. Nat. Sci. Ed.* **2006**, *20*, 85–88.
15. Przybysz, A.; Popek, R.; Stankiewicz-Kosyl, M.; Zhu, C.; Małecka-Przybysz, M.; Maulidyawati, T.; Mikowska, K.; Deluga, D.; Griżuk, K.; Sokalski-Wieczorek, J.; et al. Where trees cannot grow—Particulate matter accumulation by urban meadows. *Sci. Total Environ.* **2021**, *785*, 147310. [CrossRef] [PubMed]
16. Terzaghi, E.; Wild, E.; Zacchello, G. Forest Filter Effect: Role of leaves in capturing/releasing air particulate matter and its associated PAHs. *Atmos. Environ.* **2013**, *74*, 378–384. [CrossRef]
17. McDonald, A.G.; Bealey, W.J.; Fowler, D. Quantifying the effect of urban tree planting on concentrations and depositions of PM_{10} in two UK conurbations. *Atmos. Environ.* **2007**, *41*, 8455–8467. [CrossRef]
18. Gao, J.H.; Wang, D.M.; Zhao, L.; Wang, D.D. Journal of Beijing Forestry University Airborne Dust Detainment by Different Plant Leaves: Taking Beijing as An Example. *J. Beijing For. Univ.* **2007**, *29*, 94–99.
19. Selmi, W.; Weber, C.; Rivière, E. Air pollution removal by trees in public green spaces in Strasbourg city, France. *Urban For. Urban Green.* **2016**, *17*, 192–201. [CrossRef]
20. Li, L.; Wang, X.F.; Zuo, Z. Monthly and seasonal variation of PM_{10} concentration and its relationships withmeteorological factors in Shapoutou of Ningxia. *Shengtaixue Zazhi* **2019**, *38*, 1175–1181.
21. Zheng, G.; Li, P. Resuspension of settled atmospheric particulate matter on plant leaves determined by wind and leaf surface characteristics. *Environ. Sci. Pollut. R.* **2019**, *26*, 19606–19614. [CrossRef]
22. Li, H. *Analysis between the Air Pollution Status and Meteorogical Conditions in Shijiazhuang*; Hebei University of Science and Technology: Shijiazhuang, China, 2015; Volume 67.
23. Yang, L.X.; Wang, D.C.; Cheng, S.H. Influence of meteorological conditions and particulate matter on visual range impairment in Jinan, China. *Sci. Total Environ.* **2007**, *383*, 164–173. [CrossRef] [PubMed]
24. Jim, C.Y.; Chen, W.Y. Assessing the ecosystem service of air pollutant removal by urban trees in Guangzhou (China). *J. Environ. Manag.* **2008**, *88*, 665–676. [CrossRef] [PubMed]
25. Li, M.; Zhang, W.J.; Fu, H.X.; Li, H.B. Air quality and its relationship with meteorological conditions in Jinan. *Environ. Sci. Technol.* **2018**, *41*, 130–139.
26. China Meteorological Administration (CMA). 2020. 2021. Available online: http://www.cma.gov.cn/ (accessed on 9 November 2020).
27. Gao, T.; Qiu, L.; Hammer, M. The Importance of Temporal and Spatial Vegetation Structure Information in Biotope Mapping Schemes: A Case Study in Helsingborg, Sweden. *Environ. Manag.* **2012**, *49*, 459–472. [CrossRef] [PubMed]
28. Liu, X.H.; Yu, X.X.; Zhang, Z.M.; Liu, M.M.; Ruan, S.Q.C. Characteristics of PM_{10} and $PM_{2.5}$ pollution in forest belt and its relationship with meteorological conditions. *Chin. J. Ecol.* **2014**, *33*, 1715–1721.
29. Dawson, J.P.; Adams, P.J.; Pandis, S.N. Sensitivity of $PM_{2.5}$ to climate in the Eastern US: A modeling case study. *Atmos. Chem. Phys.* **2007**, *7*, 4295–4309. [CrossRef]
30. Huffman, J.A.; Prenni, A.J.; DeMott, P.J. High concentrations of biological aerosol particles and ice nuclei during and after rain. *Atmos. Chem. Phys.* **2013**, *13*, 6151–6164. [CrossRef]
31. Freer-Smith, P.H.; El-Khatib, A.A.; Taylor, G. Capture of Particulate Pollution by Trees: A Comparison of Species Typical of Semi-Arid Areas (Ficus Nitida and Eucalyptus Globulus) with European and North American Species. *Water Air Soil Pollut.* **2004**, *155*, 173–187. [CrossRef]
32. Tomašević, M.; Vukmirović, Z.; Rajšić, S. Characterization of trace metal particles deposited on some deciduous tree leaves in an urban area. *Chemosphere* **2005**, *61*, 753–760. [CrossRef]
33. Kretinin, V.M.; Selyanina, Z.M. Dust retention by tree and shrub leaves and its accumulation in light chestnut soils under forest shelterbelts. *Eurasian Soil Sci.* **2006**, *39*, 334–338. [CrossRef]
34. Mu, L.; Deng, F.; Tian, L. Peak expiratory flow, breath rate and blood pressure in adults with changes in particulate matter air pollution during the Beijing Olympics: A panel study. *Environ. Res.* **2014**, *133*, 4–11. [CrossRef]
35. Zhang, F.; Cheng, H.; Wang, Z. Fine particles ($PM_{2.5}$) at a CAWNET background site in Central China: Chemical compositions, seasonal variations and regional pollution events. *Atmos. Environ.* **2014**, *86*, 193–202. [CrossRef]

36. Li, M. *Spatial and Temporal Characteristics and Influencing Factors of Haze Pollution in Henan Province*; Chongqing University: Chongqing, China, 2017; p. 55.
37. Leonard, R.J.; McArthur, C.; Hochuli, D.F. Particulate matter deposition on roadside plants and the importance of leaf trait combinations. *Urban For. Urban Green.* **2016**, *20*, 249–253. [CrossRef]
38. Wu, Z.W.; Ma, X.; Yang, X. Uneven Distribution of $PM_{2.5}$ and Residential Area Planning. *Appl. Mech. Mater.* **2013**, *361–363*, 966–972. [CrossRef]
39. Song, Y.; Tang, X.Y.; Xie, S.D. Source apportionment of $PM_{2.5}$ in Beijing in 2004. *J. Hazard. Mater.* **2007**, *146*, 124–130. [CrossRef] [PubMed]
40. Chan, Y.C.; Simpson, R.W.; Mctainsh, G.H.; Vowles, P.D. Characterisation of chemical species in $PM_{2.5}$ and PM_{10} aerosols in Brisbane, Australia. *Atmospheric Environment* **1997**, *31*, 3773–3785. [CrossRef]
41. Brechtel, F.J.; Kreidenweis, S.M. Predicting particle critical supersaturation from hygroscopic growth measurements in the humidified TDMA. Part I: Theory and sensitivity studies. *J. Atmos. Sci.* **2000**, *57*, 1854–1871. [CrossRef]
42. Zhao, C.X.; Wang, Y.J.; Wang, Y.Q.; Zhang, H.L. Interactions between fine particulate matter ($PM_{2.5}$) and vegetation: A review. *Chin. J. Ecol.* **2013**, *32*, 2203–2210.
43. Chang, Y.M.; Yan, P.B.; Yang, J. Suggestions on Tree Species Selection for Urban Greening in Beijing to Control $PM_{2.5}$ Pollution. *Chinese Landscape Architecture* **2015**, *1*, 15.
44. Hwang, H.J.; Yook, S.J.; Ahn, K.H. Experimental investigation of submicron and ultrafine soot particle removal by tree leaves. *Atmos. Environ.* **2011**, *45*, 6987–6994. [CrossRef]
45. Zhao, S.T.; Li, X.Y.; Li, Y.M. The Characteristics of deposition of airborne particulate matters with different size on certain plants. *Ecol. Environ. Sci.* **2014**, *23*, 271–276.
46. Huang, C.; Lin, M.; Khlystov, A. The Effects of Leaf Area Density Variation on the Particle Collection Efficiency in the Size Range of Ultrafine Particles (UFP). *Environ. Sci. Technol.* **2013**, *47*, 11607–11615. [CrossRef]
47. Sehmel, G.A. Particle and gas dry deposition: A review. *Atmos. Environ.* **1980**, *14*, 983–1011. [CrossRef]
48. Łukowski, A.; Popek, R.; Karolewski, P. Particulate matter on foliage of *Betula pendula*, *Quercus robur*, and *Tilia cordata*: Deposition and ecophysiology. *Environ. Sci. Pollut. Res.* **2020**, *27*, 10296–10307. [CrossRef] [PubMed]
49. Setyan, A.; Zhang, Q.; Merkel, M. Characterization of submicron particles influenced by mixed biogenic and anthropogenic emissions using high-resolution aerosol mass spectrometry: Results from CARES. *Atmos. Chem. Phys.* **2012**, *12*, 8131–8156. [CrossRef]
50. Lv, L.Y.; Li, H.Y.; Yang, J.N. The temporal-spatial variation characteristics and influencing factors of absorbing air particulate matters by plants: A review. *Chin. J. Ecol.* **2016**, *35*, 524–533.
51. Fenn, M.E.; Bytnerowicz, A. Summer throughfall and winter deposition in the San Bernardino Mountains in Southern California. *Atmos. Environ.* **1997**, *31*, 673–683. [CrossRef]
52. Haynes, A.; Popek, R.; Boles, M.; Paton-Walsh, C.; Robinson, S. Roadside Moss Turfs in South East Australia Capture More Particulate Matter Along an Urban Gradient than a Common Native Tree Species. *Atmosphere* **2019**, *10*, 224. [CrossRef]

forests

MDPI

Article

The Role of Trees in Winter Air Purification on Children's Routes to School

Adrian Hoppa [1], Daria Sikorska [1,*], Arkadiusz Przybysz [2], Marta Melon [1] and Piotr Sikorski [1]

[1] Department of Remote Sensing and Environmental Assessment, Institute of Environmental Engineering, Warsaw University of Life Sciences-SGGW, Nowoursynowska 159 Str., 02-776 Warsaw, Poland; adrian_hoppa@sggw.edu.pl (A.H.); marta_melon@sggw.edu.pl (M.M.); piotr_sikorski@sggw.edu.pl (P.S.)

[2] Institute of Horticultural Sciences, Warsaw University of Life Sciences-SGGW, Nowoursynowska 159, 02-776 Warsaw, Poland; arkadiusz_przybysz@sggw.edu.pl

* Correspondence: daria_sikorska@sggw.edu.pl; Tel.: +48-2259-35393

Abstract: Air pollution is now considered to be the world's largest environmental health threat, accounting for millions of deaths globally each year. The social group that is particularly exposed to the harmful effects of air pollution is children. Their vulnerability results from higher breathing frequency and being subject to concentration peaks just above the ground. The negative effects of ambient particulate matter also depend on the time of exposure. A daily route to school can constitute an important component of children's physical activity, but air pollution can pose a threat to their health. Numerous studies have proved that high loads of PM can be effectively reduced by vegetation. Little is known, however, on whether vegetation can also reduce PM during leaf dormancy. In this study we investigated the role of trees in air purification during the leafless period in children's routes to selected schools located in Warsaw during winter. The results obtained show a weak impact of the tree canopy in winter.

Keywords: air pollution; particulate matter; WAI; urban trees; brown leaf area index

Citation: Hoppa, A.; Sikorska, D.; Przybysz, A.; Melon, M.; Sikorski, P. The Role of Trees in Winter Air Purification on Children's Routes to School. *Forests* **2022**, *13*, 40. https://doi.org/10.3390/f13010040

Academic Editors: Radu-Daniel Pintilii and Diego Varga

Received: 8 December 2021
Accepted: 23 December 2021
Published: 1 January 2022

1. Introduction

Air pollution is now considered to be the world's largest environmental health threat, accounting for millions of deaths globally each year [1–3]. The main component of air pollution is particulate matter (PM), which can be emitted to the atmosphere directly (primary PM), or can be formed as a result of chemical reactions (secondary PM) [4]. Apart from natural sources of PM, the anthropogenic PM emissions include primarily fuel combustion and manufacturing processes [5,6]. Recent years have shown that the highest annual average concentrations of PM_{10} and $PM_{2.5}$ in Europe occur in central and eastern European countries, mainly in Poland [7].

Among various air pollutants, particulate matter (PM), because of their small particle size, is the most harmful and most representative pollutant [8,9] and its major toxicological effects on human health and the environment have been observed for decades [10]. PM has been associated with an increased risk of respiratory health outcomes among children [11,12] and an increased risk of cardiovascular diseases, including heart failure and myocardial infarction, hypertension and stroke [13]. Children are a particularly vulnerable group to the effects of PM [14], as they are more active, breathe proportionately more air than adults, their respiratory systems are still developing, and they spend more time outdoors, inhaling the highest PM concentrations just above the ground. Children growing up in the most polluted areas reveal significant lung function deficits [15] and studies show an increased incidence of ADHD [16] and developing allergies [17]. The locations particularly important in terms of risks of children's exposure to air pollution are their routes to school. Walking to school takes up more than 50% of children's active time, compared to about 20% spent at school, 10% at home and 1% in green areas [18].

Concerns over the health and well-being of the city residents, particularly the most vulnerable groups, make it necessary to take appropriate measures to shield them from exposure to harmful PM. As a significant share of the children's daily activities is their commute to school, the proper design of their routes can support a friendly and healthy environment and reduce the negative effects of air pollution on their journey [19]. The selection of proper plant species in these areas can have a positive effect on the ambient air quality, particularly where PM levels are exceeded [20]. Positive relations between the presence of greenery and their beneficial role in ensuring children's health and well-being have been long investigated. Children who live near urban green areas have better lung capacity [21], while street trees have been proven to be beneficial for childhood asthma prevention [22]. Growing up in greener neighborhoods may also be beneficial for brain development and cognitive functions [23]. Children who grew up in environments with the lowest levels of green were 55% more likely to develop mental disorders [24].

Factors that influence the ability of plants to accumulate pollutants are the location and structure of greenery, morphological characteristics of plants forming the plant community and environmental conditions [25]. Plants do not have the ability to escape from a contaminated site, and therefore they have evolved mechanisms that allow them to survive in a contaminated environment. This ability is the basis of phytoremediation technology, which involves using plants to trap pollutants and, under certain conditions, break them down. Phytoremediation uses selected species of trees, shrubs and climbers that are able to accumulate on their leaves PM harmful to human health, thus supporting the process of air purification from pollutants [26]. Research shows that the presence of plants near buildings can have a positive impact on well-being, as well as physical and mental health [27]. However, during the winter, when air emissions are particularly high [28], there is a lack of comprehensive research on the role of trees in the leafless season, which makes the role of plants in the winter season unclear. The aim of this study was to determine the extent of particulate matter pollution that children are exposed to on their routes to school during the leafless season, and to investigate how the adjacent trees can reduce exposure to high PM concentrations.

2. Materials and Methods

The study area is located in Warsaw, Poland's capital and the largest city with a population of 1.79 million [29]. The average annual temperature in Warsaw is about 9.3 °C, while the yearly precipitation is about 695 mm. Despite being a relatively green city, with the vegetation cover exceeding 50% [30], Warsaw is characterized by the phenomenon of urban heat island [31]. The character of pollution in Warsaw is typical of large urban agglomerations, $PM_{2.5}$, PM_{10}, nitrogen and carbon oxides and sulfur dioxide being the dominant pollutants [32]. In the very center of Warsaw and in densely populated districts located outside the city center, high levels of PM have been noted. The analysis of PM concentrations in the winter season showed that the main emission sources are of anthropogenic origin (energy production based on coal and biomass combustion). In the warm season, the pollutants mainly originate from local emission sources, mainly urban traffic and transportation [33]. Warsaw launched the program to improve air quality and has been continuously encouraging and subsidizing the removal of black-smoke-belching stoves; however, the share of existing ones still remains at a high level, being unevenly distributed throughout the city (Figure 1).

Figure 1. Location of the selected schools and the examined routes and study plots. Examined schools (**A**)—no. 319 (Ursynów district), (**B**)—no. 218 (Wawer district), (**C**)—no. 385 (Wesoła district), (**D**)—no. 70 (Mokotów district).

In this study we investigated the pollution levels during a daily school commute of children on their way to school. We took into consideration primary schools, due to the fact that children aged 6–14 are more susceptible to high pollutant concentration levels, but they are also most likely to commute to school on foot. Due to regionalization of primary education in Poland, those children are more likely to attend the nearest school. Out of 320 elementary schools in Warsaw [34], we selected four that were located outside the strict city center, so that the mean annual pollution levels did not differ significantly (Table 1). We selected schools representing various possible pollution levels originating from municipal emissions and similar conditions in terms of possible traffic (proximity to larger roads). The schools were selected in pairs—one group near a cluster of black-smoke-belching stoves (B and D) and one beyond such a cluster (A and C) (Table 1). One pair (B and C) was selected in a location where the share of tree-covered areas in the neighbourhood (500 m buffer zone) was high (over 30%) and the other schools (A and D) were characterized by a lower share of tree-covered areas in their immediate vicinity (Table 1). For the assessment of the share of vegetated surfaces and traffic conditions, we used BDOT (Database of Topographic Objects for Poland), which is the most fundamental source of information on the location of topographic objects in Poland [35].

Table 1. Characteristics of investigated school locations and neighborhood in a 500 m buffer zone (source: BDOT 10k) and average annual concentrations of $PM_{2.5}$ and PM_{10} [36] and locations of black-smoke-belching stoves—heating furnaces [37].

School	District	Share of Tree Covered Area in a 500 m Buffer Zone(%)	Mean Annual Concentration of $PM_{2,5}$ (μg/m^3)	Mean Annual Concentration of PM_{10} (μg/m^3)	Number of Black-Smoke-Belching Stoves in 1 Km Buffer
A	Ursynów	21.1	20.5	26.8	0
B	Wawer	36.6	19.6	25.4	8
C	Wesoła	57.2	19.1	24.5	0
D	Mokotów	0.7	22.1	29.0	7

We inventoried all walking routes from the schools' entrance within a buffer of 400 m that we observed to be frequently used by children as their home-school routes (after initial observations conducted during one workday at each of the selected schools). In each of the locations, we chose 3 to 4 of the main routes most frequently used by the pupils.

Along the routes, we took regular measurements in plots located every 20 m. In each plot we identified winter vegetation density (LAI) and measured PM concentrations. In our research we took into account the locations under the tree canopy and beyond the tree cover. In cases when there was another walking route crossing the path, or where the plot was located at the edge of the tree canopy, the plot was rejected to avoid the edge effect.

At all locations we measured $PM_{2.5}$ and PM_{10} concentrations in each of the study plots in December 2020 on three windless days (nearest local weather station indicating winds below 0.2 m/s) at weekly intervals. Measurements for each day were made in the morning (8–10:00), representing highest concentration rates typical for peak hours (possible increased loads due to traffic) and in the afternoon (12–14 PM) hours at the time when children were travelling from school, but also when the PM concentrations recorded could be lower and originate further from the heating sources.

We measured $PM_{2.5}$ and PM_{10} concentrations with the Dust Air device [38] at 140 cm, corresponding to the height of the primary school pupils. For each school we recorded a series of measurements in 3 days, every week, meaning there were 3 repetitions per each school, concerning both morning and afternoon measurements. Each series lasted 60 s with a 10 s interval between the measurements. In order to determine the relation between vegetation density and PM loads we measured Leaf Area Index (LAI) along the selected home-school routes using the SS1-COM-R4 Complete System with Radio Link [39]. The measurements were taken along the walking routes on their right side in the direction of walking, the study plots were located 1.5 m away from the side of the pavement. As the LAI measurement is being calculated per 1 m^2, to ensure we captured the diversity of vegetation density we took 3 adjacent measurements and averaged the score per plot. The LAI meter is most commonly used to determine the density of canopy; however, during the vegetation dormancy season it can be effectively used for the assessment of the density of branches, which could allow for the deposition of PM and also a sheltering of the walking routes from high loads of pollutants. LAI measured during leafless season is referred to as Wooden Area Index (WAI), which provides information on the density of woody shoots and leaves remaining in the winter [40].

Statistical Analysis

We analysed the data on PM concentration and vegetation density in Statistica 10 software. We tested the relationship between WAI and PM concentration levels with Pearson's correlation (after having confirmed the data is normally distributed with Kolmogorov–Smirnov test). We used one-way ANOVA for comparing the data at $p < 0.05$ significance level.

3. Results

3.1. Particulate Matter Content of the Studied Routes to Schools

Ambient air PM concentrations recorded along routes to school in wintertime were very much associated with the location of the school in terms of the proximity to individual household heating emitters (Table 1). Irrespective of the PM fraction and the time of day when measurements were made, significantly higher PM concentrations were recorded at schools B and D, which were surrounded by more emission sources (Figure 2). Concentrations at these locations exceeded the average acceptable level for the 24-h period (PM_{10} 50 $\mu g/m^3$). Concentrations recorded in the afternoon were significantly higher than in the morning during rush hours (Figure 2).

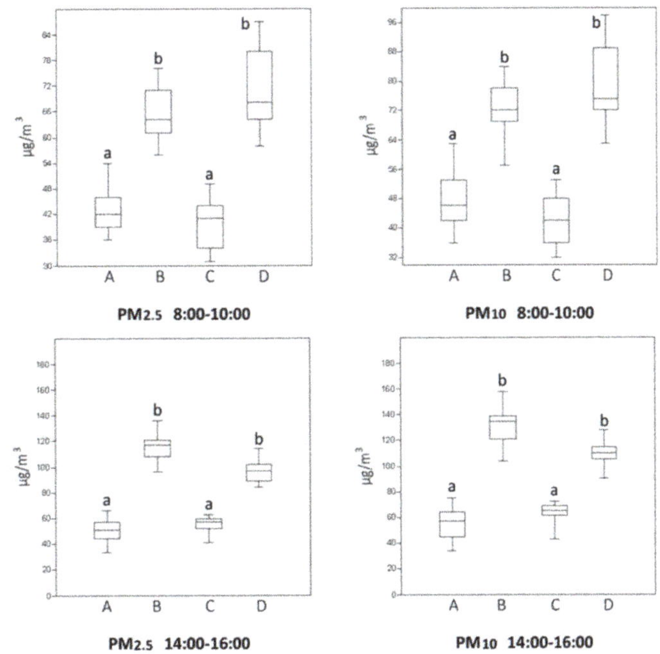

Figure 2. Mean $PM_{2.5}$ and PM_{10} (µg/m^3) concentrations of in the morning and evening on the studied sections of the road to schools from all measurement points. a, b—homogeneous groups, significant differences at $p < 0.05$. Letters A–D refer to school's symbols.

3.2. *Vegetation Density along the School Routes in Winter*

There were no statistical differences in the amount of WAI along the studied routes to schools. The average index ranged from 0.2 to 0.6, with high spatial variability (Figure 3). The highest mean WAI values were recorded in the surroundings of schools A and C (Figure 3), where the proportion of forests and woodlands was the highest (Table 1).

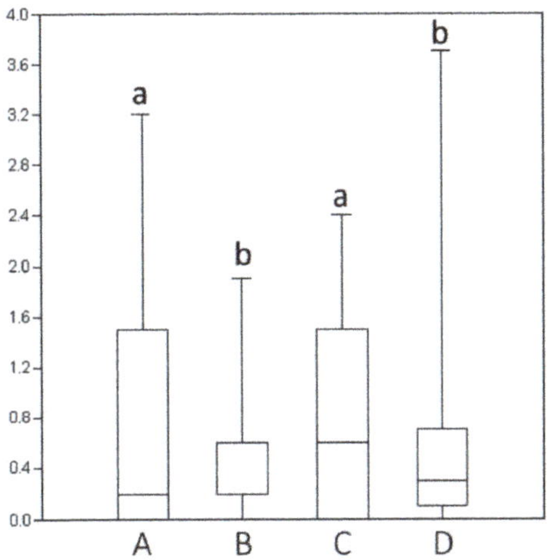

Figure 3. Comparison of the WAI along the studied routes to schools. a, b—homogeneous groups at $p < 0.05$.

3.3. Relation between PM Concentration and Vegetation Density in Winter

We found an ambiguous negative relationship between the greenery and PM concentrations during the vegetation dormancy period (Figures 4–7). On one hand we found a significant effect (Figures 4–7) in the morning hours for schools A and D (Figures 4 and 7), in sites with a low proportion of trees in their neighbourhood (Table 1). The relationship between WAI and PM concentration was positive, WAI related to a local increase in both $PM_{2.5}$ and PM_{10} pollution (Figures 4 and 7).

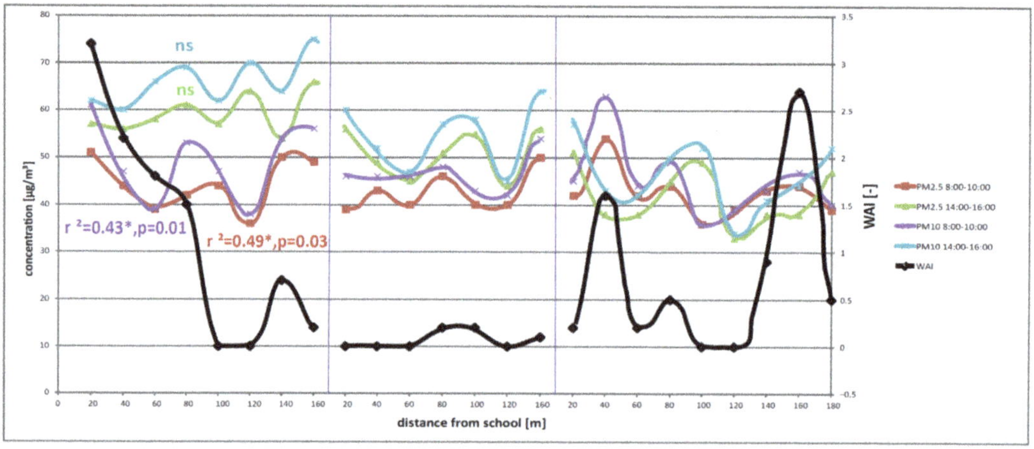

Figure 4. WAI distribution and an average PM concentration from 8:00 to 10:00 and from 14:00 to 16:00 in study plots along walking paths to school A (low share of trees, low number of local heating emission sources). ns—no significance.

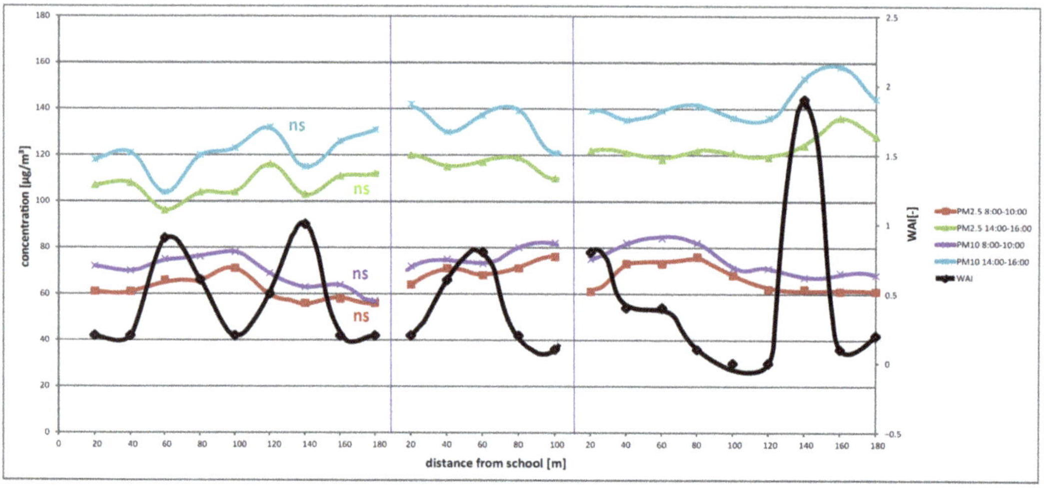

Figure 5. WAI distribution and an average PM concentration from 8:00 to 10:00 and from 14:00 to 16:00 in study plots along walking paths to school B (high share of trees, high number of local heating emission sources). ns—no significance.

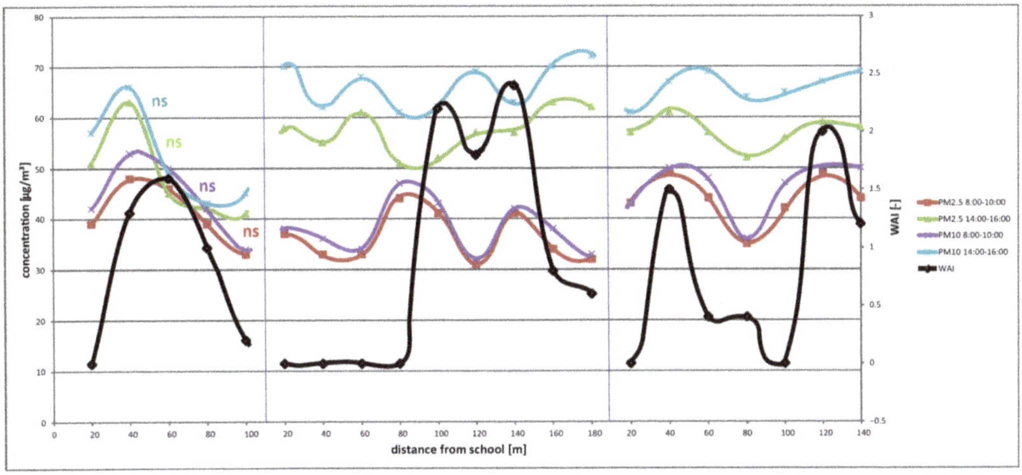

Figure 6. WAI distribution and an average PM concentration from 8:00 to 10:00 and from 14:00 to 16:00 in study plots along walking paths to school C (very high share of trees, low number of local heating emission sources). ns—no significance.

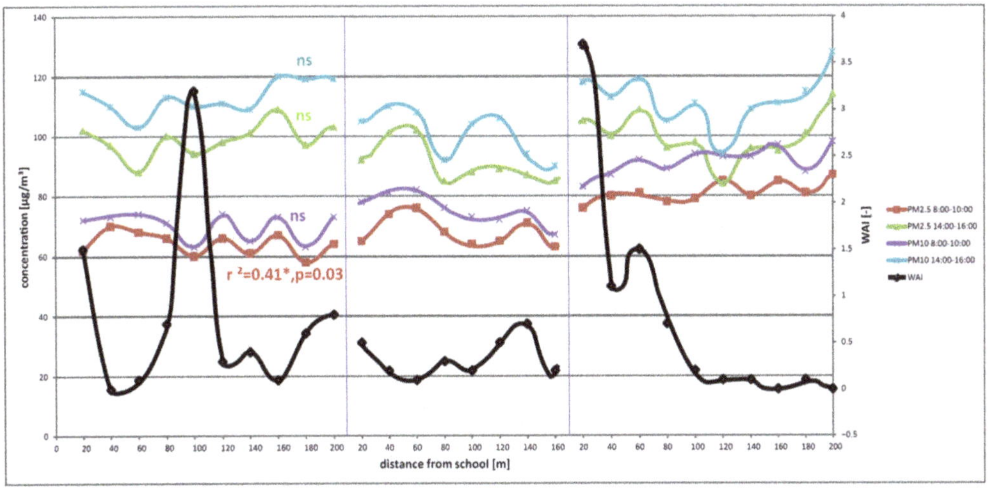

Figure 7. WAI distribution and an average PM concentration from 8:00 to 10:00 and from 14:00 to 16:00 in study plots along walking paths to school D (low share of trees, high number of local heating emission sources). ns—no significance.

4. Discussion

In Warsaw, similarly to many other European cities, where energy coal-based heating systems are still common and the transformation towards clean energy sources is still ineffective, the majority of PM is produced in winter and it is related to heating activities [41,42]. Undoubtedly, drawing out from fossil fuels in the residential sector is essential for reducing PM pollution and thereby improving the city residents health status [41]. Fundamentally improving air quality requires deep decarbonisation of the energy system, as well as more synergistic pathways to simultaneously address air pollution and global climate change [43]. However, the lack of local solutions related to CHP emissions [42] results in very localised problems. A study of routes to schools in Warsaw found that proximity to local CHP emitters had the strongest impact on pollution on the way to schools (Figure 2). Given that the children spend more than 50% of their active time commuting to school [18], exposure to pollutants can have a critical effect on their health, which can impede pulmonary function development [1], leading to asthma development [2] and susceptibility to otitis media [3].

An example of possible actions aimed at mitigating high loads of PM is the introduction of more greenery into cities. Models on the reduction of PM dispersion by trees in cities show promising results [44], indicating that the size, distribution and species composition of vegetation play a key role in PM reduction [45]. Urban trees contribute to improving air quality and can be used in national air protection strategies to reduce air pollutant concentrations [46]. The ability to effectively capture PM is an important factor in selecting optimal plant species to be used in urban greening [47]. Particularly promising might be the evergreen plants, which keep their leaves throughout the seasons. This means that they can be used for air purification in the winter season [48]. However, current studies have not proved their efficiency in PM removal.

The presence of greenery close to schools is mostly associated with their beneficial aesthetical as well as educational role for the young generation as even the sole visibility of vegetation outside the window was linked to improved performance of school pupils, not to mention the educational and aesthetic value [19,49].

Models predicting efficiency of greening interventions have been shown to poorly capture the seasonal variability of greening, even when the main source PM is an increased residential heating in winter [50]. There is a lack of research on the role of trees in the leafless state in reducing high loads of PM, which impedes our understanding of its impact on human health and well-being. Additionally, effectiveness of vegetation and application of phytoremediation methods is mostly criticized to be effective only during the vegetation season, while some cases are reported where dense vegetation in the winter can lead to creating local concentrations of PM [51,52]. Moreover, the trees can reduce the air flow and in some cases contribute to the local increase of pollutants, due to changes in their dispersal mode [46]. Our results suggest that such situations are possible during the winter (Figures 4 and 7). However, the influence of tree density in the leafless state expressed by WAI on the formation of local pollutant concentrations is questionable (Figures 4 and 7). The investigated walking routes in places with the highest density of trees tended to stimulate PM concentrations. On the other hand, the tree stands of comparable density did not show such peaks (Figures 4 and 7). Trees, characterized by a high foliage density, can act as a barrier preventing pollutant dispersal and be used as a biological filter [25,53]. The number of locations studied does not allow us to unambiguously resolve whether WAI is significantly related to the local accumulation of pollutants. Studies performed during the vegetation season state tend to argue that pollution decreases with the density of trees [54]. A sufficiently large green area with a well-chosen species composition can be a viable way to improve air quality and in some cases even reduce PM pollution to acceptable levels [55]. In this work, the surroundings of schools B and C were the greenest (Figure 3), but this did not translate into significant reductions in PM concentrations measured at height and along the children's route to school (Figure 2). It seems that in order to effectively filter PM from the air in wintertime, urban greenery should be as numerous and dense as possible while maintaining porosity that guarantees air movement. In this way, two unfavourable phenomena, local PM stagnation and uncontrolled transport of pollutants to potentially clean locations, may be limited.

The role of trees during dormancy season in air purification processes requires further research [54]. Potentially, the rough surface of branches and the remaining withered leaves of some species, accompanied by a few evergreen and coniferous plants, could potentially have some positive effect on reducing PM loads which could be deposited and trapped on their surface, accompanied by the sheltering effect, allowing less pollutant to reach the paths used by the children. There are no studies, however, which could confirm that phenomenon and assess its scale and extent. If it is confirmed that the occurrence of locally increased PM concentrations is a frequent phenomenon, further steps should be taken to counteract this, especially in places such as children's routes to school. Coniferous trees seem to be useful in this respect. These species also absorb PM during the winter months when air quality is poorer. Due to their smaller leaves, a larger wax layer on their needles and more complex shoot structures, they have a high capacity to capture pollutants from the air [56]. Species such as black pine or common yew [57], as well as climbers, could be grown along streets and be more effective in PM removal [58]. Undoubtedly greening interventions should always meet social approval, and the aesthetic function plays a primary role in species selection and the greening solution used. Therefore, plant species proved to be most efficient in PM reduction might not necessarily be those most meeting public preferences. However, the numerous studies, included those performed in Warsaw, show a growing approval for innovative greening interventions, showing that the public is willing to accept other-than-traditional forms of greenery, if they are supported by economic or ecological benefits [59,60].

The results we obtained did not show a positive role of trees in air purification in any of the locations. This could have been caused by the difficult environmental conditions along the roads, the discontinuity of greenery, and the relatively small proportion of trees in the area (Table 1). However, the results showing the negative impact of tree canopies during the winter season should not discourage the use of greenery, as the benefits outweigh the effects of exposure [61]. We hope to encourage further research and search for solutions in identifying these negative positions.

We suggest that areas next to schools and roads to schools should be considered to require special attention, as all locations studied were exposed to above-normal concentrations of particulate air pollution (Figure 2). Areas along school roads should be greened continuously [54] to allow ventilation. Species selection and planting structure should also take into account the most difficult winter heating period in Northern Europe, especially near local pollutant emitters.

5. Conclusions

- Children attending schools located near thermal heating emitters were twice as exposed to high PM concentrations.
- Tree planting on the way to school in the winter season showed no positive effect on air quality along the analysed sections of the road to the school, and locally even increased PM concentrations.
- Research on how to avoid local concentrations of pollutants should be expanded by modifying the structure of greenery and the share of evergreen species to increase the effectiveness of plants in the winter season.
- We propose to treat roads to schools as special zones of street greenery.

Author Contributions: Conceptualization, D.S. and A.H.; methodology, D.S., P.S. and A.H.; software, P.S.; validation, D.S., P.S. and A.H.; formal analysis, A.H., D.S., P.S. and A.P.; investigation, A.H.; resources, P.S.; data curation, D.S., P.S. and A.H.; writing—original draft preparation, A.H.; writing—review and editing, P.S., D.S. and A.P.; visualization, M.M.; supervision, P.S., D.S. and A.P.; project administration, D.S. and A.H.; funding acquisition, D.S. All authors have read and agreed to the published version of the manuscript.

Funding: This research was funded by the National Science Centre (Poland) grant no. 2020/39/B/HS4/03240.

Informed Consent Statement: Not applicable.

Data Availability Statement: Not applicable.

Conflicts of Interest: The authors declare no conflict of interest. The funders had no role in the design of the study; in the collection, analyses, or interpretation of data; in the writing of the manuscript, or in the decision to publish the results.

References

1. Horak, F.; Studnicka, M.; Gartner, C.; Spengler, J.; Tauber, E.; Urbanek, R.; Veiter, A.; Frischer, T. Particulate matter and lung function growth in children: A 3-yr follow-up study in Austrian schoolchildren. *Eur. Respir. J.* **2002**, *19*, 838–845. [CrossRef]
2. Holst, G.J.; Pedersen, C.B.; Thygesen, M.; Brandt, J.; Geels, C.; Bønløkke, J.H.; Sigsgaard, T. Air pollution and family related determinants of asthma onset and persistent wheezing in children: Nationwide case-control study. *Br. Med. J.* **2020**, *370*, 2791. [CrossRef]
3. Park, M.; Han, J.; Park, J.; Jang, M.; Park, M. Particular matter influences the incidence of acute otitis media in children. *Sci. Rep.* **2021**, *11*, 19730. [CrossRef]
4. Degórska, A.; Śnieżek, T. Skład pyłu zawieszonego na Stacji Puszcza Borecka z uwzględnieniem oddziaływania antropogenicznych i naturalnych źródeł emisji. *Monit. Środ. Przyr.* **2012**, *13*, 31–40.
5. Shandilya, K.; Khare, M. Particulate Matter: Sources, Emission Rates and Health Effects; Environmental and Agricultural Researcher Biographical Sketches and Research Summaries. In *Environmental Research Advances*; Cacioppo, L.T., Ed.; Nova: Toledo, Spain, 2014; Volume 1, pp. 305–306.
6. Juda-Rezler, K.; Toczko, B. (Eds.) *Pyły Drobne w Atmosferze. Kompendium Wiedzy o Zanieczyszczeniu Powietrza Pyłem Zawieszonym w Polsce*; Główny Inspektorat Ochrony Środowiska: Warszawa, Poland, 2016.
7. Kobza, J.; Geremek, M.; Dul, L. Characteristics of air quality and sources affecting high levels of PM10 and PM2.5 in Poland, Upper Silesia urban area. *Environ. Monit. Assess.* **2018**, *190*, 515. [CrossRef]
8. Chen, R.; Hu, B.; Liu, Y.; Xu, J.; Guosheng, Y.; Xu, D.; Chen, C. Beyond PM2.5: The role of ultrafine particles on adverse health effects of air pollution. *Bioch. Bioph. Acta* **2016**, *1860*, 2844–2855. [CrossRef]
9. Kim, K.H.; Kabir, E.; Kabir, S. A review on the human health impact of airborne particulate matter. *Environ. Int.* **2015**, *74*, 136–143. [CrossRef]
10. Ghorani, A.; Riahi-Zanjani, B.; Balali-Mood, M. Effects of air pollution on human health and practical measures for prevention in Iran. *J. Res. Med. Sci.* **2016**, *21*, 65. [CrossRef]
11. Xu, D.; Chen, Y.; Wu, L.; He, S.; Xu, P.; Zhang, Y.; Luo, J.; Ye, X.; Chen, Z.; Wang, X.; et al. Acute effects of ambient PM2.5 on lung function among schoolchildren. *Sci. Rep.* **2020**, *10*, 4061. [CrossRef]
12. Dunea, D.; Iordache, S.; Pohoata, A. Fine particulate matter in urban environments: A trigger of respiratory symptoms in sensitive children. *Int. J. Environ. Res. Public Health* **2016**, *13*, 1246. [CrossRef]
13. Krzeszkowiak, J.; Pawlas, K. Pył zawieszony (PM2,5 oraz PM10), właściwości oraz znaczenie epidemiologiczne ekspozycji krótko i długookresowej dla chorób układu oddechowego oraz krążenia. *Med. Srod.* **2018**, *21*, 7–13. [CrossRef]
14. Kapka, L.; Wdowiak, L. Negative health effects induced by airborne. *Med. Ogólna* **2009**, *15*, 421–429.
15. Rückerl, R.; Schneider, A.; Breitner, S.; Cyrys, J.; Peters, A. Health effects of particulate air pollution: A review of epidemiological evidence. *Inhal. Toxicol.* **2011**, *23*, 555–592. [CrossRef]
16. Markevych, I.; Tesch, F.; Datzmann, T.; Romanos, M.; Schmitt, J.; Heinrich, J. Outdoor air pollution, greenspace, and incidence of ADHD: A semi-individual study. *Sci. Total Environ.* **2018**, *642*, 1362–1368. [CrossRef]
17. Czarnobilska, E.; Mazur, M. Wpływ zanieczyszczenia środowiska na występowanie chorób alergicznych u dzieci i młodzieży szkolnej w Krakowie. *Lek. Wojsk.* **2017**, *95*, 32–39.
18. Rainham, D.G.; Bates, C.J.; Blanchard, C.M.; Dummer, T.J.; Kirk, S.F.; Shearer, C.L. Spatial classification of youth physical activity patterns. *Am. J. Prev. Med.* **2012**, *42*, 87–96. [CrossRef]
19. Łaszkiewicz, E.; Sikorska, D. Children's green walk to school: An evaluation of welfare-related disparities in the visibility of greenery among children. *Environ. Sci. Policy* **2020**, *110*, 1–13. [CrossRef]
20. Szczepanowska, H.B. Drzewa w mieście—Zielony kapitał wartości i usług ekosystemowych. *Człowiek I Sr.* **2015**, *39*, 5–28.
21. Fuertes, E.; Markevych, I.; Thomas, R.; Boyd, A.; Granell, R.; Mahmoud, O.; Heinrich, J.; Garcia-Aymerich, J.; Roda, C.; Henderson, J.; et al. Residential greenspace and lung function up to 24 years of age: The ALSPAC birth cohort. *Environ. Int.* **2020**, *140*, 105749. [CrossRef]
22. Yu, H.; Zhou, Y.; Wang, R.; Qian, Z.; Knibbs, L.D.; Jalaludin, B.; Schootman, M.; McMillin, S.E.; Howard, S.W.; Lin, L.-Z.; et al. Associations between trees and grass presence with childhood asthma prevalence using deep learning image segmentation and a novel green view index. *Environ. Poll.* **2021**, *286*, 117582. [CrossRef]
23. Dadvand, P.; Pujol, J.; Macià, D.; Martínez-Vilavella, G.; Blanco-Hinojo, L.; Mortamais, M.; Alvarez-Pedrerol, M.; Fenoll, R.; Esnaola, M.; Dalmau-Bueno, A.; et al. The Association between Lifelong Greenspace Exposure and 3-Dimensional Brain Magnetic Resonance Imaging in Barcelona Schoolchildren. *Environ. Health Perspect.* **2018**, *126*, 027012. [CrossRef]
24. Engemann, K.; Pedersen, C.B.; Arge, L.; Tsirogiannis, C.; Mortensen, P.B.; Svenning, J. Residential green space in childhood is associated with lower risk of psychiatric disorders from adolescence into adulthood. *Proc. Natl. Acad. Sci. USA* **2019**, *116*, 5188–5193. [CrossRef]
25. Przybysz, A.; Wińska-Krysiak, M.; Malecka Przybysz, M.; Stankiewicz-Kosyl, M.; Skwara, M.; Kłos, A.; Kowalczyk, S.; Jarocka, K.; Sikorski, P. Urban wastelands: On the frontline between air pollution sources and residential areas. *Sci. Total Environ.* **2020**, *721*, 13795. [CrossRef] [PubMed]
26. Popek, R. Aktualny stan wiedzy na temat zanieczyszczenia powietrza pyłem zawieszonym i jego fitoremediacji przy pomocy roślin drzewiastych. In *Badania i Rozwój Młodych Naukowców w Polsce—Ochrona Środowiska*; Nyćkowiak, J., Leśny, J., Eds.; Młodzi Naukowcy: Poznań, Poland, 2017; pp. 126–131.

27. Twohig-Bennett, C.; Jones, A. The health benefits of the great outdoors: A systematic review and meta-analysis of greenspace exposure and health outcomes. *Environ. Res.* **2018**, *166*, 628–637. [CrossRef]
28. Letter, C.; Jäger, G. Simulating the potential of trees to reduce particulate matter pollution in urban areas throughout the year. *Environ. Dev. Sustain.* **2020**, *22*, 4311–4321. [CrossRef]
29. Bank Danych Lokalnych: GUS. Available online: https://bdl.stat.gov.pl/BDL (accessed on 12 December 2021).
30. Sikorska, D.; Łaszkiewicz, E.; Krauze, K.; Sikorski, P. The role of informal green spaces in reducing inequalities in urban green space availability to children and seniors. *Environ. Sci. Policy* **2020**, *108*, 144–154. [CrossRef]
31. Kuchcik, M. Mortality and thermal environment (UTCI) in Poland—Long-term, multi-city study. *Int. J. Biometeor.* **2021**, *65*, 1529–1541. [CrossRef]
32. Majewski, G.; Czechowski, P.O.; Badyda, B.; Brandyk, A. Effect of air pollution on visibility in urban conditions. Warsaw case study. *Environ. Prot. Eng.* **2014**, *40*, 47–64. [CrossRef]
33. Majewski, G.; Rogula-Kozłowska, W. The elemental composition and origin of fine ambient particles in the largest Polish conurbation: First results from the short-term winter campaign. *Theor. Appl. Climatol.* **2016**, *125*, 79–92. [CrossRef]
34. Szkoły Podstawowe w Warszawa—Mapa Polski. Available online: http://mapa.edu.pl/szkoly-podstawowe/warszawa_20571.htm (accessed on 12 December 2021).
35. Feltynowski, M.; Kronenberg, J.; Bergier, T.; Kabisch, N.; Łaszkiewicz, E.; Strohbach, M.W. Challenges of urban green space management in the face of using inadequate data. *Urban. For. Urban. Green.* **2018**, *31*, 56–66. [CrossRef]
36. Urząd Statystyczny w Warszawie. Ranking Dzielnic Warszawy pod Względem Atrakcyjności Warunków Życia. Available online: https://warszawa.stat.gov.pl/publikacje-i-foldery/warunki-zycia/ranking-dzielnic-warszawy-pod-wzgledem-atrakcyjnosci-warunkow-zycia,1,2.html (accessed on 12 December 2021).
37. Miasto Jest Nasze. Kopciuchy Miejskie w Dzielnicach. Available online: http://smog.miastojestnasze.org/index.html (accessed on 12 December 2021).
38. Pyłomierz Osobisty Dust Air. Available online: https://dustair.gig.eu/ (accessed on 12 December 2021).
39. SS1 SunScan Canopy Analysis System—Delta-T Devices. Available online: https://delta-t.co.uk/product/sunscan (accessed on 12 December 2021).
40. Breda, N.J.J. Ground-based measurements of leaf area index: A review of methods, instruments and current controversies. *J. Exper. Bot.* **2003**, *54*, 2403–2417. [CrossRef]
41. Zhang, Q.; Zheng, Y.; Tong, D.; Shao, M.; Wang, S.; Zhang, Y.; Xu, X.; Wang, J.; He, H.; Liu, W.; et al. Drivers of improved $PM_{2.5}$ air quality in China from 2013 to 2017. *Proc. Natl. Acad. Sci. USA* **2019**, *116*, 24463–24469. [CrossRef] [PubMed]
42. Juda-Rezler, K.; Reizer, M.; Maciejewska, K.; Błaszczak, B.; Klejnowski, K. Characterization of atmospheric PM2.5 sources at a Central European urban background site. *Sci. Total Environ.* **2020**, *713*, 136729. [CrossRef]
43. Xi, L.; Shaojun, Z.; Jia, X.; Yunjie, W.; Wenhui, C.; Dian, D.; Ye, W.; Shuxiao, W.; Lei, D.; Jiming, H. Progress of air pollution control in China and its challenges and opportunities in the ecological civilization era. *Engineering* **2020**, *6*, 1423–1431. [CrossRef]
44. Nowak, D.J.; Hirabayashi, S.; Bodine, A.; Hoehn, R. Modeled $PM_{2.5}$ removal by trees in ten U.S. cities and associated health effects. *Environ. Poll.* **2013**, *178*, 395–402. [CrossRef]
45. Kim, S.; Lee, S.; Hwang, K.; An, K. Exploring Sustainable Street Tree Planting Patterns to Be Resistant against Fine Particles (PM2.5). *Sustainability* **2017**, *9*, 1709. [CrossRef]
46. Riondato, E.; Pilla, F.; Sarkar Basu, A.; Basu, B. Investigating the effect of trees on urban quality in Dublin by combining air monitoring with i-Tree Eco model. *Sustain. Cities Soc.* **2020**, *61*, 102356. [CrossRef]
47. Gratani, L.; Crecente, M.F.; Varone, L. Long-Term monitoring of metal pollution by urban trees. *Atmos. Environ.* **2008**, *42*, 8273–8277. [CrossRef]
48. Jin, E.J.; Yoon, J.H.; Bae, E.J.; Jeong, B.R.; Yong, S.H.; Choi, M.S. Particulate Matter Removal Ability of Ten Evergreen Trees Planted in Korea Urban Greening. *Forests* **2021**, *12*, 438. [CrossRef]
49. Dadvand, P.; Nieuwenhuijsen, M.J.; Esnaola, M.; Forns, J.; Basagaña, X.; Alvarez-Pedrerol, M.; Rivas, I.; López-Vicente, M.; De Castro Pascual, M.; Su, J.; et al. Green spaces and cognitive development in primary schoolchildren. *Proc. Natl. Acad. Sci. USA* **2015**, *112*, 7937–7942. [CrossRef]
50. Pace, R.; Grote, R. Deposition and resuspension mechanisms into and from tree canopies: A study modeling particle removal of conifers and broadleaves in different cities. *Front. For. Glob. Change* **2020**, *3*, 26. [CrossRef]
51. Eisenman, T.S.; Jariwala, S.P.; Lovasi, G.S. Urban trees and asthma: A call for epidemiological research. *Lancet Resp. Med.* **2019**, *7*, 19–20. [CrossRef]
52. Tong, Z.; Whitlow, T.H.; MacRae, P.F.; Landers, A.J.; Harada, Y. Quantifying the effect of vegetation on near-road air quality using brief campaigns. *Environ. Poll.* **2015**, *201*, 141–149. [CrossRef] [PubMed]
53. Hong, B.; Lin, B.; Qin, H. Numerical investigation on the coupled effects of building-tree arrangements on fine particulate matter ($PM_{2.5}$) dispersion in housing blocks. *Sustain. Cities Soc.* **2017**, *34*, 358–370. [CrossRef]
54. Wang, X.; Teng, M.; Huang, C.; Zhou, Z.; Chen, X.; Xiang, Y. Canopy density effects on particulate matter attenuation coefficients in street canyons during summer in the Wuhan metropolitan area. *Atmos. Environ.* **2020**, *240*, 117739. [CrossRef]
55. Yang, J.; Chang, Y.; Yan, P. Ranking the suitability of common urban tree species for controlling $PM_{2.5}$ pollution. Atmos. *Poll. Res.* **2015**, *6*, 267–277. [CrossRef]

56. Przybysz, A.; Nersisyan, G.; Gawroński, S.W. Removal of particulate matter and trace elements from ambient air by urban greenery in the winter season. *Environ. Sci. Poll. Res.* **2019**, *26*, 473–482. [CrossRef]
57. Przybysz, A.; Sæbø, A.; Hanslin, H.M.; Gawroński, S.W. Accumulation of particulate matter and trace elements on vegetation as affected by pollution level, rainfall and the passage of time. *Sci. Total Environ.* **2014**, *481*, 360–369. [CrossRef]
58. Dzierżanowski, K.; Popek, R.; Gawrońska, H.; Sæbø, A.; Gawroński, S.W. Deposition of particulate matter of different size fractions on leaf surfaces and in waxes of urban forest species. *Int. J. Phytoremed.* **2011**, *13*, 1037–1046. [CrossRef]
59. Sikorski, P.; Wińska-Krysiak, M.; Chormański, J.; Krauze, K.; Kubacka, K.; Sikorska, D. Low-maintenance green tram tracks as a socially acceptable solution to greening a city. *Urban For. Urban Green.* **2018**, *35*, 148–164. [CrossRef]
60. Sikorska, D.; Macegoniuk, S.; Łaszkiewicz, E.; Sikorski, P. Energy crops in urban parks as a promising alternative to traditional lawns–Perceptions and a cost-benefit analysis. *Urban For. Urban Green.* **2020**, *49*, 126579. [CrossRef]
61. Tainio, M. Burden of disease caused by local transport in Warsaw, Poland. *J. Transp. Health* **2015**, *2*, 423–433. [CrossRef] [PubMed]

forests MDPI

Article

Exploring National Park Visitors' Judgements from Social Media: The Case Study of Plitvice Lakes National Park

Carlotta Sergiacomi [1,*], Dijana Vuletić [2], Alessandro Paletto [3], Elena Barbierato [1] and Claudio Fagarazzi [1]

1 Department of Agriculture, Food, Environment and Forestry, University of Florence, p.le delle Cascine 18, 50144 Florence, Italy; elena.barbierato@unifi.it (E.B.); claudio.fagarazzi@unifi.it (C.F.)
2 Division for International Scientific Cooperation in Southeast Europe, Croatian Forest Research Institute, Cvjetno naselje 41, HR-10450 Jastrebarsko, Croatia; dijanav@sumins.hr
3 Research Center for Forestry and Wood, Council for Agricultural Research and Economics (CREA), p.zza Nicolini 6, 38123 Trento, Italy; alessandro.paletto@crea.gov.it
* Correspondence: carlotta.sergiacomi@unifi.it

Abstract: This study aims to conduct a survey of visitor reviews of the Plitvice Lakes National Park in Croatia to detect strengths and weaknesses of the park. In total, 15,673 reviews written in the period between 2007 and 2021 were scraped from the social media platform TripAdvisor. The research applies a comprehensive combination of multidimensional scaling, sentiment analysis, and natural language processing approaches to a sample area of international naturalistic interest. Analyzing the opinions of visitors, the authors identify: the main topics of interest related to the management of the park; and the strengths and weaknesses on the basis of definitely positive and decidedly negative reviews, respectively. The tested methodology is easily applicable for the analysis of different naturalistic contexts and protected areas, even in different countries, thanks to the use of translated reviews. The results obtained show that visitors to protected natural areas are not only interested in naturalistic and landscape aspects but also in issues such as accessibility and management of routes and visits.

Keywords: forest recreation; protected area management; text mining; natural language processing; sentiment analysis; multidimensional scaling method; web scraping; customer satisfaction; TripAdvisor reviews

Citation: Sergiacomi, C.; Vuletić, D.; Paletto, A.; Barbierato, E.; Fagarazzi, C. Exploring National Park Visitors' Judgements from Social Media: The Case Study of Plitvice Lakes National Park. *Forests* **2022**, *13*, 717. https://doi.org/10.3390/f13050717

Academic Editor: Radu-Daniel Pintilii

Received: 4 April 2022
Accepted: 29 April 2022
Published: 3 May 2022

1. Introduction

In the last decades, technological advances applied to the tourism sector have radically changed the way information is produced and consulted [1]. Tourists can access an increasing number of sources of knowledge and have many channels available to share their opinions on experiences and places. When the experiences are shared online, they help to define a concrete image of tourist destinations and to shape the decisions of future visitors [2,3]. In particular, social media platforms offer a space to freely share experiences and make judgements [4,5] through the so-called user-generated contents (UGC) [6–8]. For this reason, these platforms are becoming increasingly important both in the planning of destinations and in the definition of management priorities for places of tourist interest [9–12]. Social media can be considered as a rich source of news within which users create, circulate, and consult such information to mutually update each other on products, services, personages, and other objects of interest [13]. They are interactive platforms where individuals or larger communities share UGCs and include, among others, blogs, forums, or social networks [14]. Some social media are of general interest (e.g., Facebook or Twitter), while others are focused on more specific topics (e.g., professional networking on LinkedIn); some of them deal with media sharing (e.g., YouTube or Flickr), while others allow you to provide reviews on products and services (e.g., Google My Business or TripAdvisor).

In this study, TripAdvisor was chosen among the many available social media, because it is the largest travel website in the world, operating in 45 countries around the world [11]. It has more than 400 million visitors visiting every month [15] and more than 450 million reviews and opinions which concern more than seven million accommodations, restaurants, and attractions [16]. Besides, TripAdvisor is available in 28 languages [17]. TripAdvisor reviews are a source of information characterized by several positive aspects, including being free and easily accessible and covering a considerable number of years [3]. In addition to reviews, users can also publish other information, such as the country of provenance and the purpose of the trip. Therefore, user reviews on TripAdvisor combine textual comments (i.e., reviews) with concise ratings (i.e., bubbles). Although recent studies have shown that textual comments receive a lower priority than synthetic evaluations [18], it should be highlighted that users may have different priorities [19] that cannot be fully explained in choosing between one and five bubbles. Therefore, it becomes essential to develop tools which allow more information to be extrapolated from the textual component of the reviews.

The massive amounts of unstructured data that are continuously generated on the Internet necessarily require the use of automated procedures for this kind of data analysis [1,7,12]. Social media analytics is receiving increasing attention from companies in many sectors, because they try to analyze the large amount of data collected through different methods [6,20,21]. Content analysis (CA) is one of the available techniques for extrapolating and analyzing the text contents which is widely used in the tourism research field [11]. Sentiment analysis (SA) approach is part of the CA field, and it is a valid option to process this type of data automatically. SA uses computational linguistics and natural language processing (NLP) to analyze the text and identify the polarity of the judgements contained within it [1,8,16]. Another technique for analyzing unstructured textual data is that of multidimensional scaling (MDS), the main purpose of which is that of a better graphical visualization of the data in order to facilitate the understanding of the text structure [22]. In the international literature, the applications of MDS in tourism studies are numerous [23,24]. MDS is usually associated with cluster analysis, a particular application of which is text clustering [6].

Today, it is essential for the tourist community to identify destinations that provide them with meaningful experiences in natural contexts. In this way, protected forest areas and forested landscapes turn out to be popular destinations thanks to the multitude of natural values that take place within them [25]. In Croatia, this type of destination is well represented by national parks, which correspond to the second-highest level in the scale of protected areas (Law on Nature Protection, OG 88/13, 15/18, 14/19, 127/19). One of the most famous and visited national parks in Croatia is Plitvice Lakes National Park (PLNP). The choice of this well-known park was guided: on one side, by the need to validate a new methodology with a case study for which a great deal of information was already available on the activities and management problems with which to compare the final results; on the other side, by the fact that that social media data prove to be a better proxy of tourist visits in reference to the most popular parks [5].

To the best of our knowledge, no previous studies have focused on visitors' experiences for PLNP. The present study tried to fill this gap in the literature by conducting an in-depth analysis of TripAdvisor tourists' reviews on PLNP, by applying a comprehensive method of text mining and natural language processing techniques.

In particular, this study aims to answer the following research questions.

- RQ1. How to collect and investigate textual data by social media platform to investigate the preferences of users of protected areas?
- RQ2. How to extrapolate and analyze the management issues of greatest interest to visitors who choose protected areas as their destination?
- RQ3. How to identify the strengths and weaknesses of the management of protected areas from the point of view of visitors?

The management of protected forest areas as a potential tourist destination is particularly demanding. This complexity is due to the trade-off between the conservation of natural ecosystems and the promotion of tourist visits for economic reasons [26,27]. Therefore, it is particularly useful to define a flexible methodology for the analysis of the management of protected areas that considers the point of view of visitors. In the present study, the answers to the research questions will allow PLNP managers to monitor the satisfaction of local and international users and plan activities aimed at improving the quality of visits to the park.

The remainder of the paper is organized into the following five sections. Foremost, Section 2 provides a literature review on the analysis of nature-based tourism using MDS and NLP tools. After that, the methodology used is illustrated in Section 3. Section 4 shows the results, while Section 5 discusses the findings. Finally, Section 6 analyzes the limitations of the study and provides suggestions for useful application and future research directions.

2. Literature Review

2.1. Nature-Based Tourism

Nowadays, it is widely recognized that some segments of the tourism sector can be considered a "clean industry" and part of the Green Economy [28]. In particular, nature-based tourism is a growing key sector of this industry [26,29,30] which seeks to respond to a growing consumer demand for a return to nature [3,25]. This need is well explained by the fact that nature is capable of generating human well-being from a physical and psychological point of view. [20,25,31–34]. Moreover, natural areas are a place of refuge for biodiversity, in addition to providing restorative surroundings for people [26,31]. The establishment of protected areas created to conserve biodiversity and esthetic value of landscapes is one of the main pillars of nature-based tourism [29,30]. Thus, protected areas and nature-based tourism represent fundamental access for people to cultural ecosystem services [25,35,36]. Particularly, national parks are characterized by a high level of biodiversity protection among protected areas and, at the same time, provide tourism opportunities [5,26,27,37]. Thus, national parks play a very important role also in the tourism sector. For this reason, it is essential to analyze the factors that attract visitors and make visits to protected areas pleasant. Both internal components (e.g., expectations for places and activities) and external components related to tourism management (e.g., accessibility, means of transportations, etc.) strongly influence visitors' perception of the natural landscape [3]. Consequently, the management of nature-based tourism services must take into account the diversified opinions that visitors have towards nature in general and recreational activities in particular [38]. Therefore, it has become fundamental to evaluate how people perceive their recreational experiences in this type of protected area [8].

2.2. Content Analysis

Content analysis (CA) is a research tool to be adopted in order to identify some particular words or more general concepts within qualitative textual data [2,39] or to extrapolate homogeneous units of meaning from a complex text. Traditionally, CA involved human subjective interpretation by researchers, which has now been replaced by automated procedures and sophisticated software [4]. One of the possible approaches of CA is sentiment analysis (SA), which is also an important component of text mining. Text mining is an interdisciplinary field which draws on information retrieval, data mining, machine learning, statistics, and computational linguistics [40]. Valid overviews on SA were produced by Ma et al. and Alaei et al., to which reference should be made for further information [1,9]. In these contributions, the authors reconstruct the main historical stages that characterized the evolution of the SA and outline its most recent features and applications. Nonetheless, it can be synthetically said that the main purpose of SA is to distinguish between positive, negative, or neutral opinions [1,12,16]. Natural language processing (NLP) is one of the available tools for SA, but its application on UGC from social media in landscape design, and planning research is still in a preliminary stage [21,41]. In the text analysis, MDS is a particularly valid automated computer algorithm. MDS is a data visualization technique based on the proximity of words and their spatial representation [23,24]. Another type of machine learning algorithm usually associated with MDS is that of cluster analysis, which is usually applied to transform unstructured word sets into structured clusters [21].

Social media analytics—in particular, SA—has been applied to social media in numerous tourism-related research fields [6,39]. The most investigated fields are food and wine tourism [19,39,41,42], hospitality [9,11,43,44], areas of interest or events in cities [4,16,45–47], and natural spaces with special regard to urban parks [20,21,31–33,48]. Conversely, national parks and nature reserves [3,6,8,25,27] are a field still not much investigated [8].

2.3. Nature-Based Tourism and Ccontent Analysis

According to the European Landscape Convention [49], landscape assessment processes should take into consideration public perception of places [50]. To evaluate visitors' perception towards natural destinations, traditional methods, such as in situ questionnaires, in-depth interviews, and focus groups, have long been employed. These techniques are usually time- and resource-consuming, in addition to not allowing the collection of results on a large scale or comparisons over time [3,6,8,27,32,50]. On the other hand, the development of modern tools for web analysis allows us to overcome all of these shortcomings. In the recent literature, numerous research contributions have used CA methods to analyze nature-based tourism destinations, but there are still few contributions that investigate the usability of the various social media platforms in relation to visits to protected areas [3].

Stoleriu et al. explores 226 online TripAdvisor reviews on Danube Delta through an automated CA in order to identify and quantify the main dimensions of visitors' experiences and memories [3]. Their results showed that managerial aspects linked to visit organisation (e.g., trip itinerary and visit duration) were more prominent themes in the tourists' reviews compared to the site characteristics. One of the main limitations of the study in relation to the use of TripAdvisor reviews is the lack of demographic and socioeconomic information of visitors. For this reason, it would be necessary to integrate this type of analysis with surveys that make it possible to evaluate the preferences of visitors based on their characteristics.

Two other recent studies [8,27] conducted SA in some national parks of South Africa. Hausmann et al. used SA and NLP techniques to analyze the content of image captions in 33,213 English posts published on Instagram relating to four national parks in South Africa [8]. The authors identified the main emotional components and the keywords formed by both a single word and a pair of adjacent words that recurred most in the posts. The results showed that the polarity of sentiment about national parks expressed by visitors on social media is generally positive, with a minor expression of negative feelings. This is significant to highlight the social role that national parks assume, favoring the development of positive interactions with nature and, therefore, well-being in visitors. Those authors found that visitors tend to idealize certain places or features of national parks and give them symbolic meaning. This meaning is what makes visiting experiences worth sharing and promoting. Among the problems identified by those authors in using this method, there are: on the one hand, the potential lack of representation of the sample of visitors who publish reviews; on the other hand, the use of an unconventional language (e.g., abbreviations, slang, emojis, etc.) which can make the use of automatic computational systems less effective. In almost the same area, Mangachena and Pickering conducted an analysis of 10,292 English tweets on Twitter about seven South African national parks [27]. Even in this case, they mostly found positive feelings and opinions related to the nature-based experience. Those authors identified a particular interest from visitors regarding specific events, such as commemorations related to the history of the park or discoveries of naturalistic interest. Furthermore, according to previous studies [8], some authors recognized that the use of concise texts, shortened words, and special characters (e.g., hashtags and emoticons), typical of social networks such as Instagram and Twitter, may also complicate text analysis of tourists' reviews [20].

Recently, Niezgoda and Nowacki investigated visitors' opinions towards one of the most visited protected areas in Poland, Tatra National Park [25]. Those authors elaborated a composite methodology made by text mining, NLP, and coding opinion procedures to process the data obtained from 624 English reviews published on TripAdvisor. The authors were interested in identifying the main reasons that led visitors to live experiences in the nature park and whether these were mainly related to the themes of ecological awareness and nature protection. The results of their study showed that the most active forms of entertainment (e.g., hiking, taking photos, mountain climbing) are the main motivation for visiting places in the open air. Those authors also highlight that in order to conduct this type of analysis it is necessary to assume that the reviews contain the elements considered most important by visitors, but it would be advisable to deepen the themes identified with more detailed surveys.

One of the latest applications of CA to national parks is that of Mirzaalian and Halpenny. In their study on Jasper National Park, they analyzed 17,224 English reviews on TripAdvisor [6]. In addition, that study analyzed destination loyalty statements using a keyword clustering approach. Among the main categories of visitor favorite destinations can be found waterfalls and lakes. Those authors acknowledge that one of the biggest limitations of this study is that the analysis did not concern some meaningful management aspects (e.g., transportation or outdoor activities).

3. Materials and Methods

The combination of several tools has made it possible to obtain different types of results that can be useful to the managers of the study area. On the one hand, the strengths and weaknesses of the PLNP from the visitor's point of view stemmed from the NLP technique (i.e., rapid automatic keyword extraction) based on SA scores. On the other hand, the MDS and cluster analyses were carried out to identify the topics most dealt with in the reviews released by PLNP visitors on TripAdvisor.

The different steps of the method used are summarized and described in a procedure flowchart (Figure 1).

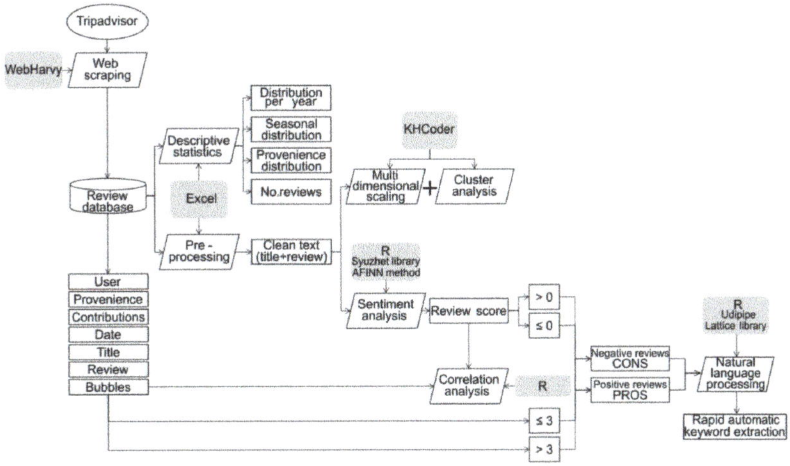

Figure 1. Flowchart of the research procedure.

3.1. Study Area

Plitvice Lakes National Park (PLNP) is one of the most famous and visited national parks in Croatia. PLNP is located in the mountainous central part of the nation and is part of the Dinaric karst area. PLNP is the oldest protected area (designated 8 April 1949) and the biggest national park (29,685.15 ha) in Croatia. The park mainly consists of forest areas, which represent about 81% of the total territory, with a complex system of lakes connected with waterfalls. The PLNP is well known for the rich biodiversity of its 296 square kilometers of forests. It is managed by the Plitvice Lakes National Park Public Institution (PLNPPI), founded by the Republic of Croatia and placed under the jurisdiction of the Ministry of the Environment and Energy (MEE). In addition, Plitvice is the only Croatian national park that is on the UNESCO World Heritage list (1979) as natural heritage and is entirely identified as a Natura 2000 site. Despite the large area of the park, only a small part of it represents the point of major tourist interest [37]. It is a lake system which includes 16 main lakes characteristic for their waterfalls, to which are added several other smaller lakes [51]. The park's finances derive entirely from the entrance tickets and hospitality services, including four hotels (380 accommodation units and 820 beds), two camping sites (2850 parking spaces for campers), seven restaurants, and eight other small park facilities (just under 3000 seats) [52]. The income of these activities is used for management and investments within the park area [37].

PLNP is one of the most visited natural sites in Central Europe and in the Mediterranean region [53]. The park's official statistics report a significant growth in the number of visitors per year, from 850,000 registered in 2007 to about 1.75 million in 2018. More than 80% of visitors visit the park in the period from May to September. The months of the greatest peak are July and August, when approximately 335,000 and 385,000 visitors were registered in 2017, with daily averages of about 10,800 and 12,400 visitors and reaching the maximum with over 16,000 visitors in a single day (August 2017). Consequently, the Park is often congested, causing both considerable discontent in the opinion of some visitors but above all putting safety procedures at risk and causing negative ecological impacts for the natural systems of the park [53].

3.2. Data Collection

Reviews relating to "Plitvice Lakes National Park" were scraped between October and November 2021 from the dedicated website on TripAdvisor (https://www.Tripadvisor.com/Attraction_Review-g303827-d554038-Reviews-Plitvice_Lakes_National_Park-Plitvice_Lakes_National_Park_Central_Croatia.html accessed on 26 September 2021).

WebHarvy software was used to scrape the reviews and obtain the following information:

- User data: name, origin, number of contributions (review number);
- Review data: date (month and year), travel purpose, number of bubbles (i.e., summary judgement), title and text of the review (i.e., extended judgement).

The software utilized is a visual web scraper that uses no script or code to scrape data. The program allows you to access the URL address of interest and to select the items that you want to collect. Thanks to the potential of the tool used, it was possible to carry out the immediate translation of the reviews and their respective titles by referring to the Google Translate plug-in. In this way, all of the reviews of all available languages were translated into English and used for subsequent analyses.

The study did not collect other types of socio-demographic information such as the age, occupation, and educational level of visitors. This is due to the fact that TripAdvisor profiles do not contain this kind of data [3]. The only personal information that TripAdvisor users commonly share is their country of origin. These data could be useful for analyzing the origin of visitor flows to the PLNP.

3.3. Multidimensional Scaling Method and Cluster Analysis

MDS and cluster analysis allow us to explore possible combinations or groups of words that share similar appearance patterns [22]. In particular, text clustering is a textual data mining method which converts the original sentences in a term-document-matrix using different feature extraction techniques [6,54]. In this way, it is possible to deduce the main elements perceived by the users (e.g., reviewers), which should be taken into consideration for an effective and rational management of the protected areas. The ease of analysis application and result interpretation are among the main advantages of the MDS [23,24].

The elaborations were carried out using KH Coder 3 software [25,39,54,55]. The KH Coder software combines two fundamental approaches of computer-based text analysis: the correlational approach, which consists in automatically extracting words from a text and analyzing them statistically; and the dictionary-based approach, which establishes coding rules for the different elements that form the text (e.g., sentences or groups of words) [55]. In order to identify the clusters of words, the Ward's minimum variance method or Ward's hierarchical clustering method was applied, as previously carried out by Barbierato et al. [39]. The Ward's method is a procedure that initially generates in clusters containing single objects. These clusters are gradually aggregated in such a way as to create clusters with the highest number of objects possible, but ensuring that the variance within each cluster is minimized [56]. The Ward's method was applied within the so-called Sammon space, which allows one to maintain a certain distance between words, preventing them from being excessively crowded and overlapping, giving more readable results [57]. Furthermore, among the options to define the distance, the cosine similarity coefficient was chosen, which is considered an efficient option in the presence of long documents (e.g., reviews) which contain, as in our case study, numerous words with an important frequency in each document [57]. A frequency threshold of 1500 terms was adopted on the basis of the term frequency–document frequency graph (i.e., TF–DF) (Figure 2a) in order to include exclusively the most representative terms that appear in several reviews. Based on the agglomeration graph (Figure 2b), it was chosen to generate seven clusters of 60 words each. For further information on the method, refer to the KH Coder software manual [57].

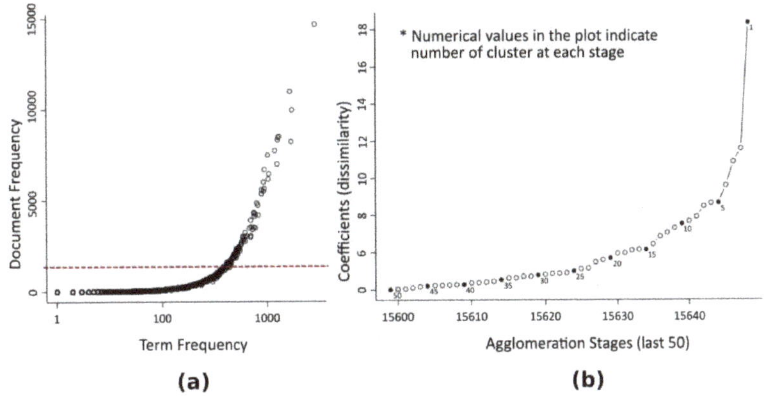

Figure 2. MDS model parameters for Plitvice Lakes National Park: TF–DF (**a**) and agglomeration graph (**b**).

3.4. Sentiment Analysis

Sentiment analysis (SA) research is driven by the importance of understanding consumer judgement [9]. In particular, SA can be used to understand consumer attitudes towards particular products, services, or places [16]. SA determines the positive or negative polarity of each relevant word in the text. Moreover, SA calculates a score based on a predefined lexicon contained within a library [39]. It should be specified that this score is not set on a reference scale between a predetermined minimum and maximum. The sentiment score varies both in reference to the text length and to the specific words contained therein. The only fixed references are the scores assigned to the individual words within the lexicon to be adopted. In the present study, the "syuzhet" library of R software was chosen, as it was applied in previous research that analyzed reviews on TripAdvisor [12,27,39]. The AFINN lexicon [58] was applied at the "syuzhet" library. Negative words and slang are commonly used in reviews on social networks (e.g., TripAdvisor). The AFINN lexicon is considered a valid option for evaluating this type of comment [59]. Furthermore, SA is widely applied to the analysis of quality perception through TripAdvisor reviews for heritage sites and natural parks [45] and urban green areas [16]. For a more in-depth analysis of the procedure used by the software, please refer to Barbierato et al. [39].

3.5. Natural Language Processing

Natural language processing (NLP) is a technology that combines computer science and linguistics in order to interpret written texts [39]. In this study, the strengths and weaknesses of the PLNP were identified using a NLP procedure. The rapid automatic keyword extraction (RAKE) procedure is a method for extrapolating multi-word keywords from documents [60]. Candidate keywords are obtained by partitioning text through stop words (e.g., and, the, of, etc.) and phrase delimiters (e.g., ; and ,) and assigning a score to each candidate multiple keyword. Only double-word keyword candidates are searched in this study. Each of the two words that constitute the candidate keyword obtains a score that is given by the ratio between the number of times each single word co-occurs with the other word of the candidate keyword and the total frequency with which it appears by itself. The final RAKE score for the entire candidate keyword is the sum of the scores of each of the two words that form the candidate keyword [61]. The procedure was carried out through the "udpipe" library [61] of R software [62], considering only adjectives and nouns. Furthermore, only the first 20 keywords as a sequence of two adjacent words—defined as bi-grams—are considered, and a frequency threshold of 6 was adopted. In addition, the "lemma" option was chosen instead of "token". Through the lemmatization process, it is possible to group the different forms in which a word can be presented (e.g., singular and plural) in a single common voice. In this way, the various forms of the same reference word are counted as a single lemma, assuming a greater weight.

The analysis of definitely positive (bubbles > 3 and sentiment score > 0) and decidedly negative (bubbles $\leq$ 3 and sentiment score $\leq$ 0) reviews allowed us to identify strengths and weaknesses of the PLNP based on the visitor's judgement.

4. Results

4.1. Data Collection and Sample Description

Overall, 15,673 online reviews were automatically retrieved from the online review website TripAdvisor. The downloaded reviews date back to the period between 2007 and 2021.

Figure 3 shows the trend in the number of reviews registered on TripAdvisor for PLNP. This trend is considered to be related to the interest of visitors. The graph shows an important growth until 2015, followed by a slight decrease until 2019. In 2020, there is a significant drop (−88% compared to the previous year) due to the international and national restrictions on travel as a consequence of the COVID-19 pandemic.

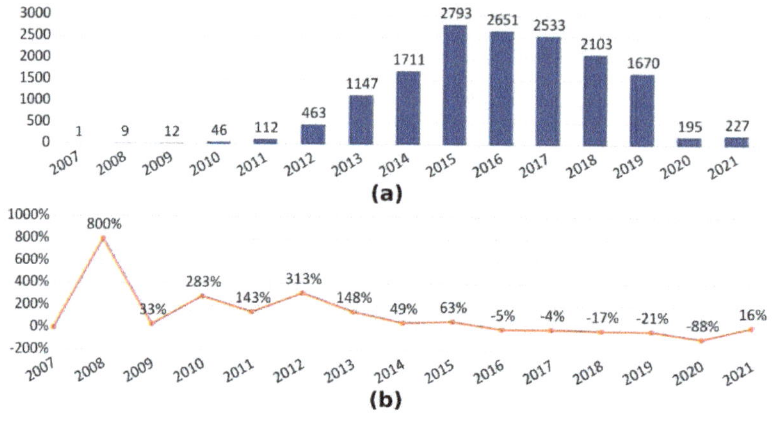

Figure 3. Frequency of reviews per year (**a**) and annual percentage growth rate of reviews (**b**).

The monthly and seasonal distribution of reviews (Figure 4) is consistent with the dynamics of visitor flows that have been analyzed in the current PLNP management plan [52]. The graph shows that in the summer—with special regard to August—the maximum peak is recorded. Instead, an intermediate influx of visitors is recorded on average in spring and autumn, even if the month of September still seems to be influenced by the importance of the summer flow. Winter is the season of least interest for visitors, as confirmed by the low number of revisions.

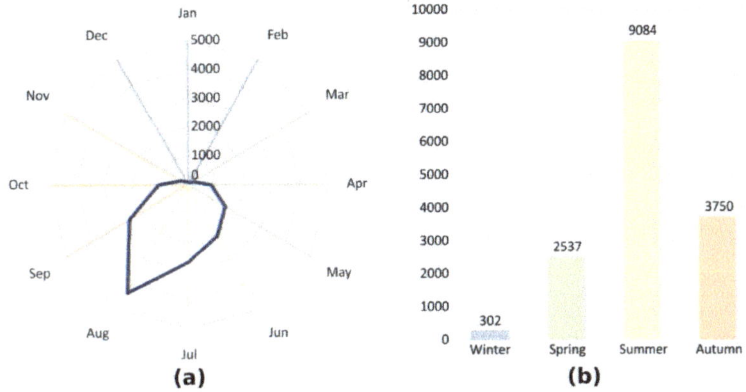

Figure 4. Monthly (**a**) and seasonal (**b**) distribution of reviews (average value for the period 2007–2021).

As regards the origin of PLNP visitors, Figure 5 shows that most of the visitors come from European countries. In particular, the largest flows are recorded from Italy, the United Kingdom, and France.

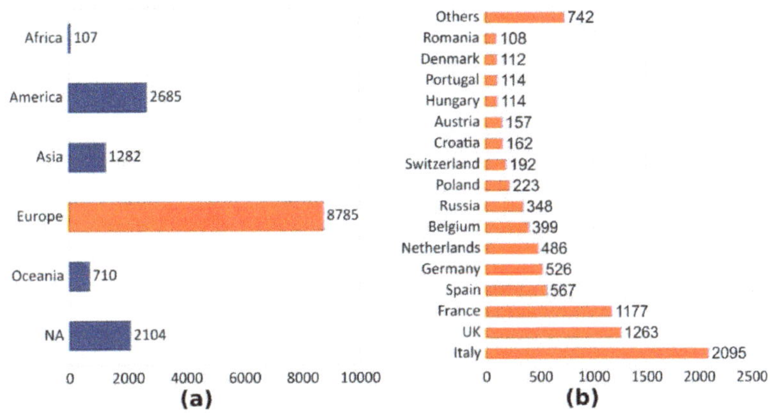

Figure 5. Provenance of the reviewers by continents (**a**) and from exclusively EU countries (**b**) (reference period 2007–2021).

4.2. Multidimensional Scaling Method and Cluster Analysis

The diagram derived from the MDS method shows seven clusters of words differentiated by color [54]. The results are in Figure 6. Cluster 1 (i.e., turquoise bubbles) concerns the principal elements that characterized PLNP landscape (i.e., "park", "lake", "waterfall") which are commonly associated with positive judgements ("beautiful"). Cluster 2 (i.e., yellow bubbles) is related to the theme of accessibility, including: the possible means of transport to access and/or visit the park (i.e., "boat", "bus", "train", "car"); the organization into "route(s)" divided by length in terms of "hour(s)"; and the real entrance to the park, which concerns different activities, such as "parking" and the purchase of the "ticket". Cluster 3 (i.e., violet bubbles) is a hybrid set of aspects that characterize the park, emphasizing the beauty of the site on the one hand, using terms such as "nice" and "good", and the disadvantages related to overcrowding in the summer months of the high season, expressed by adjectives such as "many", "long", and "lot". Clusters 4 (i.e., red bubbles) and 6 (i.e., orange bubbles) contain the main favorable appreciations thus synthesizable: "great", "worth", "wonderful", "natural" connected to "nature", "beauty", and "experience" for Cluster 4; "stunning", "amazing", "clear", and "different" (in the positive sense of "different" landscapes and sceneries) relating in general to the "Croatia(n)" "national" park of "Plitvice" for Cluster 6. All of the positive adjectives of the Clusters 4 and 6 are also related to the nearest central terms of the Cluster 1. Cluster 5 (i.e., blue bubbles) contains the most negative elements, referring to the main problems related to the PLNP management: the presence of "crowd" and "queue(s)" in many different "point(s)", "path(s)", and "way(s)" of the area. Finally, Cluster 7 (green bubbles) represents a small deepening of the nearby Cluster 2 themes, recovering the theme of the fruition through the use of words such as "walk", "trip", and "tour". In this cluster, some information about the division in the "upper" and "lower" districts of the park are included.

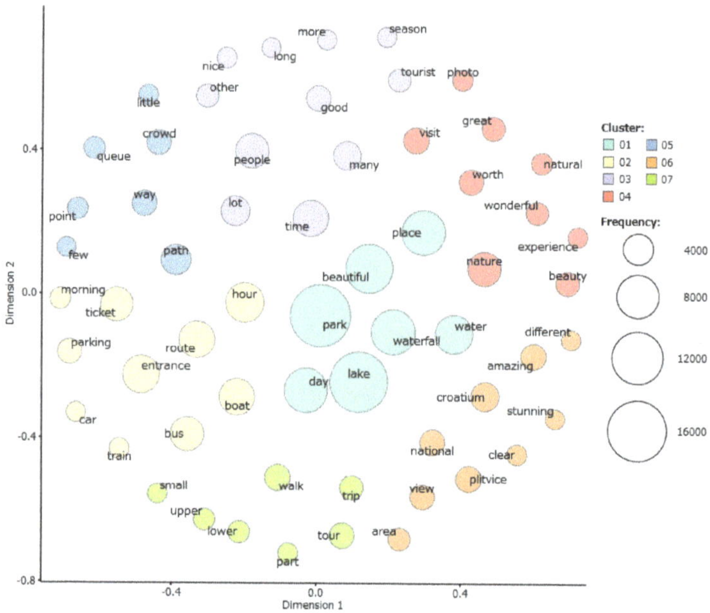

Figure 6. Multidimensional scaling method and cluster analysis results for Plitvice Lakes National Park.

These results make it possible to identify the issues (i.e., the seven clusters) related to the PLNP management that are of greatest interest to visitors. The issues thus identified would be useful if applied to guide a participatory planning of the park in which samples of visitors were also involved.

4.3. Sentiment Analysis

The results of the SA are shown in Table 1. The reviews for PLNP are basically positive (mean value of 9.16) and the dispersion is relatively symmetrical (1st Qu. = 5; 3rd Qu. = 13). In fact, the mean value is shifted upwards, as the group of reviews designated with five bubbles represents over 78% of the total reviews (15,673). The SA results show that mean and median values tend to increase with the increment in the number of bubbles (i.e., short judgement).

Table 1. Sentiment analysis scores for Plitvice Lakes National Park.

Bubbles	No. Reviews	Min.	1st Qu.	Median	Mean	3rd Qu.	Max.
●○○○○	210	−27	−3	0	0.40	4	23
●●○○○	228	−19	−1	3	3.04	7	36
●●●○○	641	−14	2	6	5.96	10	27
●●●●○	2317	−15	4	8	8.10	11	40
●●●●●	12,277	−16	6	9	9.79	13	72
Total	15,673	−27	5	9	9.16	13	72

The non-normal distribution of the SA scores was visually verified through normal quantile plots, histograms, and box plots for each group related to the five review ratings (i.e., bubbles) (see Appendix A: Figures A1–A3). Furthermore, the Shapiro–Wilks test was performed for the groups of Bubbles 1, 2, 3, and 4 (in R, the Shapiro–Wilks test cannot be performed on sets of more than 5000 units). The *p*-value of all four groups (min $< 2.2 \times 10^{-16}$; max = 0.002) showed that the data do not follow a normal distribution. For this reason, the non-parametric Kruskal–Wallis test was applied to verify the correspondence between the SA scores and the bubbles assigned by the reviewers themselves.

The results confirmed the hypothesis of a statistically significant difference between the groups of bubbles in relation to the dependent variable of SA scores ($K = 848.91$; *p*-value $< 2.2 \times 10^{-16}$; $\alpha = 0.05$). In addition, a pairwise comparison using the non-parametric Mann–Whitney U test was conducted to highlight where the statistically significant differences between groups of bubbles are [34]. Although the differences within each pair of groups are statistically significant (Table 2), according to Barbierato et al. [39] the complete database was divided only into two sub-databases in order to simplify the data analysis: one definitely positive (bubbles > 3 and sentiment score > 0) and one decidedly negative (bubbles ≤ 3 and sentiment score ≤ 0), which were used separately in NLP analyses.

Table 2. Mann–Whitney U test ($\alpha = 0.05$) results for Plitvice Lakes National Park.

Pair of Groups of Bubbles		*W*	*p*-Value
●○○○○	●●○○○	17,998	6.934×10^{-6}
●○○○○	●●●○○	33,963	$<2.2 \times 10^{-16}$
●○○○○	●●●●○	86,522	$<2.2 \times 10^{-16}$
●○○○○	●●●●●	348,787	$<2.2 \times 10^{-16}$
●●○○○	●●●○○	52,868	5.059×10^{-10}
●●○○○	●●●●○	141,873	$<2.2 \times 10^{-16}$
●●○○○	●●●●●	584,911	$<2.2 \times 10^{-16}$
●●●○○	●●●●○	589,860	1.306×10^{-15}
●●●○○	●●●●●	2,551,728	$<2.2 \times 10^{-16}$
●●●●○	●●●●●	11,975,881	$<2.2 \times 10^{-16}$

4.4. Natural Language Processing: The RAKE Analysis

The RAKE analysis was applied to the two sub-databases obtained dividing positive from negative reviews considering the SA scores. The double-word keywords most frequently encountered in TripAdvisor reviews for PLNP were identified by the RAKE analysis (Figure 7). The most cited characteristics can be identified both in the definitely positive reviews, to be interpreted as the main strengths, and in the decidedly negative reviews, to be read as the most critical weaknesses. Definitely positive RAKE analysis results (Figure 7a)—deriving from the sub-database containing the reviews with bubbles > 3 and sentiment score > 0—show that the natural heritage and landscape elements are the most appreciated aspects of the PLNP. In particular, the "UNESCO" designation is considered as an extremely positive characteristic, as highlighted by three keywords: "UNESCO heritage", "UNESCO site", and "UNESCO list". The negative results—deriving from the sub-database containing the reviews with bubbles ≤ 3 and sentiment score ≤ 0—show that the main weaknesses are represented by the phenomenon of crowding ("many people"), because the presence of a "mass tourism" during the "high season" is the cause of complex management problems, such as "traffic jam" and "endless queue" (Figure 7b). In addition to "long (waiting) time", there are also complaints about the organization of "parking lot" and the "high price" of the entrance ticket.

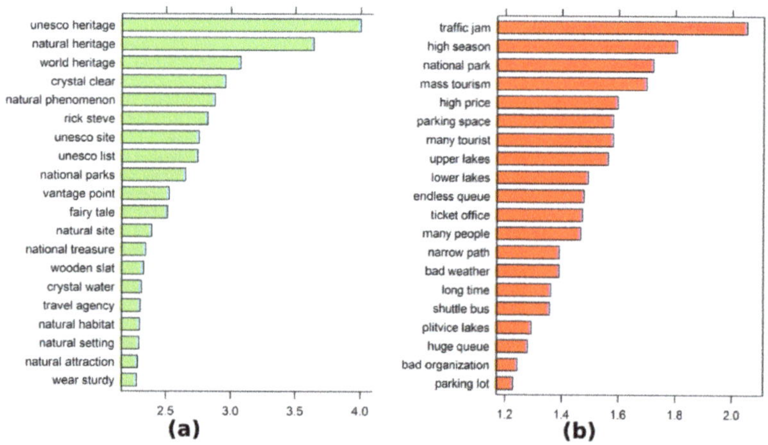

Figure 7. RAKE analysis for positive (**a**) and negative (**b**) reviews for Plitvice Lakes National Park.

5. Discussion

5.1. Answers to Research Questions

The importance of the PLNP at national and international levels is now recognized (Figures 3 and 5). The descriptive statistics highlighted the recurring seasonal trend of visits (Figure 4). This trend has made it essential to implement strategies to redistribute tourist pressure acting on the protected area in a more balanced way.

Regarding the first research question (RQ1), the research has shown that efficient tools exist as an alternative to manual coding (e.g., the software WebHarvy) to collect extensive data relating to lengthy textual reviews (e.g., TripAdvisor online platform). Moreover, the combination of CA with MDS method and cluster analysis turned out to be exhaustive to analyze visitors' preferences and perception for areas of naturalistic interest. First of all, these techniques make it possible to identify the most important symbols and attributes that characterize national parks in accordance with the visitors' opinions. The SA results (Table 1) confirm that national parks and, in general, nature-based experiences arouse positive sentiments in visitors, as already found in other studies [6,8].

MDS methods and cluster analysis are valid instruments to investigate the principal management issues from visitors' point of view (RQ2). The seven clusters identified by this study can help guide a participatory discussion on the issues that visitors consider most important for the reality of PLNP. As stated by Hausmann et al., visitors to national parks tend to idealize some particular places in their destinations, assigning them meanings that make those places worth visiting [8]. In fact, some of the naturalistic and landscape aspects of the PLNP (Cluster 1, 4, and 6, Figure 6) assume a symbolic meaning that almost exclusively attracts the interest of visitors. The most recurring element is the complex aquatic ecosystem of lakes and waterfalls. Also Mirzaalian and Halpenny have identified this type of water elements as one of the main categories of destinations preferred by visitors and a recurring element in the reviews of naturalistic sites [6]. On the one hand, the water system represents the most important naturalistic attraction of the PLNP, but it is also the place where visitors flock the most, representing the fulcrum of tourist organizational problems. In this way, interest in high landscape and environmental or historical values of other areas of the park is excluded a priori. The most evident example is that of the large forest area which is not mentioned at all in any clusters. Other relevant aspects identified are those of accessibility and management of paths and visitors (Clusters 2, 5, and 7, Figure 6). The results obtained show that visitors are aware of and interested in discussing and expressing opinions on organizational issues related to the fruition of places, as already found by Stoleriu et al. [3]. In particular, words like "route" (Cluster 2), "experience" (Cluster 4), "path" (Cluster 5), and "walk" (Cluster 7) emphasize the attention of visitors towards active experiences (e.g., hiking or nature photography). Other studies have also identified these activities as being of great interest in the outdoor visits [25]. In addition, the organizational capacity and the entertainment activities promoted by a tourist destination is an indispensable experiential factor for all those who do not have naturalness as their primary interest [25]. In any case, the most relevant management aspect identified is the management of visitor flows and the problem of overcrowding (Cluster 3 and 5, Figure 6), which was also found by the RAKE analysis.

About the third research question (RQ3), NLP techniques proved to be fundamental to highlight strengths and weaknesses that characterize the image of PLNP. These techniques are of greater interest to identify the negative aspects to be solved and improved rather than the positive aspects to maintain and enhance. The problem of overcrowding is already widely recognized by the Plitvice Lakes National Park Management Plan 2019–2028 [50], which talks about the dissatisfaction of visitors (e.g., due to numerous encounters on the trails or impossibility of taking good photos of pristine landscapes) and the countless organizational problems (e.g., the overcoming of the physical capability of means of transport such as buses and boats or the inability to find parking) detected in the high season [53]. Visitor congestion caused by the crowds of visitors and the consequent recreational conflicts are recurring themes also in other studies focused on the use of protected areas of international interest [21,25,63]. Only a small part of the PLNP's surface represents the main focal point [37], with the "upper lake(s)" and "lower lake(s)" zones (see Figures 6 and 7), where the majority of visits are concentrated [51]. This means that an organizational and promotional effort could be conducted to make the other parts of the park more attractive with activities and guided tours. In fact, the organization of specific events, preferably connected to naturalistic aspects, are of particular interest and attract a large number of visitors as found by Mangachena and Pickering [27].

The automated text analysis processes on social media can provide park managers useful information relating to environment and organizational perception of visitors [27] with a view to collaborative and participatory planning.

5.2. Theoretical Implications

This study makes significant theoretical contributions in the management of areas of naturalistic interest. Firstly, the research demonstrates the flexibility and effectiveness in using an automated approach to obtain information from a large amount of content generated by visitors. From a methodological point of view, the web scraper software applied, WebHarvy, proved to be a valid alternative to manual coding tools. One of the most important innovations of this study is the use of reviews in different languages. In fact, the automatic translation procedure made it possible to use a large number of reviews compared to previous studies that only used reviews written in English [6,8,11,16,25,27,33,39]. Secondly, this study answers a series of research questions regarding the users' judgement on the management of areas of naturalistic interest. In fact, it was possible to identify the topics most cited in visitor reviews, give an order of importance to their discussion, and summarize those that are considered the most important strengths and weaknesses. The study made it possible to extend the use of text mining and NLP techniques already widely applied in other research topics related to tourism in general [9,19,39,44,45] but less explored [8] in nature-based tourism [6,25,27].

Finally, the use of this innovative technique for a well-known study area of international interest (i.e., Plitvice Lakes National Park) allowed to validate the effectiveness of the tool, finding results in accordance with previous knowledge. This step will permit extending the use of the method to other less investigated areas of naturalistic interest, being able to contribute substantially to the identification of key management factors.

5.3. Practical Implications

The results show that social media analysis can be very validly applied to the nature-based tourism field [8]. In particular, these techniques can help decision makers and managers to interpret the online image of national parks constructed by visitors [3,8]. CA—with special regard to SA—effectively identifies negative trends in online reviews, making the tourism operators of national parks capable of being proactive and developing targeted strategies [9]. On the one hand, the method adopted makes it possible to monitor the perception of visitors' recreational experiences in order to plan attractive and well-organized tourist activities. On the other hand, the need to create protected areas and implement conservation and enhancement strategies within them would be supported by similar results [8,53]. In fact, the results of this study demonstrate the high interest and involvement that visitors have towards these very popular tourist destinations. Furthermore, starting from the results obtained, social media could be used by tourism actors (e.g., park managers, tour operators, etc.) to communicate their strategies and marketing proposals to consumers [6]. In particular, for the PLNP both the topics of greatest interest treated by visitors in their reviews and the less contemplated elements are identified, thanks to the use of the methodology adopted. Particularly, the forest ecosystem is not taken into consideration by the visitor reviews, while it would represent the largest percentage of the park area. In line with what has been identified in the current Management Plan [52], it becomes essential to enrich the program of visits with activities that encourage the exploration of all areas of the park. For example, experiences of great interest [25], such as group excursions or guided naturalistic visits, could generate greater appreciation for the complexity of the park's natural systems other than the aquatic ones already widely known. Given the importance attached by visitors to events and special occasions, a further solution to improve the management of the PLNP could be to organize theme-days, highly appreciated by visitors to national parks [27], in order to attract tourists even in less crowded periods, for example, during the winter season, and, therefore, reduce the pressure of the summer season. The PLNP managers could monitor the effectiveness in the proposal of the new visiting programs and events by repeating in the future an analysis of the TripAdvisor reviews with the method adopted in this study in order to search for the presence or absence of the "forests" theme among the interests of visitors.

Thus, in general, from a managerial point of view, these findings can help PLNP managers to better understand visitors' preferences. Furthermore, in this way, managers can more consciously decide which aspects to devote more attention to and how to best redistribute investments to ensure visitor satisfaction.

5.4. Limitations and Future Research

Through the use of social media, it is possible to involve visitors in a first level of participation for protected natural resource management, that of information gathering. In fact, it is extremely complex to include visitors in the subsequent steps of the process, first of all, because it would be necessary to involve very large samples to be representative for the entire population and, secondly, because it is difficult to find simple and adequate channels to contact and interview so many people. Conversely, one of the most relevant advantages is due to the opportunity to carry out investigations on very large samples at extremely low costs. It is also true that other social media (e.g., Instagram and Twitter) allow analysis on a larger scale [8,27], even if they reported some difficulties in processing much shorter texts with a definitely lower amount of information [27].

In the present study, in order to obtain a consistent sample (15,673 online reviews) it was decided to use TripAdvisor reviews on the PLNP issued over a long period (2007–2021). Future research could investigate shorter periods of time to analyze the evolutionary dynamics of the park as well as the effectiveness of the different management strategies used over the years. Furthermore, it must be said that the analysis was restricted to a single Croatian National Park, even if it is the best known (i.e., PLNP). A further study could be, for example, that of a broader analysis of the overall network of national parks that would make it possible to systematize the monitoring and management of protected areas based on a shared investigation effort. It should also be noted that the study presents some biases related to the habits of people in the use of social media. In fact, it has been demonstrated that social media are mostly used among younger people [8,32], which highlights the fact that the analyzed sample is not representative of some categories of people (i.e., children and elderly). The absence of socio-demographic information from TripAdvisor users does not allow for more extensive surveys on the characteristics of the sample [3], while it would be advisable to analyze the preferences of visitors based on their personal characteristics through subsequent in-depth surveys. In fact, it has not been forgotten that the combination of current and traditional survey methods certainly allows the carrying out of very extensive investigations but also allows one to deepen some aspects of the issue in detail [3]. Likewise, it is assumed that all reviews analyzed come from honest opinions of visitors. However, this assumption may not be true, as fake reviews are not uncommon, and it is likely that some of them were included in the sample used in this as well as other sector studies [19]. Since that of natural areas, and in particular of national parks, is a topic not yet particularly deepened in the CA field [3], it could be useful to develop a recreational dictionary specific for national parks that can improve the accuracy of the analysis of the text thanks to the reference to specific terms for the description of the perception of natural environments [8]. Finally, future research could exploit the information available relating to the country of provenance in order to investigate the different preferences expressed by visitors from diverse geographic clusters [27], which have not been investigated in this study.

Despite the above-mentioned limitations, it is believed that the research conducted can be a reliable and useful starting point in the context of tourism analyses to deepen the opinions of the users of the areas of naturalistic interest and extrapolate from their reviews important information for better planning of management activities.

6. Conclusions

The present study investigated the strengths and weaknesses of the PLNP through a large sample of visitor reviews. The results demonstrated the flexibility and effectiveness of applying the developed method to unstructured textual data of online reviews. The present study contributes to fill a research gap in visitor perception analysis for natural areas. The management of the forest area of the PLNP is complex, as it must combine the conservation of natural ecosystems and the tourist destination promotion. In other words, the management must consider the trade-off between the tourism-recreation function and other ecosystem services. The combined use of different and complementary techniques allowed us to develop two research branches in parallel. In the first, the sentiment analysis scores were used to implement a natural language processing technique (i.e., RAKE analysis) from which the strengths and weaknesses of the PLNP have been extrapolated from the visitors' point of view. In the second, the multidimensional scaling method and cluster analysis were used to identify the key topics covered in visitors' reviews. In accordance with the latter result, it might be appropriate to involve visitors in a more in-depth investigation so as to collect visitors' opinions on the priorities defined by the park managers. Despite the limitations encountered, the social media data analysis turns out to be an exhaustive investigation method capable of providing useful information. On the one hand, theoretical advantages can be achieved, contributing in the field of research to the definition of increasingly in-depth and efficient survey tools, and, on the other hand, it is possible to obtain practical information to be provided to the figures who deal with the management and planning related to protected natural areas.

Author Contributions: Conceptualization, A.P., C.F., C.S. and D.V.; methodology, A.P., C.F., C.S., D.V. and E.B.; formal analysis, A.P., C.S. and E.B.; investigation, C.S.; data curation, A.P. and C.S.; writing—original draft preparation C.S.; writing—review and editing, A.P., C.F., C.S. and D.V.; visualization, C.S.; supervision, A.P., C.F. and D.V. All authors have read and agreed to the published version of the manuscript.

Funding: This research received no external funding.

Data Availability Statement: Data available from authors upon request.

Conflicts of Interest: The authors declare no conflict of interest.

Appendix A

The non-normal distribution of the sentiment analysis scores was visually verified in the following graphs.

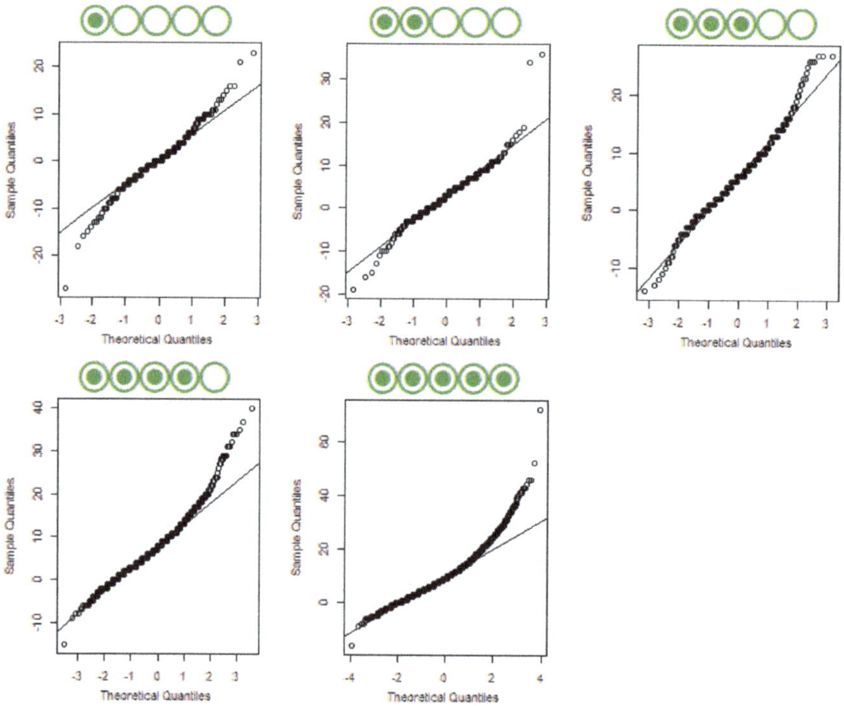

Figure A1. Quantile-quantile plots of the variable "score" for the five groups of bubbles.

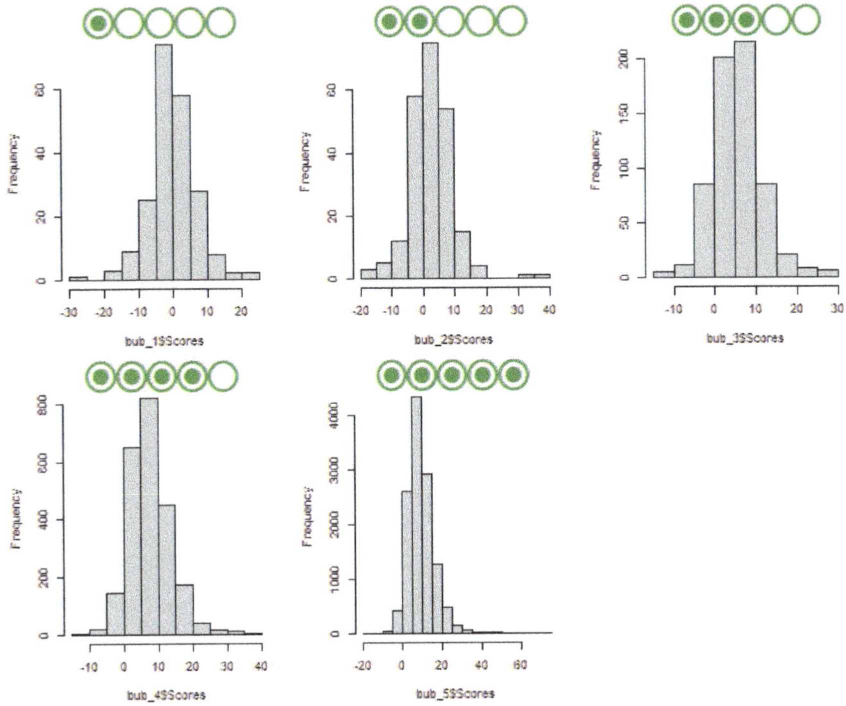

Figure A2. Histograms of the variable "score" for the five groups of bubbles.

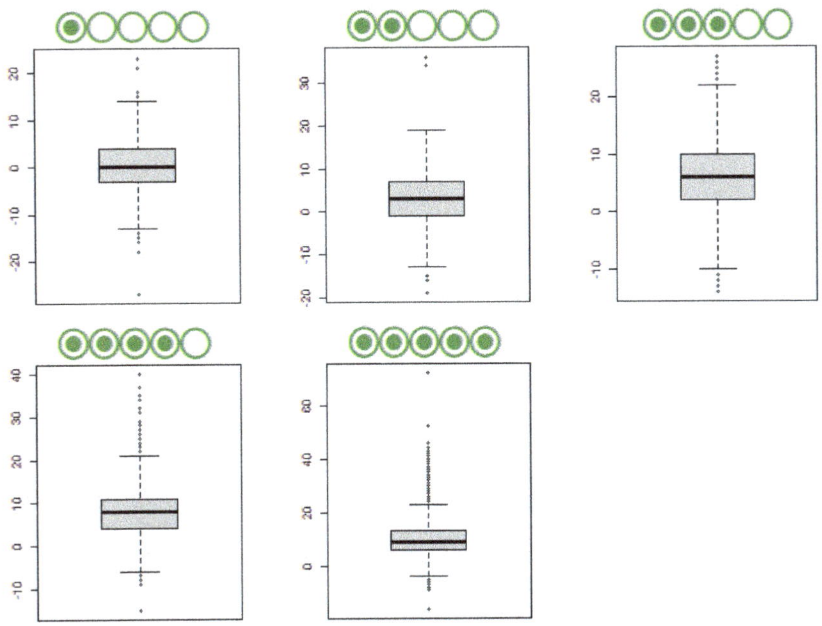

Figure A3. Box plots of the variable "score" for the five groups of bubbles.

References

1. Alaei, A.R.; Becken, S.; Stantic, B. Sentiment Analysis in Tourism: Capitalizing on Big Data. *J. Travel Res.* **2019**, *58*, 175–191. [CrossRef]
2. Zhang, Y.; Cole, S.T. Dimensions of lodging guest satisfaction among guests with mobility challenges: A mixed-method analysis of web-based texts. *Tour. Manag.* **2016**, *53*, 13–27. [CrossRef]
3. Stoleriu, O.M.; Brochado, A.; Rusu, A.; Lupu, C. Analyses of Visitors' Experiences in a Natural World Heritage Site Based on TripAdvisor Reviews. *Visit. Stud.* **2019**, *22*, 192–212. [CrossRef]
4. Lai, L.S.L.; To, W.M. Content analysis of social media: A grounded theory approach. *J. Electron. Commer. Res.* **2015**, *16*, 138–152.
5. Tenkanen, H.; Di Minin, E.; Heikinheimo, V.; Hausmann, A.; Herbst, M.; Kajala, L.; Toivonen, T. Instagram, Flickr, or Twitter: Assessing the usability of social media data for visitor monitoring in protected areas. *Sci. Rep.* **2017**, *7*, 1–11. [CrossRef]
6. Mirzaalian, F.; Halpenny, E. Exploring destination loyalty: Application of social media analytics in a nature-based tourism setting. *J. Destin. Mark. Manag.* **2021**, *20*, 100598. [CrossRef]
7. Yang, M.; Han, C. Revealing industry challenge and business response to Covid-19: A text mining approach. *Int. J. Contemp. Hosp. Manag.* **2020**, *33*, 1230–1248. [CrossRef]
8. Hausmann, A.; Toivonen, T.; Fink, C.; Heikinheimo, V.; Kulkarni, R.; Tenkanen, H.; Di Minin, E. Understanding sentiment of national park visitors from social media data. *People Nat.* **2020**, *2*, 750–760. [CrossRef]
9. Ma, E.; Cheng, M.; Hsiao, A. Sentiment analysis—A review and agenda for future research in hospitality contexts. *Int. J. Contemp. Hosp. Manag.* **2018**, *30*, 3287–3308. [CrossRef]
10. Xiang, Z.; Du, Q.; Ma, Y.; Fan, W. A comparative analysis of major online review platforms: Implications for social media analytics in hospitality and tourism. *Tour. Manag.* **2017**, *58*, 51–65. [CrossRef]
11. Yu, Y.; Li, X.; Jai, T.M. (Catherine) The impact of green experience on customer satisfaction: Evidence from TripAdvisor. *Int. J. Contemp. Hosp. Manag.* **2017**, *29*, 1340–1361. [CrossRef]
12. Valdivia, A.; Luzón, M.V.; Herrera, F. Sentiment Analysis in TripAdvisor. *IEEE Intell. Syst.* **2017**, *32*, 72–77. [CrossRef]
13. Blackshaw, P.; Nazzaro, M. *Consumer Generated Media (CGM) 101: Word-of-Mouth in the Age of Web-Fortified Consumer*; Nielsen, A., Ed.; BuzzMetrics White Pap: Diepoldsau, Switzerland, 2006.
14. Khan, M.F.; Jan, A. Social Media and Social Media Marketing: A Literature Review. *IOSR J. Bus. Manag.* **2015**, *17*, 12–15. [CrossRef]
15. Filieri, R.; Acikgoz, F.; Ndou, V.; Dwivedi, Y. Is TripAdvisor still relevant? The influence of review credibility, review usefulness, and ease of use on consumers' continuance intention. *Int. J. Contemp. Hosp. Manag.* **2021**, *33*, 199–223. [CrossRef]
16. Ghahramani, M.; Galle, N.J.; Duarte, F.; Ratti, C.; Pilla, F. Leveraging artificial intelligence to analyze citizens' opinions on urban green space. *City Environ. Interact.* **2021**, *10*, 100058. [CrossRef]
17. Tripadvisor Tripadvisor Investor Relations. Available online: https://ir.tripadvisor.com/ (accessed on 11 January 2022).
18. Aicher, J.; Asiimwe, F.; Batchuluun, B.; Hauschild, M.; Martina, Z.; Egger, R. *Information and Communication Technologies in Tourism 2016*; Springer: Cham, Switzerland, 2016; pp. 369–382. [CrossRef]
19. Pezenka, I.; Weismayer, C. Which factors influence locals' and visitors' overall restaurant evaluations? *Int. J. Contemp. Hosp. Manag.* **2020**, *32*, 2793–2812. [CrossRef]
20. Roberts, H.; Resch, B.; Sadler, J.; Chapman, L.; Petutschnig, A.; Zimmer, S. Investigating the emotional responses of individuals to urban green space using twitter data: A critical comparison of three different methods of sentiment analysis. *Urban Plan.* **2018**, *3*, 21–33. [CrossRef]
21. Huai, S.; Van de Voorde, T. Which environmental features contribute to positive and negative perceptions of urban parks? A cross-cultural comparison using online reviews and Natural Language Processing methods. *Landsc. Urban Plan.* **2022**, *218*, 104307. [CrossRef]
22. Borg, I.; Groenen, P.J.F.; Mair, P. *Applied Multidimensional Scaling and Unfolding*, 2nd ed.; Springer: Cham, Switzerland, 2018; ISBN 978-3-319-73470-5.
23. Marcussen, C. Multidimensional scaling in tourism literature. *Tour. Manag. Perspect.* **2014**, *12*, 31–40. [CrossRef]
24. Chhetri, P.; Arrowsmith, C.; Jackson, M. Determining hiking experiences in nature-based tourist destinations. *Tour. Manag.* **2004**, *25*, 31–43. [CrossRef]
25. Niezgoda, A.; Nowacki, M. Experiencing nature: Physical activity, beauty and tension in Tatra National Park-analysis of tripadvisor reviews. *Sustainability* **2020**, *12*, 601. [CrossRef]
26. Kaffashi, S.; Radam, A.; Shamsudin, M.N.; Yacob, M.R.; Nordin, N.H. Ecological conservation, ecotourism, and sustainable management: The case of Penang National Park. *Forests* **2015**, *6*, 2345–2370. [CrossRef]
27. Mangachena, J.R.; Pickering, C.M. Implications of social media discourse for managing national parks in South Africa. *J. Environ. Manag.* **2021**, *285*, 112159. [CrossRef] [PubMed]
28. Duffy, R. Nature-Based Tourism and Neoliberalism: Concealing Contradictions. *Tour. Geogr.* **2015**, *17*, 529–543. [CrossRef]
29. Balmford, A.; Beresford, J.; Green, J.; Naidoo, R.; Walpole, M.; Manica, A. A global perspective on trends in nature-based tourism. *PLoS Biol.* **2009**, *7*, e1000144. [CrossRef] [PubMed]
30. Dedeke, A.N. Creating sustainable tourism ventures in protected areas: An actor-network theory analysis. *Tour. Manag.* **2017**, *61*, 161–172. [CrossRef]
31. Schwartz, A.J.; Dodds, P.S.; O'Neil-Dunne, J.P.M.; Danforth, C.M.; Ricketts, T.H. Visitors to urban greenspace have higher sentiment and lower negativity on Twitter. *People Nat.* **2019**, *1*, 476–485. [CrossRef]

32. Kovacs-Györi, A.; Ristea, A.; Kolcsar, R.; Resch, B.; Crivellari, A.; Blaschke, T. Beyond spatial proximity-classifying parks and their visitors in london based on spatiotemporal and sentiment analysis of twitter data. *ISPRS Int. J. Geo-Inf.* **2018**, *7*, 378. [CrossRef]
33. Plunz, R.A.; Zhou, Y.; Carrasco Vintimilla, M.I.; Mckeown, K.; Yu, T.; Uguccioni, L.; Sutto, M.P. Twitter sentiment in New York City parks as measure of well-being. *Landsc. Urban Plan.* **2019**, *189*, 235–246. [CrossRef]
34. Zhang, T.; Zhang, W.; Meng, H.; Zhang, Z. Analyzing visitors' preferences and evaluation of satisfaction based on different attributes, with forest trails in the Akasawa National Recreational Forest, Central Japan. *Forests* **2019**, *10*, 431. [CrossRef]
35. MEA Millenium Ecosystem Assessment (MA). *Ecosystems and Human Well-Being: Synthesis*; Island Press: Washington, DC, USA, 2005.
36. Hausmann, A.; Slotow, R.; Burns, J.K.; Di Minin, E. The ecosystem service of sense of place: Benefits for human well-being and biodiversity conservation. *Environ. Conserv.* **2016**, *43*, 117–127. [CrossRef]
37. Vurnek, M.; Brozinčević, A.; Čulinović, K.; Novosel, A. Challenges in the Management of Plitvice Lakes National Park, Republic of Croatia. In *National Parks—Management and Conservation*; Suratman, M.N., Ed.; IntechOpen: London, UK, 2018; pp. 55–72.
38. Winter, P.L.; Selin, S.; Cerveny, L.; Bricker, K. Outdoor recreation, nature-based tourism, and sustainability. *Sustainability* **2020**, *12*, 81. [CrossRef]
39. Barbierato, E.; Bernetti, I.; Capecchi, I. Analyzing TripAdvisor reviews of wine tours: An approach based on text mining and sentiment analysis. *Int. J. Wine Bus. Res.* **2021**, *1751*, 1062. [CrossRef]
40. Gupta, V.; Lehal, G.S. A Survey of Text Mining Techniques and Applications. *J. Emerg. Technol. Web Intell.* **2009**, *1*, 60–76. [CrossRef]
41. Gan, Q.; Ferns, B.H.; Yu, Y.; Jin, L. A Text Mining and Multidimensional Sentiment Analysis of Online Restaurant Reviews. *J. Qual. Assur. Hosp. Tour.* **2017**, *18*, 465–492. [CrossRef]
42. Zuheros, C.; Martínez-Cámara, E.; Herrera-Viedma, E.; Herrera, F. Sentiment Analysis based Multi-Person Multi-criteria Decision Making methodology using natural language processing and deep learning for smarter decision aid. Case study of restaurant choice using TripAdvisor reviews. *Inf. Fusion* **2021**, *68*, 22–36. [CrossRef]
43. Li, H.; Liu, Y.; Tan, C.W.; Hu, F. Comprehending customer satisfaction with hotels: Data analysis of consumer-generated reviews. *Int. J. Contemp. Hosp. Manag.* **2020**, *32*, 1713–1735. [CrossRef]
44. Serrano, L.; Ariza-Montes, A.; Nader, M.; Sianes, A.; Law, R. Exploring preferences and sustainable attitudes of Airbnb green users in the review comments and ratings: A text mining approach. *J. Sustain. Tour.* **2021**, *29*, 1134–1152. [CrossRef]
45. Lee, J.; Benjamin, S.; Childs, M. Unpacking the Emotions behind TripAdvisor Travel Reviews: The Case Study of Gatlinburg, Tennessee. *Int. J. Hosp. Tour. Adm.* **2020**, *23*, 347–364. [CrossRef]
46. Grandi, R.; Neri, F. Sentiment Analysis and City Branding. In *New Trends in Databases and Information Systems*; Springer: Cham, Switzerland, 2014; Volume 241, pp. 339–349. ISBN 9783319018621.
47. Pearce, P.L.; Wu, M.Y. Entertaining International Tourists: An Empirical Study of an Iconic Site in China. *J. Hosp. Tour. Res.* **2015**, *42*, 772–792. [CrossRef]
48. Song, Y.; Fernandez, J.; Wang, T. Understanding perceived site qualities and experiences of urban public spaces: A case study of social media reviews in Bryant Park, New York city. *Sustainability* **2020**, *12*, 36. [CrossRef]
49. Council of Europe. *European Landscape Convention*; Council of Europe: Florence, Italy, 2000.
50. Koblet, O.; Purves, R.S. From online texts to Landscape Character Assessment: Collecting and analysing first-person landscape perception computationally. *Landsc. Urban Plan.* **2020**, *197*, 103757. [CrossRef]
51. Mandić, A. Protected area management effectiveness and COVID-19: The case of Plitvice Lakes National Park, Croatia. *J. Outdoor Recreat. Tour.* **2021**, 100397. [CrossRef]
52. *Plitvice Lakes National Park Management Plan 2019–2028*; Plitvice Lakes National Park Public Institution: Zagreb, Croatia, 2019; ISBN 9789534850350.
53. McCool, S.F.; Eagles, P.F.J.; Skunca, O.; Vukadin, V.; Besancon, C.; Novosel, A. Integrating Marketing and Management Planning for Outstanding Visitor Experiences in a Turbulent Era: The Case of Plitvice Lakes National Park. In *Mediterranean Protected Areas in the Era of Overtourism*; Springer: Cham, Switzerland, 2021; pp. 221–240. [CrossRef]
54. Nattuthurai, P. Content Analysis of Dark Net Academic Journals from 2010–2017 Using KH Coder. *ACET J. Comput. Educ. Res.* **2017**, *12*, 1–10.
55. Hirahara, S. Evaluation of a structure providing cultural ecosystem services in forest recreation: Quantitative text analysis of essays by participants. *Forests* **2021**, *12*, 1546. [CrossRef]
56. Ward, J.H. Hierarchical Grouping to Optimize an Objective Function. *J. Am. Stat. Assoc.* **1963**, *58*, 236–244. [CrossRef]
57. Higuchi, K. *KH Coder 3 Reference Manual*; Ritsumeikan University: Kyoto, Japan, 2016; p. 99.
58. Nielsen, F.Å. AFINN Word Database an Affective Lexicon. Available online: https://www2.imm.dtu.dk/pubdb/pubs/6010-full.html (accessed on 17 December 2021).
59. Naldi, M. A review of sentiment computation methods with R packages. *arXiv* **2019**, arXiv:1901.08319.
60. Rose, S.; Engel, D.; Cramer, N.; Cowley, W. Automatic keyword extraction from individual document. In *Text Mining: Applications and Theory*; John Wiley & Sons Ltd.: Hoboken, NJ, USA, 2010; pp. 1–277. ISBN 9780470749821.
61. Wijffels, J. *udpipe: Tokenization, Parts of Speech Tagging, Lemmatization and Dependency Parsing with the "UDPipe" "NLP" Toolkit*; R Foundation for Statistical Computing: Vienna, Austria, 2021.

62. R Core Team. *R: A Language and Environment for Statistical Computing*; R Foundation for Statistical Computing: Vienna, Austria, 2021.
63. Mirzaalian, F.; Halpenny, E. Multi-Dimensional Explorations into Visitors' Experience Sharing through TripAdvisor Using Social Media Analytics: An Investigation on Jasper National Park. In Proceedings of the TTRA Canada 2018 Conference, Halifax, NS, Canada, 25–28 September 2018.

forests

MDPI

Article

Habitat Characteristics of *Magnolia* Based on Spatial Analysis: Landscape Protection to Conserve Endemic and Endangered *Magnolia sulawesiana* Brambach, Noot., and Culmsee

Julianus Kinho [1], Diah Irawati Dwi Arini [1], Lutfy Abdulah [1], Ruliyana Susanti [1], Arif Irawan [2], Mira Yulianti [1], Subarudi Subarudi [3], Rinaldi Imanuddin [1,*], Marfuah Wardani [4], Denny Denny [4], Titi Kalima [1], Asef Kurniyawan Hardjana [5], Adi Susilo [1], Ika Heriansyah [4] and Ana Tampang [6]

1 Research Center for Ecology and Ethnobiology, National Research and Innovation Agency, Cibinong 16911, West Java, Indonesia; julianus.kinho@brin.go.id (J.K.); irawati.diah@gmail.com (D.I.D.A.); lutf008@brin.go.id (L.A.); ruli001@brin.go.id (R.S.); mira001@brin.go.id (M.Y.); titi_kalima@yahoo.co.id (T.K.); adis013@brin.go.id (A.S.)

2 Institution for Instrument Standard Implementation on Environment and Forestry of Manado, Ministry of Environment and Forestry, Manado 95119, North Sulawesi, Indonesia; arif_net23@yahoo.com

3 Center for Instrument Standardization on Disaster and Climate Change Resilience, Bogor 16610, West Java, Indonesia; rudi.subarudi@yahoo.co.id

4 Research Center for Plant Conservation, Botanic Gardens, and Forestry, National Research and Innovation Agency, Bogor 16610, West Java, Indonesia; marf002@brin.go.id (M.W.); dennybppn@gmail.com (D.D.); ika.heriansyah@brin.go.id (I.H.)

5 Center for Standardization Instrument on Environment Quality, Ministry of Environment and Forestry, Samarinda 75119, East Kalimantan, Indonesia; akhardjana78@gmail.com

6 Faculty of Forestry, University of Papua, Manokwari 98314, West Papua, Indonesia; a.tampang@unipa.ac.id

* Correspondence: rinaldiimanuddin@yahoo.com; Tel.: +62-812-1809-663

Citation: Kinho, J.; Arini, D.I.D.; Abdulah, L.; Susanti, R.; Irawan, A.; Yulianti, M.; Subarudi, S.; Imanuddin, R.; Wardani, M.; Denny, D.; et al. Habitat Characteristics of *Magnolia* Based on Spatial Analysis: Landscape Protection to Conserve Endemic and Endangered *Magnolia sulawesiana* Brambach, Noot., and Culmsee. *Forests* **2022**, *13*, 802. https://doi.org/10.3390/f13050802

Academic Editors: Radu-Daniel Pintilii, Diego Varga and Luis Diaz-Balteiro

Received: 14 February 2022
Accepted: 17 May 2022
Published: 20 May 2022

Abstract: Based on habitat preferences, in this study, we investigated the spatial distribution of the *Magnolia* genus in the northern part of Sulawesi. Habitat characteristics, especially temperature, precipitation, and topography, were determined using spatial analysis. The temperature and precipitation datasets were obtained from WorldClim BIO Variables V1, and topographical data were obtained from the Google Earth Engine. Data collection began in 2008–2009 and was completed in 2019–2020. In total, we analyzed 786 waypoints. The genus distribution was then predicted based on the most suitable habitat characteristics and mapped spatially. This study confirmed that *Magnolia* spp. distribution is affected by the annual temperature range, precipitation seasonality, and elevation. We discovered endemic and endangered species, *Magnolia sulawesiana* Brambach, Noot., and Culmsee, that were previously distributed exclusively in the central part of Sulawesi. Five waypoints of the endemic species were found in the conservation area of the Gunung Ambang Nature Reserve and on the border of Bogani Nani Wartabone Nation Park. In general, *M. sulawesiana* is distributed at higher elevations than other *Magnolia* species. This study provides a scientific basis for forest officers to develop in-situ and ex-situ conservation strategies and landscape protection measures to maintain the sustainable use of the genus, especially the sustainability of endemic species.

Keywords: endemic; landscape protection; conservation; *Magnolia sulawesiana*; habitat characteristic; spatial distribution

1. Introduction

Magnolia (Fam. Magnoliaceae) is a plant genus that consists of more than 300 species [1–5]. This genus has a wide distribution in subtropical and tropical Asia and America [2,4,5]. The Magnolia genus includes evergreen and deciduous trees and shrubs [2,4], and many species are prominently used as ornamental plants, timber, medicinal raw materials, cosmetics, and essential oils [2,3,6–10]. Despite Magnolia's crucial uses, the assessment in [2] using the International Union for Conservation of Nature (IUCN) criteria resulted in 147 species of *Magnolia* being categorized as threatened (Critically Endangered, Endangered, and Vulnerable) due to various threats such as continued deforestation, habitat destruction, and over-harvesting.

In Indonesia, there are 28 species of *Magnolia* distributed in Sumatra, Java, Lesser Sunda Islands, Sulawesi, Moluccas, and Papua [2,7,10]. Among these species, one species was categorized as Threatened, i.e., *Magnolia sulawesiana* Brambach, Noot., and Culmsee (Endangered); one as Near Threatened (*M. borneensis*); five under Least Concern; and the rest under Data Deficiency. Indonesia is one of the countries with the least amount of information on *Magnolia,* especially for the threatened taxa [11].

The endangered *M. sulawesiana* is an endemic species that grows naturally only in three locations within the mountain range in the central part of Sulawesi [12]. Considering the increasing rate of forest-cover loss in Sulawesi, which was 10.98% between 2000 and 2007 [13], as well as the species' current red-list status as Endangered [2], it is crucial to find this endemic species in other areas of Sulawesi. The central parts of Sulawesi, including the Central and West Sulawesi Province, where *M. sulawesiana* was found, face deforestation rates of 0.68 and 0.84%, respectively [13]. Deforestation is not the only threat to *Magnolia;* overharvesting, poor fruiting, and low natural regeneration [11] also add pressure to the endemic species' vulnerability in the wild.

In Indonesia, overharvesting might become a real threat since Magnolia species are commercially traded. This is especially true for *M. sulawesiana* because it is challenging to distinguish *M. sulawesiana* wood from other Magnolia woods on the market. Magnolia species have a long historical connection with the Minahasa tribe, one of the tribes in the northern part of Sulawesi. The *Magnolia* woods are known as *Cempaka* or *Wasian* and were used as material to construct Woloan, a traditional Minahasa wooden house [14–16]. In the 1970s, when forest concession rightsholder companies began to operate in the production forest, *Magnolia* wood became prominent because of its good quality. In response to the high demand for *Magnolia* wood, local communities started to plant *Magnolia* species [17,18]. Today, the Minahasa district is known to have the largest community plantation forest containing *Magnolia* species among the areas in Sulawesi [15,19,20].

There is also *Magnolia cubensis* in Cuba, which is a highly endangered and endemic species that requires conservation measures. The need for conservation action was determined based on the findings of studies on the influence of habitat fragmentation on the species' population structure and genetic diversity [21]. In Western Mexico, *Magnolia granbarrancae, M. pugana, M. talpana,* and *M. vallartensis* are also critically endangered species because their extent of occurrence (EOO) and area of occupancy (AOO) are below the limits set by the IUCN, and they also have a low genetic diversity [22].

The spatial distribution of plant species is not the result of a random event but is influenced by environmental variables, especially climate and topography characteristics [23,24], as well as soil, temperature, hydrology, and spatial constraints, which affect plant distribution [25–29]. Information on the distribution of *Magnolia* spp. in North Sulawesi is crucial for conservation strategies and landscape protection. Understanding the habitat preferences and suitability of *Magnolia* spp. is also important to determine the species' functions in its surrounding community, including the associated animals [30]. An assessment conducted in China using climate and terrain variables demonstrated differing habitat preferences among *Magnolia* species [8]. Despite its importance for conservation strategies, this type of assessment has never been used for *Magnolia* species in Indonesia.

This study aims to identify the distribution of *Magnolia* species in the northern part of Sulawesi, including the endemic *M. sulawesiana.* We distinguished the habitat preferences of *Magnolia* and combined them with spatial data to estimate the potential species distribution. The discovery of *M. sulawesiana* distribution in the northern part of Sulawesi will lead to a new record of this species' distribution in the Wallacea bioregion [29], which features high species endemism but is still poorly understood.

2. Materials and Methods

2.1. Study Area and Targeted Species

This study was conducted in the northern part of Sulawesi Islands, Indonesia, covering an area of 13,892 km^2 (Figure 1). Most of the topography in this area features hills and mountains that are more than 1000 m above sea level (m asl), with steep contour intervals of less than 12.5 m.

The following Magnolia species [5,19] are known to exist in North Sulawesi: *Magnolia tsiampacca, M. tsiampacca* var. *tsiampacca, M. vrieseana, M. lilliifera, M. champaca,* and *M. candollei.* The endemic *M. sulawesiana*, which closely resembles *M. tsiampacca* var. *tsiampacca* [12], is the most common species in the study areas. Even though it is challenging to distinguish all species in the field, *M. sulawesiana* can be easily differentiated by its golden leaf color. Local people sometimes refer to *M. sulawesiana* as "gold Cempaka Wasian" because of its leaf color. In all study areas, we specifically distinguished *M. sulawesiana* and grouped other Magnolia species that were found as *Magnolia* spp.

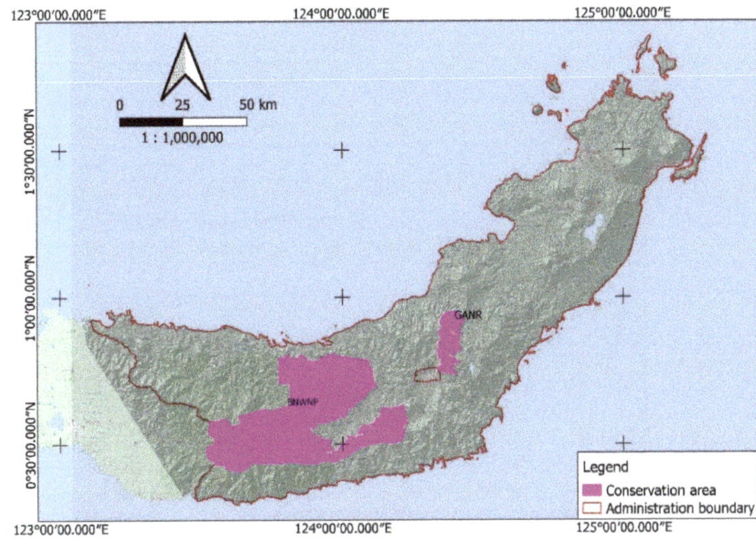

Figure 1. Study area in the northern part of Sulawesi: The purple color represents conservation areas of Bogani Nani Wartabone National Park (BNWNP) and the Gunung Ambang Nature Reserve (GANR).

A preliminary study was conducted to locate the *Magnolia* species in North Sulawesi by collecting information from the district forestry office, local herbarium data, local people, and the literature [19,31,32]. All information was subsequently mapped to produce a survey map. A field survey and ground check were then conducted twice to record the presence or absence of the genus. The first survey was conducted in 2008–2009, concentrating in the western part, and the second in 2019–2020 for the eastern part of North Sulawesi. In all locations in which *Magnolia* trees (diameter at breast height > 15 cm) were found, we recorded the geographical positioning system coordinates as waypoint data. Associated species found around *Magnolia* spp. were also recorded. We recorded 786 waypoints for *Magnolia* spp., 5 of which were for the endemic species *M. sulawesiana.*

2.2. Habitat Characteristics

Habitat characteristics were analyzed using a hierarchical approach with parameters for the criteria, indicators, and verifiers based on those in [8] as presented in Table 1. A flowchart of this methodology used in the present study is presented in Figure 2. The first parameter used was temperature, which was divided into several indicators as shown in Table 2 [33]. We also included land surface temperature (LST), which was derived from the MODIS_LST dataset. This dataset provides daily surface temperature information with a spatial resolution of 1 km [34]. The temperature data were collected from the WorldClim BIO Variables V1 dataset [35].

Table 1. Parameters analyzed in each data category.

Temperature	Precipitation	Topography
AMT	AP	E (Elevation)
DRT	PDM	S (slope)
MinT	PWM	A (Aspect)
MaxT	PCQ	
TAR	PWQ	
TS	PDQ	
LST	PWEQ	
Isot	PS	

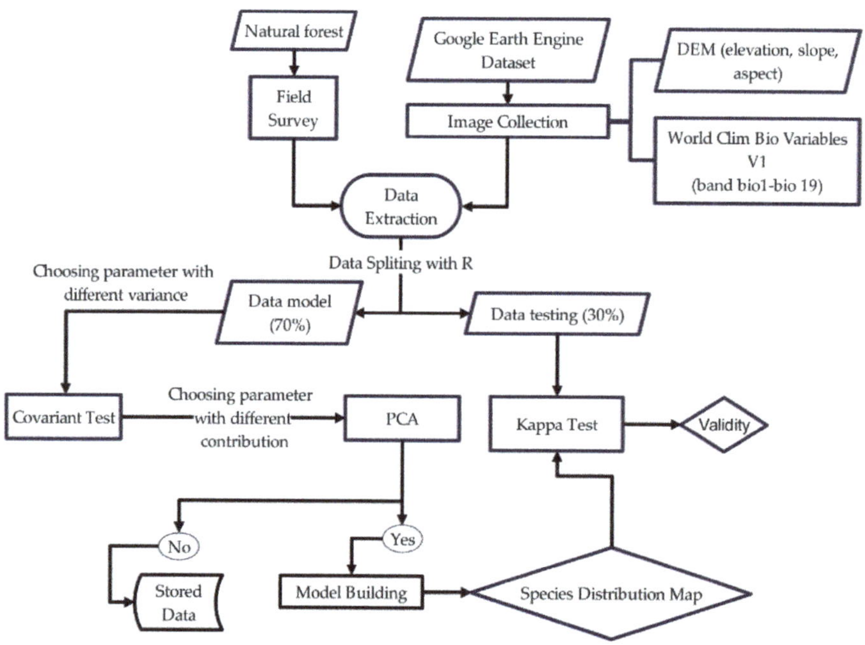

Figure 2. Flowchart of the methodology used to identify *Magnolia* and *M. sulawesiana* distribution.

Table 2. Classification of *Magnolia* spp. distribution prediction based on covariance model analysis.

Class	Description
1	Not a natural distribution area because the parameter values are out of range.
2	Possible distribution area of *Magnolia* spp. Because one of the parameter values is within range.
3	Possible distribution area of *Magnolia* spp. Because two of the parameter values are within range (elevation and temperature annual range).
4	Possible distribution area of *Magnolia* spp. Because two of the parameter values are within range (elevation and precipitation seasonality).
5	Possible distribution area of *Magnolia* spp. because two of the parameter values are within range (temperature annual range and precipitation seasonality).
6	High possible distribution area because all parameter values are within range.

Notes:

- Annual mean temperature (AMT) is the average temperature each year based on the ratio of energy obtained by the ecosystem and the duration of the covered times.
- Diurnal range temperature (DRT) records the daily temperature fluctuations obtained from the maximum and minimum daily temperature differences in a period.
- Maximum temperature (MaxT) is an indicator that states the highest daily temperature in the month with the hottest temperature in each season of the year.
- Minimum temperature (MinT) is the lowest temperature in the coldest month, which varies from year to year. As a result, the lowest temperature will be average.
- Temperature annual range (TAR) is an indicator that states the differences between MaxT and MinT.
- Temperature seasonality (TS) is a periodic temperature indicator of the daily average and deviation of temperature.
- Annual precipitation (AP) expressed in mm/year is the average yearly rainfall in the observation period.
- Precipitation of the driest month (PDM) and wettest month (PWM) record the average amount of rain in the driest month and rainy season, respectively.
- Precipitation of the coldest quarter (PCQ), the warmest quarter (PWQ), the driest quarter (PDQ), and the wettest quarter (PWEQ) indicate the average daily rain in the coldest quarter of a year, the warmest quarter of a year, the dry season quarter of a year, and the highest rainy quarter of a year, respectively. In Indonesia, PWQ and PWEQ are difficult to distinguish between since there are only two seasons, the rainy and dry seasons, between which the temperature differences are very narrow. Precipitation data were also collected from the WorldClim BIO Variables V1 dataset [35].
- Isothermality (Isot) is a thermodynamic process in which the temperature of a system remains constant.
- Aspect (A) is the compass direction or azimuth that a terrain surface faces.

2.3. Data Analysis

Magnolia spp. distribution was predicted based on the most suitable habitat characteristics. Since one of the *Magnolia* spp. found in North Sulawesi was *M. sulawesiana*, which is an endemic and endangered species [12,32,36], we predicted the suitable habitat characteristics for this particular species separately for conservation purposes.

For the spatial analysis, a covariance model was used to determine the species dependency [37] and to increase the prediction accuracy based on the species distribution [30,31]. The hypotheses used in the covariance test for each indicator were as follows:

Hypothesis H0. *There will be differences in variance for each indicator where the species grow (p-value $< \alpha = 5\%$).*

Hypothesis H1. *There are no differences in variance for each indicator where the species grow (p-value $> \alpha = 5\%$).*

The value interval of each indicator was calculated to determine the pattern of each waypoint. The contribution of each parameter and quadrant was then determined using a principal component analysis and a bi-plot contribution graph. To avoid the possibility of errors in classification, the required contribution should be close to 100%.

The classification of species distribution patterns depends on the indicator for the number of species present, where a higher score indicates a greater possibility for a species to be present. All the parameter data were overlayed with a spatial raster. The classes were then developed based on the presence/absence of tree species. If there is an n-parameter with the presence/absence of *Magnolia* spp. trees, then the number of classes becomes 2n if the species is present. A higher value of n indicates a higher possibility of *Magnolia* spp.

being present and vice versa. The interval between the highest and lowest n-value then indicates variations in the probability (Table 2).

Distribution classification testing was conducted using separated data testing based on the method in [38]. An accuracy test was then completed using the Kappa value [39], which was able to express accuracy in the classification [38]. The comparison matrix is presented in Table 3.

Table 3. Confusion matrix of the validation distribution test.

		Reference				
	Class	**1**	**2**	**..**	**m**	**Σ**
Map	1	p_{11}	p_{12}	$p_{1.}$	p_{1m}	$p_{1.}$
	2	p_{21}	p_{22}	$p_{2.}$	p_{2m}	$p_{2.}$
	..	$p_{\cdot 1}$	$p_{.2}$	$p_{..}$	$p_{.m}$	$p_{..}$
	n	p_{n1}	p_{n2}	$p_{n.}$	p_{nm}	$p_{n.}$
	Σ	$p_{\cdot 1}$	$p_{\cdot 2}$	$p_{\cdot n}$	$p_{\cdot m}$	1

$$p_0 = \sum_{i=1}^{m} pii \tag{1}$$

Here, p_0 is the probability of accuracy, which determines whether it is appropriate to locate the point in the polygon that indicates the presence of *M. sulawesiana* or *Magnolia* spp. and vice versa for the point that indicates that the polygon does not contain *M. sulawesiana* or *Magnolia* spp.

$$p_e = \sum_{i=1}^{m} pi.\ p.i \tag{2}$$

Here, p_e is the chance of error, which determines whether the point in the polygon that indicates that there is no *Magnolia* spp. or *M. sulawesiana* can be precisely determined and vice versa for the point that determines whether the polygon does not contain *Magnolia* spp. or *M. sulawesiana* when they exist in the polygon. The Kappa coefficient (κ) and standard error (σ_κ) were then measured using the following equation:

3. Results

3.1. Habitat Characteristics

M. sulawesiana and other Magnolias grow under a similar diurnal temperature range, land surface temperature, and isothermality but have different ranges for the rest of the indicators (Figure 3 and Table 4). *Magnolia* spp. grow at an annual average temperature of between 20 and 26.4 °C, with an average of 23 °C, while *M. sulawesiana* grows at 20–25 °C, with an average of 21°C. *Magnolia* spp. grow at minimum temperatures of between 15 and 22 °C, while *M. sulawesiana* can prevail at minimum temperatures of between 15 and 20 °C. Concerning the highest temperature, *Magnolia* spp. can grow at temperatures of 25–32 °C, with 25–30 °C for *M. sulawesiana*. Based on the temperature differences between the rainy and the dry season, *Magnolia* spp. grow in areas with a temperature difference of 2.2–3.8 °C. Meanwhile, *M. sulawesiana* grows in areas with a temperature difference of 2.3–2.5 °C.

Based on the precipitation data, *Magnolia* spp. and *M. sulawesiana* have a similar range only in annual precipitation (Figure 4 and Table 5). The average annual rainfall is between 1900 and 3000 mm. *Magnolia* spp. grow in areas with rainfall of between 2060 and 3034 mm/year, with an average annual rainfall of 2400 mm. Meanwhile, *M. sulawesiana* grows in areas with rainfall of 1900–2400 mm/year and an average of 2300 mm/year.

The topographical data (Figure 5 and Table 6) show that *Magnolia* spp. grow at elevations between 400 and 800 m asl. However, some trees are found at up to 1300 m asl. Meanwhile, *M. sulawesiana* grows at an elevation of 1000–1300 m asl, with most trees growing above 1200 m asl. *Magnolia* spp. and *M. sulawesiana* grow in an overlapping topography range from 1000 to 1300 m asl.

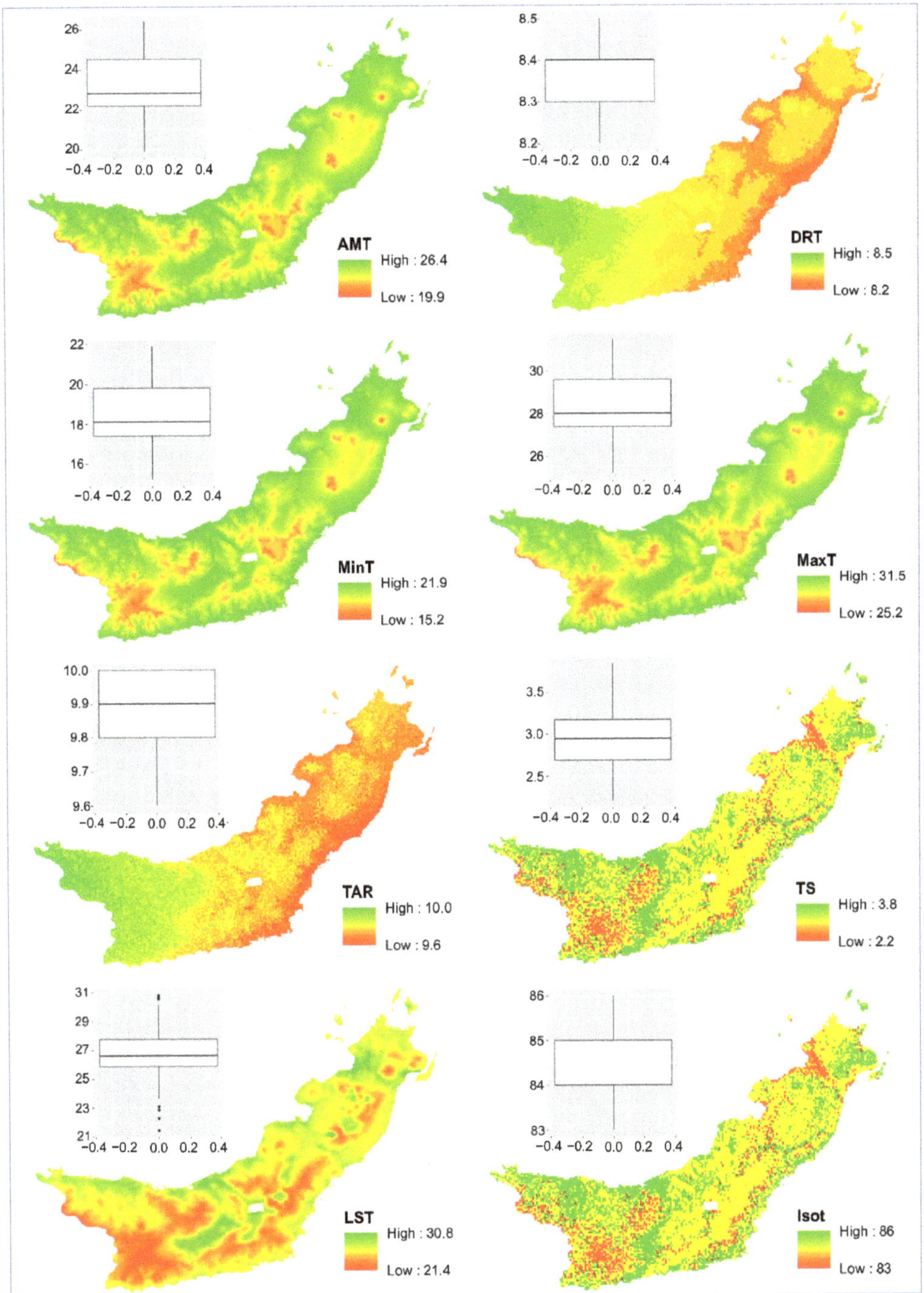

Figure 3. Temperature indicator images for *Magnolia* spp. in North Sulawesi.

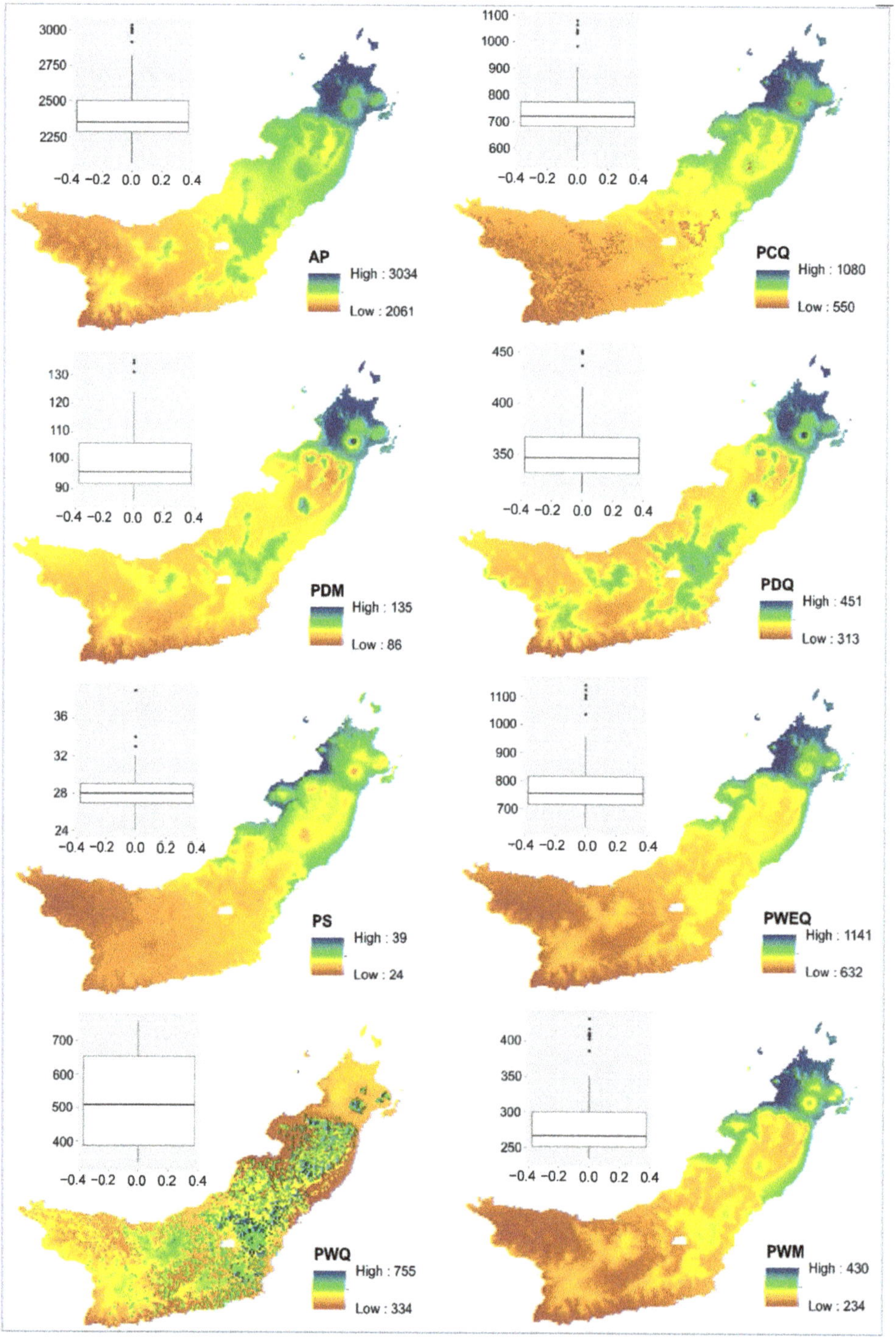

Figure 4. Precipitation indicator images for *Magnolia* spp. in North Sulawesi.

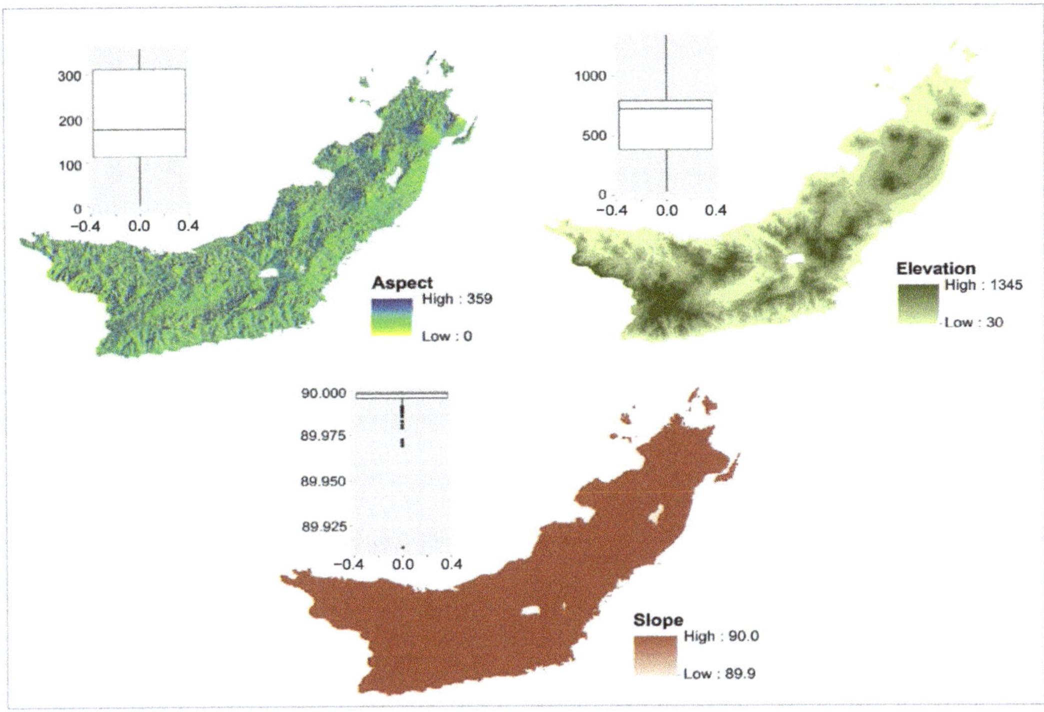

Figure 5. Topographical indicator images for *Magnolia* spp. in North Sulawesi.

The statistical analysis results (Table 6) indicate that *Magnolia* spp. can be found at an altitude of between 30 and 1345 m asl with a variation of up to 45%. Meanwhile, *M. sulawesiana* grows between 179 and 1414 m asl with a variation of 48%. The average *Magnolia* spp. and *M. sulawesiana* grow at 638 m asl and 1017 m asl, respectively. Meanwhile, both have similar preferences in terms of slope and aspect.

Table 4. Statistical descriptions for temperature indicators.

Variable	Tree Species	Average	Standard Deviation	Coefficient of Varian	Min	Max
AMT	*Magnolia* spp.	23	1.5	6.4	19.9	26.4
	M. sulawesiana	21	2.3	10.9	19.5	25.0
DRT	*Magnolia* spp.	8.4	0.06	0.8	8.2	8.5
	M. sulawesiana	8.4	0.05	0.7	8.4	8.5
MinT	*Magnolia* spp.	18.3	1.5	8.2	15.2	21.9
	M. sulawesiana	16.2	2.3	14.0	14.7	20.2
MaxT	*Magnolia* spp.	28.2	1.4	5.1	25.2	31.5
	M. sulawesiana	26.2	2.3	8.7	24.7	30.2
TAR	*Magnolia* spp.	9.9	0.09	0.9	9.6	10.0
	M. sulawesiana	10	0.0	0.0	10.0	10.0
TS	*Magnolia* spp.	2.9	0.3	10.1	2.2	3.8
	M. sulawesiana	2.4	0.1	4.7	2.3	2.5
LST	*Magnolia* spp.	26.5	1.8	6.7	21.4	30.8
	M. sulawesiana	25.3	1.9	7.4	22.6	27.4
Isot	*Magnolia* spp.	84.3	0.5	0.6	83.0	86.0
	M. sulawesiana	84.4	0.5	0.7	84.0	85.0

Table 5. Statistical descriptions for precipitation indicators.

Variable	Tree Species	Average	Standard Deviation	Coefficient of Varian	Min	Max
AP	*Magnolia* spp.	2414.0	171.7	7.1	2061.0	3034.0
	M. sulawesiana	2290.6	210.6	9.2	1915.0	2407.0
PDM	*Magnolia* spp.	101.7	11.2	11.0	86.0	135.0
	M. sulawesiana	103.4	6.5	6.3	92.0	108.0
PWM	*Magnolia* spp.	281.7	35.7	12.7	234.0	430.0
	M. sulawesiana	266.2	34.8	13.1	204.0	284.0
PCQ	*Magnolia* spp.	738.9	81.0	11.0	550.0	1080.0
	M. sulawesiana	474.2	57.2	12.1	427.0	551.0
PWQ	*Magnolia* spp.	512.8	129.0	25.2	334.0	755.0
	M. sulawesiana	553.8	74.3	13.4	421.0	592.0
PDQ	*Magnolia* spp.	358.6	32.7	9.1	313.0	451.0
	M. sulawesiana	377.2	37.2	9.9	311.0	398.0
PWEQ	*Magnolia* spp.	777.3	86.4	11.1	632.0	1141.0
	M. sulawesiana	705.6	65.3	9.3	589.0	742.0
PS	*Magnolia* spp.	27.9	2.1	7.4	24.0	39.0
	M. sulawesiana	23.6	0.5	2.3	23.0	24.0

Table 6. Statistical descriptions for topography parameter.

Variable	Tree Species	Average	Standard Deviation	Coefficient of Varian	Min	Max
Elevation	*Magnolia* spp.	638.49	288.50	45.18	30.00	1345.00
	M. sulawesiana	1017	487	47.93	179	1414
Slope	*Magnolia* spp.	89.996	0.00527	0.01	89.913	90.000
	M. sulawesiana	89.997	0.00177	0.00	89.995	89.999
Aspect	*Magnolia* spp.	202.15	106.01	52.44	0.00	359.40
	M. sulawesiana	197.1	123.9	62.84	54.4	329.0

Variation Test for Habitat Characteristics

The results of the F test showed that each category of habitat parameters featured variations among both *Magnolia* spp. and *M. sulawesiana*. The parameters of temperature annual range, precipitation seasonality, elevation, and slope showed significant differences in *Magnolia* spp. (Table 7).

The six temperature and seven precipitation indicators tested showed no differences for *Magnolia* spp. and *M. sulawesiana*, meaning that *Magnolia* spp. and *M. sulawesiana* have similar habitat preferences. Significant differences in preferences were observed in the gap between the maximum and minimum temperature (TAR) and the ratio between the standard deviations of annual rainfall (PS) (Table 7). While *Magnolia* spp. and *M. sulawesiana* have different ranges in elevation and slope, they both have the same variation in aspect, which is supported by the work in [40] showing that precipitation and annual mean temperature make critical contributions to endemic and critically endangered species in Kashmir Himalaya. Based on these results, we concluded that the habitat characteristics of *Magnolia* spp. and *M. sulawesiana* are influenced by the temperature annual range, precipitation seasonality, and elevation. This information serves as the basic information to predict the spatial distribution of *Magnolia* spp. and *M. sulawesiana*.

3.2. Species Distribution

3.2.1. Magnolia spp.

The variable contribution and bi-plot analysis of *Magnolia* spp. showed that the slope has a lower impact than the other key variables, including annual temperature range, precipitation seasonality, and elevation. In addition, the slope variable is also in the same quadrant as the annual temperature range (Figure 6). Therefore, the slope variable can be neglected when estimating the distribution of *Magnolia* spp.

The spatial distribution prediction map for *Magnolia* spp. is presented in Figure 7. There are six distribution classifications based on the range of TAR, PS, and elevation parameters (as shown in Table 2). The highest number of criteria (6, green color) represent the most suitable habitat, while the lowest number (3, yellow color) represents the least suitable habitat for *Magnolia* spp. In general, *Magnolia* spp. is spread in mountainous areas and follows the direction of the slopes with a concentric habitat pattern.

Table 7. F-test results for all habitat characteristic categories.

Categories		*p*-Value F Test
Temperature	Annual mean temperature (AMT)	0.09
	Diurnal range temperature (DRT)	0.87
	Temperature minimum (MinT)	0.11
	Temperature maximum (MaxT)	0.09
	Temperature annual range (TAR)	2.2×10^{-16} **
	Temperature seasonality (TS)	0.07
	Land surface temperature (LST)	0.72
	Isothermality	0.65
Precipitation	Annual Precipitation (AP)	0.39
	Precipitation of driest month (PDM)	0.29
	Precipitation of wettest month (PWM)	0.87
	Precipitation of coldest quarter (PCQ)	0.53
	Precipitation of warmest quarter (PWQ)	0.29
	Precipitation of driest quarter (PDQ)	0.54
	Precipitation of Wettest quarter (PWEQ)	0.63
	Precipitation Seasonality (PS)	0.02 **
Topography	Elevation (E)	0.04 **
	Slope (S)	0.04 **
	Aspect (A)	0.49

** significantly different.

3.2.2. *Magnolia sulawesiana*

The main characteristics of the *M. sulawesiana* species found in the northern part of Sulawesi are leaves with a coriaceous shiny-green top (pale greenish-brown to reddish-brown when dry) and a paler bottom (dark golden-brown to chestnut when dry). The tree bark is grey-brown, fissured, and lenticellate with a mealy texture and flakes off in large, irregular plates on older trees. Older trees feature silver-grey bark with fine longitudinal cracks (Figure 8). According to the species characteristics and identification key provided in [12], we are confident that the species found in the study area is *M. sulawesiana.*

The flowering and fruiting seasons of *M. sulawesiana* have irregular patterns in the five locations. This study identified two flowering and fruiting seasons, of February to April and August to September. The study also found that single-mother trees can have both a flowering and fruiting season (monoecious). Buds, young and mature flowers, and ripe fruits were observed together in one individual tree. We also observed very few seedlings around the mother tree, although closer observations of the reproduction strategy of this species must be conducted. Poor fruiting and low natural regeneration were reported in [11]. These findings will add to the knowledge of *M. sulawesiana* in the northern part of Sulawesi. The other location was the mountain forest of the Gunung Ambang Nature Reserve (GANR) and production forest (Figure 9), where we found four individuals of *M. sulawesiana* at an elevation of 1163–1378 m asl in a hilly primary forest dominated by Myristicaceae, Euphorbiaceae, and Calophyllaceae. Additionally, one individual was found in the slope area near Kotulidak River. The last individual found was located in the production forest at Bolaang Mongondow District (175 m asl). This production forest is part of a natural forest located near the boundaries of BNWNP. Based on the analysis of the results, this endemic species seems to have a wide distribution, from 179–1414 m asl to 1600–2200 m asl, as described in [12]. This wide range of elevation creates a greater possibility of finding this species in the mountainous area down to the lowland forest in Sulawesi.

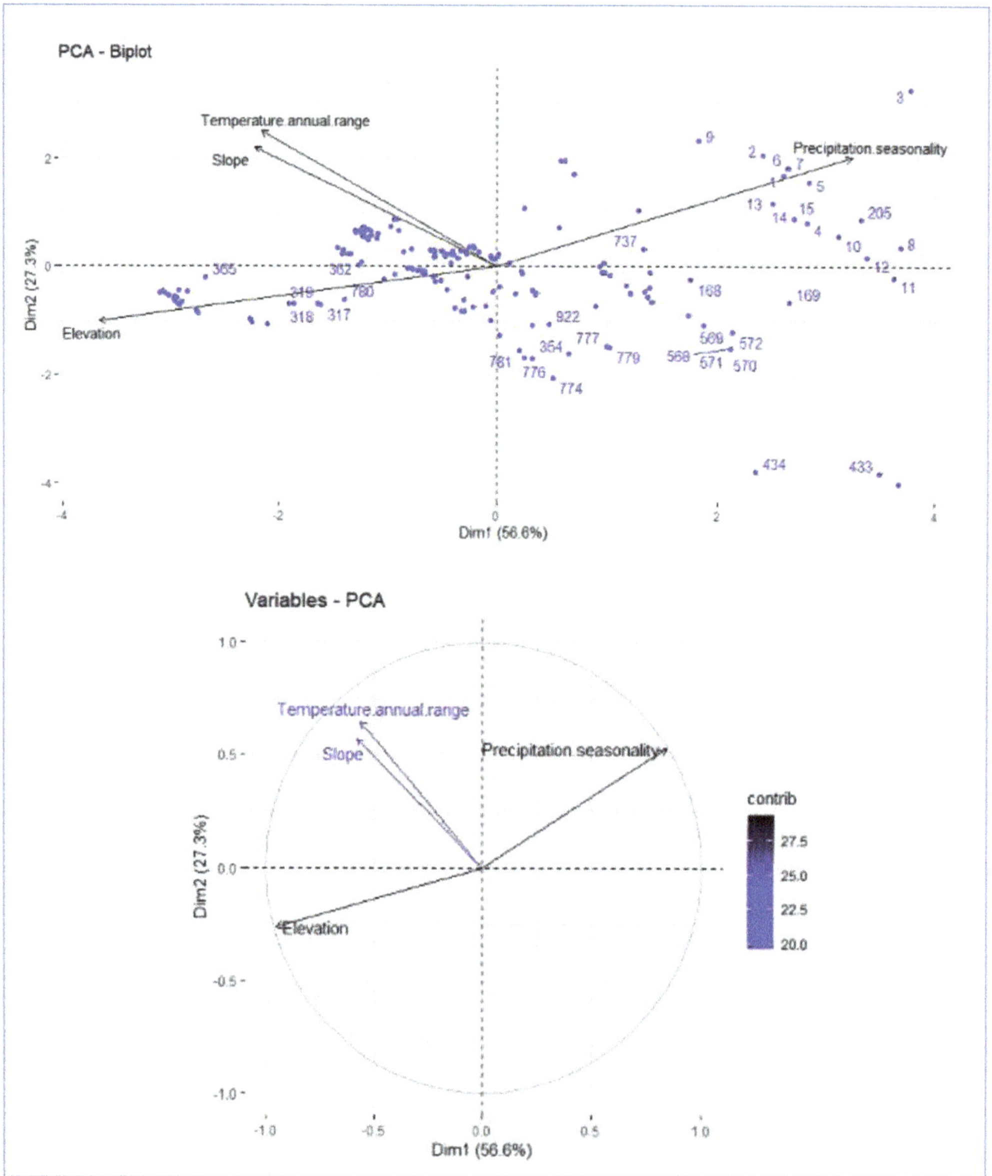

Figure 6. Variable contribution analysis and bi-plot of the annual temperature range, precipitation seasonality, elevation, and slope of *Magnolia* spp. Above: PCA—Biplot; below: Variables—PCA.

3.3. *Accuracy Test Classification*

The results of the accuracy test (Table 8) showed that the estimated kappa distribution was 100%; thus, the standard error was zero. A random selection of waypoints in the classification accuracy test was also conducted. The number of waypoints in the *Magnolia* spp. distribution test was 116. All points were of the highest class (6). Thus, the Kappa value was 100%, which means the classification accuracy is appropriate.

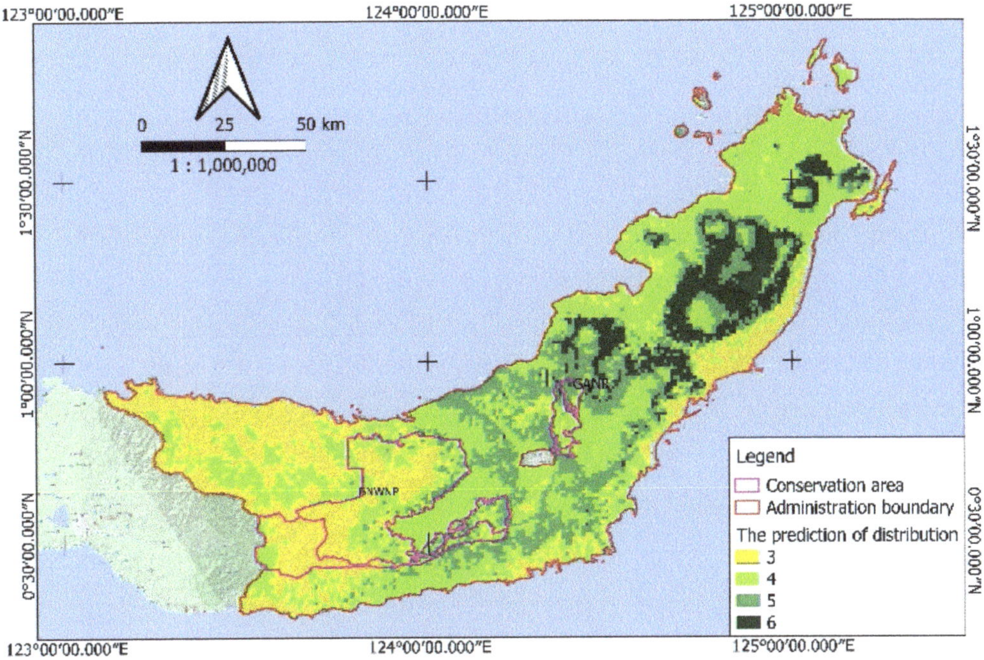

Figure 7. Distribution prediction patterns of *Magnolia* spp. in North Sulawesi, with the number representing habitat suitability, with 3 indicating the least suitable habitat and 6 representing the most suitable habitat. In the map, the darker the color (6), the more suitable the habitat for *Magnolia* spp.

Figure 8. Morphological characteristics of *M. sulawesiana* found in the northern part of Sulawesi: (**a**) adaxial leaf; (**b**) abaxial leaf; (**c**) older tree bark; (**d**) leaves on twig with an open flower; (**e**) leaves on twig with fruit; (**f**) closer look at ripe fruits.

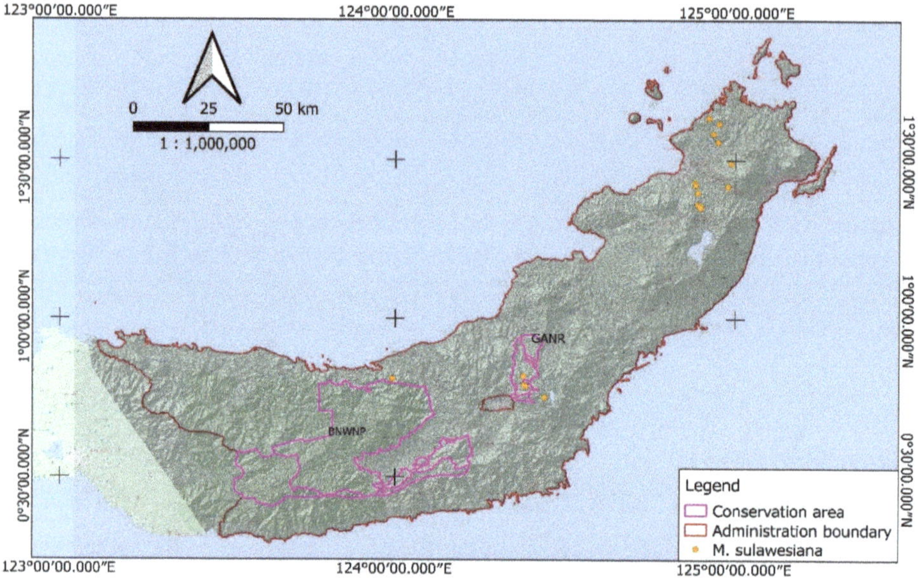

Figure 9. Distribution of *M. sulawesiana* found in GANP and the production forest at the border of BNWNP in North Sulawesi.

Table 8. Accuracy test classification results for *Magnolia* spp.

	Class	0	1	Σ	p_0	p_e	κ	σ_κ
	0	0	0	0	0	1	100%	0
Magnolia spp.	1	0	116	116				
	Σ	0	116	116				

4. Discussion

The spatial distribution of *Magnolia* in North Sulawesi Province is influenced by the annual temperature range, precipitation seasonality, and elevation. At the landscape scale in this extremely varied environment, topography and climate were reported to be significant determinants of species richness, endemic richness, and endemicity [23,24]. The influence of climate on *Magnolia* distribution demonstrates the species' vulnerability to climate changes [40]. *Magnolia* is also reported to have allopatric speciation [3]. Thus, geographical isolation [3], topography, and climatic factors [26] have led to the scarcity of this genus due to a lack of specific habitat suitability. On the other hand, changes in climatic parameters such as temperature can also lead to the distribution of a species outside of its native range, as shown in [41] for *M. grandiflora*.

Based on our results, despite the importance of climatic factors, elevation seems to have the largest influence on *Magnolia* species distribution in North Sulawesi Province. Elevation can have a 10% to 50% effect on plant distribution [27,42]. For *M. sulawesiana,* we found that this species has a wider elevation range due to the presence of outlier data. One individual was found at 175 m asl, while the rest of the individuals, including the individual recorded in [12], were found from 1100 to 2000 m asl. This is a very interesting result because the individual was identified as *M. sulawesiana* based on morphological characteristics [12] and located inside GANP, where it remains unclear if this species grows naturally. A closer study needs to be conducted to explain this phenomenon. There might be other factors that determine species distribution other than climatic and elevation factors. Other research shows that most of the endemic and endangered species of *Magnolia* are naturally found in tropical mountain forests and at high elevations. Examples include *M. schiedeana* in Mexico [43], *M. sinica* in Yunan (1339–1707 m asl) [44], *M. vovidesii* in Mexico (1520–1550 m asl) [21], and *M. granbarrancae* (1073–1215 m asl) [22].

The growing demand for *Magnolia* trees in the lumber market throughout the year has led to consequences such as the increasing rarity of this genus. To meet these needs, wood is harvested not only from *Magnolia* plantations but also from the species' natural habitat. This study demonstrated that the distribution of *Magnolia* spp. in production forests is decreasing. Without any effort to create a community development program to maintain the balance between different needs, *Magnolia* species will gradually become rare and, eventually, extinct. As a result, the existence of this genus is threatened. Scattered, small-scale *Magnolia* plantations managed by local people still exist in several areas in North Sulawesi. Examples include Rumoong Atas Village, South Minahasa District [17], and Kawatak Village, Minahasa District [18]. The plantation in Rumoong Atas village has existed for decades, and the Cempaka trees in the region were planted on inheritance land [17]. The plantations typically cover about 1–2 ha and is managed from generation to generation. While the local community in Kawatak village developed plantations under the Community Forestry Program, the local people planted several *Magnolia* species, including the endemic species, *M. sulawesiana*.

Habitat preference data will serve as the basic information for landscape management approaches to ensure the survival of the genus, including in-situ and ex-situ conservation. This approach is also expected to maintain the remaining natural population in the protected area while ensuring sustainable use through plantation and community forests. The *Magnolia* species, especially the endemic species, found inside conservation areas, need to be protected in-situ [22]. As an endemic and endangered species, *M. sulawesiana* also needs to be considered as a protected species. For this reason, conservation efforts were conducted at both the habitat and species level. The other *Magnolia* species found outside the conservation areas could also be proposed for protection to maintain their sustainability. Additionally, the area could be designated as a buffer zone. Using the same information, the ex-situ conservation of the species could be conducted through the development of plantations or community forests within the most suitable habitat preferences in collaboration with the local people. Community forest development could act as a buffer for the natural habitat of the species in the conservation area. Ex-situ conservation could also lessen the risk of extinction for threatened species and support in-situ conservation efforts [22]. The remaining forest in Sulawesi plays a crucial role as a life support system due to its geographical conditions, with extreme faults being prone to landslides. Rapid changes in land use create further difficulties for conservation efforts on this island. A spatial distribution map is crucial for the local forest district to develop landscape-scale protection [22], which is important not only for the targeted species but also for Wallace's unique wildlife and Sulawesi's fragile ecosystems more broadly [45].

To facilitate the effective implementation of conservation, especially for endemic species with unique habitats [46], further research needs to be conducted to determine the genetic diversity of all populations, and inbreeding and genetic diversity levels [22,46] could be used to determine protection priorities, especially at the landscape level. Information on the population size, phenological patterns, morphological variations [47,48], and population genetic diversity [49] in natural habitats, including the populations in the northern and central part of Sulawesi, will determine the actions that should be taken concerning conservation in natural habitats (in-situ). This conservation should involve the indigenous knowledge of "*Eluren Eng Kayobaan*" (keeping and maintaining the Earth) [50] to encourage the planting of Wasian trees on community lands.

5. Conclusions

The spatial distribution of *Magnolia* spp. is affected by climatic (temperature annual range and precipitation seasonality) and elevation variables. Among the discovered *Magnolia* spp., we located five sites featuring the endemic and endangered species,

Magnolia sulawesiana, which was previously distributed exclusively in the central part of Sulawesi. The existence of this essential species alters the paradigm of landscape protection. In this study, we provided a scientific foundation from which to develop in-situ and ex-situ conservation strategies and landscape protection measures to maintain the sustainability of endemic species. Simultaneously, it is also important to ensure the sustainable use of other *Magnolia* species. Further research is required to strengthen conservation and plantation-management practices.

Author Contributions: Each author (J.K., D.I.D.A., L.A., R.S., A.I., M.Y., S.S., R.I., M.W., D.D., T.K., A.K.H., A.S., I.H. and A.T.) has an equal role as the main contributor who equally discussed the conceptual ideas and the outline, provided critical feedback for each section, and helped shape and write the manuscript. All authors have read and agreed to the published version of the manuscript.

Funding: Part of the field work of this research was funded by ITTO Project PD 646/12 Rev.3 (F) Initiating the conservation of cempaka tree species (*Elmerrillia* sp.) through plantation development with the local community participation in North Sulawesi.

Institutional Review Board Statement: Not applicable.

Data Availability Statement: Not applicable.

Acknowledgments: We thank the ITTO for supporting this research, especially for the field data collection through the Project PD 646/12 Rev.3(F), and cooperation between ITTO and Manado Environment and Forestry Research and Development Institute (BP2LHK). We also thank Hiras Sidabutar as a project advisor; Mochlis, the former head BP2LHK, for the facilitation and encouragement and ensuring the project ran smoothly; the North Sulawesi Forestry Service; our valuable friend at Forum Cempaka for all the discussion and support; and the reviewers provided usefull comments for the paper improvement.

Conflicts of Interest: The authors declared no conflict of interset.

References

1. Figlar, R.B.; Nooteboom, H.P. Notes on Magnoliaceae IV. *Blumea Biodivers. Evol. Biogeogr. Plants* **2004**, *49*, 87–100. [CrossRef]
2. Rivers, M.; Beech, E.; Murphy, L.; Oldfield, S. *The Red List of Magnoliaceae—Revised and Extended*; Botanic Gardens Conservation International: Richmond, UK, 2016.
3. Sánchez-Velásquez, L.R.; Pineda-López, M.R.; Vásquez-Morales, S.G.; Avendaño-Yáñez, M.L. Ecology and conservation of endangered species: The case of magnolias. *Endanger. Species Threat Conserv. Futur. Res.* **2016**, 63–84.
4. Wang, Y.B.; Liu, B.B.; Nie, Z.L.; Chen, H.F.; Chen, F.J.; Figlar, R.B.; Wen, J. Major clades and a revised classification of Magnolia and Magnoliaceae based on whole plastid genome sequences via genome skimming. *J. Syst. Evol.* **2020**, *58*, 673–695. [CrossRef]
5. Plants of the World Online. Available online: http://www.plantsoftheworldonline.org/ (accessed on 16 August 2021).
6. Heyne, K. *Tumbuhan Berguna Indonesia*, 1st ed.; Yayasan Sarana Wana Jaya: Jakarta, Indonesia, 1987.
7. Rozak, A.H. Status taksonomi, distribusi dan kategori status konservasi Magnoliaceae di Indonesai. *Bul. Kebun Raya* **2012**, *15*, 81–91.
8. Song, C.; Liu, H.; Gao, J. Habitat preference and potential distribution of *Magnolia officinalis* subsp. *Officinalis* and *M. O.* SubSp. *biloba* in China. *Nat. Conserv.* **2019**, *36*, 93–111.
9. Song, C.; Liu, H. Habitat differentiation and conservation gap of *Magnolia biondii*, *M. denudata*, and *M. sprengeri* in China. *PeerJ* **2019**, *6*, e6126. [CrossRef]
10. Maesaroh, S.; Özel, Ç.A. Biotechnological approaches for improvement of Magnolia Genus Grown in Indonesia. *J. Nat. Appl. Sci.* **2021**, *4*, 186–203. [CrossRef]
11. Cires, E.; De Smet, Y.; Cuesta, C.; Goetghebeur, P.; Sharrock, S.; Gibbs, D.; Oldfield, S.; Kramer, A.; Samain, M.S. Gap analyses to support ex situ conservation of genetic diversity in *Magnolia*, a flagship group. *Biodivers. Conserv.* **2013**, *22*, 567–590. [CrossRef]
12. Brambach, F.; Nooteboom, H.P.; Culmsee, H. *Magnolia sulawesiana* described, and a key to the species of *Magnolia* (*Magnoliaceae*) occurring in Sulawesi. *Blumea J. Plant Taxon. Plant Geogr.* **2013**, *58*, 271–276. [CrossRef]
13. Supriatna, J.; Shekelle, M.; Fuad, H.A.H.; Winarni, N.L.; Dwiyahreni, A.A.; Farid, M.; Mariati, S.; Margules, C.; Prakoso, B.; Zakaria, Z. Deforestation on the Indonesian island of Sulawesi and the loss of primate habitat. *Glob. Ecol. Conserv.* **2020**, *24*, e01205. [CrossRef]
14. Sasmuko, S.A. Karakteristik kayu lokal untuk rumah woloan di Provinsi Sulawesi Utara. *Penelit. Has. Hutan* **2010**, *28*, 278–290. [CrossRef]
15. Kinho, J.; Mahfudz. *Prospek Pengembangan Cempaka di Sulawesi Utara*; Balai Penelitian Kehutanan Manado: Manado, Indonesia, 2011.
16. Kurnia Manoppo, F.; Anggereany Ester Kaunang, R.; Weol, W. The Dialogue of Christian Education and Minahasa Traditional Community in "Rumamba" Tradition in Disruptive Era: A Reconciliation or Disaster? In *Advances in Social Science, Education and Humanities Research, Proceedings of the 4th Asian Education Symposium (AES 2019), Manado, Indonesia, 14–15 August 2019*; Atlantis Press: Paris, France, 2020.

17. Mohtar, A.; Walangitan, H.D.; Katiandagho, T.M. Kontribusi hutan rakyat terhadap pendapatan rumah tangga petani Di Desa Rumoong Atas Kecamatan Tareran Kabupaten Minahasa Selatan. *Cocos* **2019**, *1*, 1–11.
18. Waisaley, N.; Thomas, A.; Nurmawan, W. Analisis potensi tegakan hutan rakyat jenis cempaka di Desa Kawatak Kecamatan Langowan Selatan. *Cocos* **2018**, *1*, 1–8.
19. Kinho, J.; Irawan, A. Studi keragaman jenis cempaka berdasarkan karakteristik morfologi di sulawesi utara. *Ekspose Has. Litbang BPK Manad.* **2011**, *6*, 61–78.
20. Arini, D.I.; Mairi, K.; Golioth, Y. Membangkitkan kembali minat masyarakat Sulawesi Utara menanam cempaka dalam skema hutan rakyat. *Policy Br. Pus. Penelit. Pengemb. Sos. Ekon. Kebijak. Perubahan Iklim* **2021**, *15*, 1–10.
21. Galván-Hernández, D.M.; Octavio-Aguilar, P.; de Jesús Bartolo-Hernández, C.; García-Montes, M.A.; Sánchez-González, A.; Ramírez-Bautista, A.; Vovides, A. Current Status of *Magnolia vovidesii* (Magnoliaceae, Magnoniales): New Data on Population Trends, Spatial Structure, and Disturbance Threats. *Trop. Conserv. Sci.* **2020**, *13*, 1–12. [CrossRef]
22. Vázquez-garcía, J.A.; Muñiz-castro, M.A.; Dahua-machoa, A.; Osorio-muñoz, E.A.; Hernandez-Vera, G.; Ortega-Pena, A.S.; de Lourdes Romo-Campos, R.; Jacobo-Pereira, C.; de Roman, N.A.; Shalisko, V. How to save endangered Magnolias? From population biology to conservation action: The case of allopatric radiation in Western Mexico. In *Endangered Plants*; Kumar, S., Ed.; Intechopen: Rijeka, Croatia, 2021.
23. Irl, S.D.H.; Harter, D.E.V.; Steinbauer, M.J.; Gallego Puyol, D.; Fernández-Palacios, J.M.; Jentsch, A.; Beierkuhnlein, C. Climate vs. topography–spatial patterns of plant species diversity and endemism on a high-elevation island. *J. Ecol.* **2015**, *103*, 1621–1633. [CrossRef]
24. Slaton, M.R. The roles of disturbance, topography and climate in determining the leading and rear edges of population range limits. *J. Biogeogr.* **2015**, *42*, 255–266. [CrossRef]
25. Chen, Y.; Yang, X.; Yang, Q.; Li, D.; Long, W.; Luo, W. Factors Affecting the Distribution Pattern of Wild Plants with Extremely Small Populations in Hainan Island, China. *PLoS ONE* **2014**, *9*, e97751. [CrossRef]
26. Yudaputra, A.; Fijridiyanto, I.Z.U.; Cropper, W.P. The potential impact of climate change on the distribution pattern of eusideroxylon zwageri (*Bornean ironwood*) in Kalimantan, Indonesia. *Biodiversitas* **2020**, *21*, 326–333. [CrossRef]
27. Usmadi, D.; Sutomo, S.; Iryadi, R.; Hanum, S.F.; Darma, I.D.P.; Wibawa, I.P.A.H. Predicting Species Distribution for True Indigo (*Indigofera tinctoria* L.) in Citarum Watershed, West Java, Indonesia. *J. Trop. Biodivers. Biotechnol.* **2021**, *6*, 65398. [CrossRef]
28. Raes, N.; Cannon, C.H.; Hijmans, R.J.; Piessens, T.; Saw, L.G.; Van Welzen, P.C.; Ferry Slik, J.W. Historical distribution of Sundaland's Dipterocarp rainforests at Quaternary glacial maxima. *Proc. Natl. Acad. Sci. USA.* **2014**, *111*, 16790–16795. [CrossRef] [PubMed]
29. Li, T.; Xiong, Q.; Luo, P.; Zhang, Y.; Gu, X.; Lin, B. Direct and indirect effects of environmental factors, spatial constraints, and functional traits on shaping the plant diversity of montane forests. *Ecol. Evol.* **2019**, *10*, 557–568. [CrossRef] [PubMed]
30. Oppel, S.; Mack, A.L. Bird assemblage and visitation pattern at fruiting *Elmerrillia tsiampaca* (Magnoliaceae) trees in Papua New Guinea. *Biotropica* **2010**, *42*, 229–235. [CrossRef]
31. Langi, Y.A.R. Model Penduga Biomassa dan Karbon Pada Tegakan Hutan Rakyat Cempaka (*Elmerrillia ovalis*) dan Wasian (*Elmerrrillia celebica*) Di Kabupaten Minahasa-Sulawesi Utara. Master Thesis, IPB University, Bogor Regency, Indonesia, 2007.
32. Govaerts, R. *World Checklist of Selected Plant Families in the Catalogue of Life*; Banki, O., Roskov, Y., Doring, M., Ower, G., Vandepitte, L., Hobern, D., Remsen, P., Schalk, R., DeWalt, M., Keping, J., et al., Eds.; The Catalogue of Life Partnership: Leiden, The Netherlands, 2019.
33. Noce, S.; Caporaso, L.; Santini, M. A new global dataset of bioclimatic indicators. *Sci. Data* **2020**, *7*, 398. [CrossRef] [PubMed]
34. LP DAAC—MOD11A1. Available online: https://lpdaac.usgs.gov/products/mod11a1v006/#citation (accessed on 14 January 2022).
35. Hijmans, R.J.; Cameron, S.E.; Parra, J.L.; Jones, P.G.; Jarvis, A. Very high resolution interpolated climate surfaces for global land areas. *Int. J. Climatol.* **2005**, *25*, 1965–1978. [CrossRef]
36. Rivers, M. Magnolia sulawesiana. Available online: https://www.iucnredlist.org/species/88981112/88981118 (accessed on 3 April 2022).
37. Choi, I. *Modeling Spatial Covariance Functions*; Purdue University: West Lafayette, IN, USA, 2014.
38. Foody, G.M. Explaining the unsuitability of the kappa coefficient in the assessment and comparison of the accuracy of thematic maps obtained by image classification. *Remote Sens. Environ.* **2020**, *239*, 111630. [CrossRef]
39. Cohen, J. A coefficient of agreement for nominal scales. *Educ. Psychol. Meas.* **1960**, *20*, 37–46. [CrossRef]
40. Hussain Mir, A.; Tyub, S.; Kamili, A.N. Ecology, distribution mapping and conservation implications of four critically endangered endemic plants of Kashmir Himalaya. *Saudi J. Biol. Sci.* **2020**, *27*, 2380–2389. [CrossRef]
41. Gruhn, J.A.; White, P.S. Magnolia grandiflora L. range expansion: A case study in a north carolina piedmont forest. *Southeast Nat.* **2011**, *10*, 275–288. [CrossRef]
42. Ahmad, I.; Verma, S.; Mushtaq, S.; Abdullah, A.; Nasser, M.; Tariq, M.; Pant, S. Saudi Journal of Biological Sciences Ecological analysis and environmental niche modelling of *Dactylorhiza hatagirea* (D. Don) Soo: A conservation approach for critically endangered medicinal orchid. *Saudi J. Biol. Sci.* **2021**, *28*, 2109–2122.
43. Vásquez-Morales, S.G.; Téllez-Valdés, O.; Del Rosario Pineda-López, M.; Sánchez-Velásquez, L.R.; Flores-Estevez, N.; Viveros-Viveros, H. Effect of climate change on the distribution of *Magnolia schiedeana*: A threatened species. *Bot. Sci.* **2014**, *92*, 575–585. [CrossRef]

44. Wang, B.; Ma, Y.; Chen, G.; Li, C.; Dao, Z.; Sun, W. Rescuing *Magnolia sinica* (Magnoliaceae), a Critically Endangered species endemic to Yunnan, China. *Oryx* **2016**, *50*, 446–449. [CrossRef]
45. Riley, E.P. Flexibility in diet and activity patterns of *Macaca tonkeana* in response to anthropogenic habitat alteration. *Int. J. Primatol.* **2007**, *28*, 107–133. [CrossRef]
46. Hernández, M.; Palmarola, A.; Testé, E.; Veltjen, E.; Asselman, P.; Larridon, I.; Samain, M.S.; González-Torres, L.R. Population structure and genetic diversity of *Magnolia cubensis* subsp. *acunae* (Magnoliaceae): Effects of habitat fragmentation and implications for conservation. *Oryx* **2020**, *54*, 451–459.
47. Coritico, F.P.; Amoroso, V.B. Threatened lycophytes and ferns in four protected areas of mindanao, philippines. *Nat. Conserv. Res.* **2020**, *5*, 78–88. [CrossRef]
48. Borah, D.; Taram, M.; Wahlsteen, E. *Begonia dicressine* (Begoniaceae): A new record for India. *Nat. Conserv. Res.* **2021**, *6*, 110–111. [CrossRef]
49. Barbosa-Silva, R.G.; Coutinho, T.S.; Vasconcelos, S.; da Silva, D.F.; Oliveira, G.; Zappi, D.C. Preliminary placement and new records of an overlooked amazonian tree, *Christiana mennegae* (Malvaceae). *PeerJ* **2021**, *9*, e12244. [CrossRef]
50. Berdame, J.; Lombogia, C.A.R. Merajut tradisi di tengah transisi: Pendidikan lingkungan hidup berbasis kearifan lokal dalam budaya mapalus suku minahasa. *Tumou Tou* **2020**, *7*, 128–142. [CrossRef]

forests

MDPI

Article

Analysis of Forest Landscape Preferences and Emotional Features of Chinese Forest Recreationists Based on Deep Learning of Geotagged Photos

Xitong Zeng [1], Yongde Zhong [1,2,3,*], Lingfan Yang [4], Juan Wei [1] and Xianglong Tang [1]

1 College of Forestry, Central South University of Forestry and Technology, Changsha 410000, China; zengxt1998@163.com (X.Z.); weijuan1004@163.com (J.W.); txl1812286068@163.com (X.T.)
2 College of Tourism, Central South University of Forestry and Technology, Changsha 410000, China
3 National Forestry and Grassland Administration State Forestry Administration Engineering Research Center for Forest Tourism, Changsha 410000, China
4 Beijing Zoo, Beijing 100000, China; yanglf@mail.bnu.edu.cn
* Correspondence: ade@csuft.edu.cn

Abstract: Forest landscape preference studies have an important role and significance for forest landscape conservation, quality improvement and utilization. However, there are few studies on objective forest landscape preferences from the perspective of plants and using photos. This study relies on Deep Learning technology to select six case sites in China and uses geotagged photos of forest landscapes posted by the forest recreationists on the "2BULU" app as research objects. The preferences of eight forest landscape scenes, including look down landscape, look forward landscape, look up landscape, single-tree-composed landscape, detailed landscape, overall landscape, forest trail landscape and intra-forest landscape, were explored. It also uses Deepsentibank to perform sentiment analysis on forest landscape photos to better understand Chinese forest recreationists' forest landscape preferences. The research results show that: (1) From the aesthetic spatial angle, people prefer the flat view, while the attention of the elevated view is relatively low. (2) From the perspective of forest scale and level, forest trail landscape has a high preference, implying that trail landscape plays an important role in forest landscape recreation. The landscape within the forest has a certain preference, while the preference of individual, detailed and overall landscape is low. (3) Although forest landscape photographs are extremely high in positive emotions and emotional states, there are also negative emotions, thus, illustrating that people's preferences can be both positive and negative.

Keywords: forest landscape; deepsentibank; deep learning; geotagged photos; sentiment analysis

Citation: Zeng, X.; Zhong, Y.; Yang, L.; Wei, J.; Tang, X. Analysis of Forest Landscape Preferences and Emotional Features of Chinese Forest Recreationists Based on Deep Learning of Geotagged Photos. *Forests* **2022**, *13*, 892. https://doi.org/10.3390/f13060892

Academic Editors: Radu-Daniel Pintilii and Diego Varga

Received: 13 May 2022
Accepted: 6 June 2022
Published: 8 June 2022

Publisher's Note: MDPI stays neutral with regard to jurisdictional claims in published maps and institutional affiliations.

1. Introduction

Forest aesthetics first originated in Germany, founded and proposed by Salisch in 1885, signaling the beginning of a focus on the natural aesthetics of forests. The increase in the dissection and discussion of forest beauty also laid the foundation for the study of forest landscapes and their preferences [1]. Forest landscape is a landscape composed of a forest ecosystem as the main body, and its research aims to reveal certain basic laws through the structure and function of forest landscape, and to implement the protection, construction, planning, restoration and management of forest landscape through scientific means on this basis [2]. Globally, forests perform a variety of functions for people and the social value of the forest environment has received a great deal of attention, especially the health function, recreational value and landscape appreciation brought by forest tourism [3–6]. People involved in forest tourism are generally inclined to prefer the plants element, the main landscape element in the forest landscape, and to care about the trees in the forest [7,8]. The aesthetic role of plants is the most important and the first to be recognized and used [9].

People have also learned from the experience of using forest play that the forest landscape is an integral part of the visual image of the forest [10]. However, there are still few studies on forest landscapes from the perspective of plants or animals [11]. Based on this, we decided to explore forest landscape preferences from a botanical point of view, to compensate, to some extent, for the lack of this panel.

China reported 6 billion forest visitors in a 4-year period (2016–2019), with an average annual growth rate of 15%. Among them, the number of forest tourism visitors reached 1.8 billion in 2019, creating a comprehensive social output value of USD 25.7 million; in the wake of the COVID-19 pandemic, forest tourism has also gained momentum and reached 84.2% of the annual visitor arrivals for 2019 [12]. As forest tourism is gradually chosen and accepted by the public, forest landscape preference studies have started to emerge gradually, focusing on human interaction studies. An early study of forest landscape preferences from a human perspective was conducted by Misgav and Amir—to study the degree of people preference for forest vegetation landscapes from a visual quality perspective [13]. Studies have shown that forest beauty does influence the number of visits to forested areas and preferences [14,15]. Investigation of forest tourism can improve the planning and management of forest resources, especially in terms of forest landscape, which is an important component of forest tourism [16,17]. Among these, forest classification has been a means of describing complex and diverse forest resources, based on vegetation features, landscape features or a combination of both, called landscapes. Forest landscape classification is important for ecological studies of forest landscapes, etc. In reviewing the relevant literature, it is not difficult to find that forest landscapes mostly appear as a certain research element of destination landscape preference studies, with few studies dedicated to them. There is also little use of forest landscape classification systems as a basis for identifying people's preferences for different categories of forest landscapes.

The advent of the Web 2.0 era has opened up more possibilities for the study of forest landscape preferences. Studies have conducted Internet-based studies on preferences for alpine forest landscapes, preferences for forest features and preferences for forest structure [18–20]. This has driven the emergence and catalyst of Travel 2.0, where travelers exchange travel-related content and engage in high levels of social interaction on the Internet. With the popularity and convenience of photography tools, travel photography is one of the indispensable behaviors in travel [21,22]. Travel experience is, again, the most important visual experience, and taking photos is also a fashionable behavior for people to share their lives and publish their experiences. Therefore, the photos shared by tourists have become the main dissemination channel for their preferences, and photo content analysis has been widely used in tourism research and is considered to play a more important role in the tourism process, with more scholars focusing on the nature behind the photographic behavior [23–26]. Some scholars have already conducted early experimental studies, in which tourists were asked to rate the content of photos and found that natural conditions, trail design and forest conditions all affect tourists' perceptions of the forest landscape and its trails [27]. Photographs are intrinsically linked to tourism, as a person must take some form of photography during their travels, and the photographs they take can reflect their unique personal motivations [24,26]. Studies have shown that there are no substantial differences in landscape preferences between visitors who post photos and participants who do not post photos [28]. Therefore, we want to explore more possibilities by using geotagged photos that people post, using the content of the photos to explore landscape preferences. In fact, travel photos shared by tourists contain not only objective information but also hidden information, so the photos posted on social media platforms can capture the emotions of tourists and the conditions of their experiences [29,30]. The visual perception of a forest environment does influence people's psychological emotions and adjusts their psychological state, and almost all studies agree that forests trigger positive emotions. Is it true that people in forest landscape environments feel only positive emotions? We wanted to explore whether other answers existed.

The field of artificial intelligence has made significant advances in computer vision (CV), image processing techniques and deep learning, offering many new possibilities and new ideas for preference, as well as for the study of travel photos. Currently, in addition to manual coding with tools, such as NVIVO and Textblob or using tools for smart tagging of photos [31–34], computer vision technology provides a better solution path for visual content analysis in tourism. Recently, Transformer and Multilayer Perceptron (MLP)-based models, such as MLP-Mixer and Vision Transformer, have started to lead the new trend because of their excellent performance in ImageNet classification tasks [35]. The spread of coding technologies, such as Python, has made interdisciplinary collaboration between the fields of tourism research and computing possible, and computer technologies, such as computer vision, image processing techniques and deep learning, have facilitated content processing, recognition and analysis of photos published on UGC platforms [36]. Based on the characteristics of deep learning, it is mainly applied to two segments of tourism research: tourist volume prediction and image content mining. The application methods are mainly divided into two types: using pre-models as is and training models by migration learning [37]. Tourism photo content mining is mainly from big data on UGC platforms for visual content analysis. Image-based tourism research is also increasingly focused on potential sentiment analysis, e.g., Deng et al. more innovatively started to use Deepsentibank's deep learning tools for tourism group imagery, as well as destination images [38,39]. In addition, several studies have used pre-trained models to classify and analyze tourism photos. Payntar et al. analyzed tourism photographs of the Cusco World Heritage Site in Peru using the ResNet model [40]. Cho et al. used deep learning techniques to classify photos on Flickr in an attempt to develop a photo classification system for tourist destinations [41]. Kim et al. used the Inception-v3 model to classify tourist photos of Seoul and used it as the basis for their study [42]. It proves that Deep Learning is accurate when mining the photo content. Although data generated by the widespread use of the Internet and social media reflect people's real preferences, there are still few articles that combine social media with computer vision algorithms in an attempt to understand individual preferences [43]. However, there have been successful uses of computer vision to characterize human–animal interactions [44], so we think the interaction between forest landscapes and humans can also be studied and analyzed by computer vision and the use of deep learning methods for the tourism segment is mostly on the large tourism categories, and there is less segmentation of the internal elements with the large categories. This literature has given us more inspiration to explore different class preferences in forest landscapes using the MLP-Mixer model that has simplicity for a large number of classifications. This will help forest managers to plan, design and manage the development of forest tourism, forest beauty and forest landscape quality in a more focused manner.

Existing studies have mostly explored preferences by using all landscape elements or images of the destination as research themes, and by default, the photos taken by people are their preferences. Thus, we use deep learning and Deepsentibank to analyze the image content with the core problem of "how to explore people's preference of forest landscape through photos posted on UGC platforms". That is to say, the exploration preference also supplements the sentiment analysis of pictures, aiming to make up for the lack of special research on forest landscape preference and try new angles and methods to discover valuable information from the photos published by people for forest tourism research. The specific objectives are: (1) Through data mining of outdoor website—2BULU (https://www.2bulu.com/ accessed on 1 December 2021)—a dataset of 15,052 photos of forest plants from six places in China, including Kanas, Gongga, Four Girls Mountains, Shennongjia, Changbai Mountain and Moganshan, was established, and through computer depth science, the photo visual content of the data was divided into eight scenes and three categories to determine the Chinese forest recreationists preference for forest landscapes on this basis. (2) Explore the emotional attitudes carried by photos using the Deepsentibank program to complement preference studies with a more objective perspective.

2. Case Sites and Datasets

2.1. Case Sites

The formation of forests is closely related to the long-term effects of the surrounding natural conditions. China is a vast country with five major climatic zones: cold temperate, temperate, warm temperate, subtropical and tropical, from north to south; precipitation generally decreases from south to north and decreases from east to west, and there are various topographies, such as high mountains, plateaus and hills, and basins. These all make the distribution of forests in different regions of China different, with obvious zonality, and also mean it has more types of forests. With the development of modern times, China has gradually increased both the importance and protection of forest cover. According to the State Forestry and Grassland Administration of China, China's forest cover has reached 23.04%, with a forest accumulation of over 17.5 billion cubic meters [45]. China is also gradually promoting forest tourism and actively fostering forest tourism products—building national forest parks, national forest trails, etc., calling for people to go to the forest to get in touch with nature. In addition, forest tourism festivals have made the social influence of forest tourism in China grow rapidly.

The vast majority of China's forest resources are concentrated in the northeast and southeast, while the vast northwest is poor in forest resources. In addition, 80% of China's population is located in the southeastern region, and tourism flows are heavily concentrated in the southeastern half of the country [46]. We selected the specific case sites considering that they have a certain number of forest recreationists and that these recreationists have uploaded a certain number of forest landscape photos for analysis in the "2BULU" app. Therefore, the majority of the case sites were in southeastern China. The difference in the level of economic development between the north and south, east and west of China limits the construction of forest recreation facilities and the accessibility of the forest for different vegetation types. Through pre-experiments and surveys, we set the case sites as Kanas, Gongga, Four Girls Mountains, Shennongjia, Changbai Mountain and Moganshan, in order to better ensure that the case sites are representative. The six selected case sites are all highly visible and influential in China, even world class, such as Shennongjia being selected as a World Biosphere Reserve Network and Changbai Mountain being selected as a United Nations "Man and Biosphere" nature reserve and an international A-class nature reserve. Based on this, we believe that the photographs of the six case sites are representative of the forest landscape preferences of Chinese forest tourists. See Figure 1 for location diagram.

2.2. Datasets

The data comes from the UGC platform "2BULU"—an app for outdoor resource sharing and community interaction—which is widely used in daily trips, travel trips, wilderness camping, etc. It has a large user base and a large amount of data, and is also a more mature platform in China for obtaining photos taken by travelers and their metadata [47]. It should be noted that the users of "2BULU" are not only outdoor travelers, but also ordinary tourists and even local residents, school students, etc. The range of users is relatively wide, so we believe that its data source tends to encompass all kinds of travelers, rather than just outdoor travelers.

We used Python tools to write script code to acquire the photo data using the keywords of the case site. Due to the large amount of data and the fact that we only needed forest photos, we used the Tencent API (Application Programming Interface) filtering port to help with the first round of filtering and selecting photos with plants in the data acquisition phase. This enables us to improve the accuracy and efficiency of data screening and reduce the difficulty and effort of processing raw data from "2BULU". After the first round of selection, we obtained a total of 35,675 photos with plants inside. We use a narrow sense of forest landscape—natural scenery with forest vegetation as the main part within the view of people at a certain point in time and space [48]—and referred to the data processing method of White et al. for analyzing forest landscapes [49]. Therefore, we conducted a second and third round of manual screening to remove photos containing a large number of people, animals, lakes and other elements, which reduces distractions and shows the true attractiveness and preference of the natural environment [49]. Finally, 15,052 photos were obtained in the final dataset.

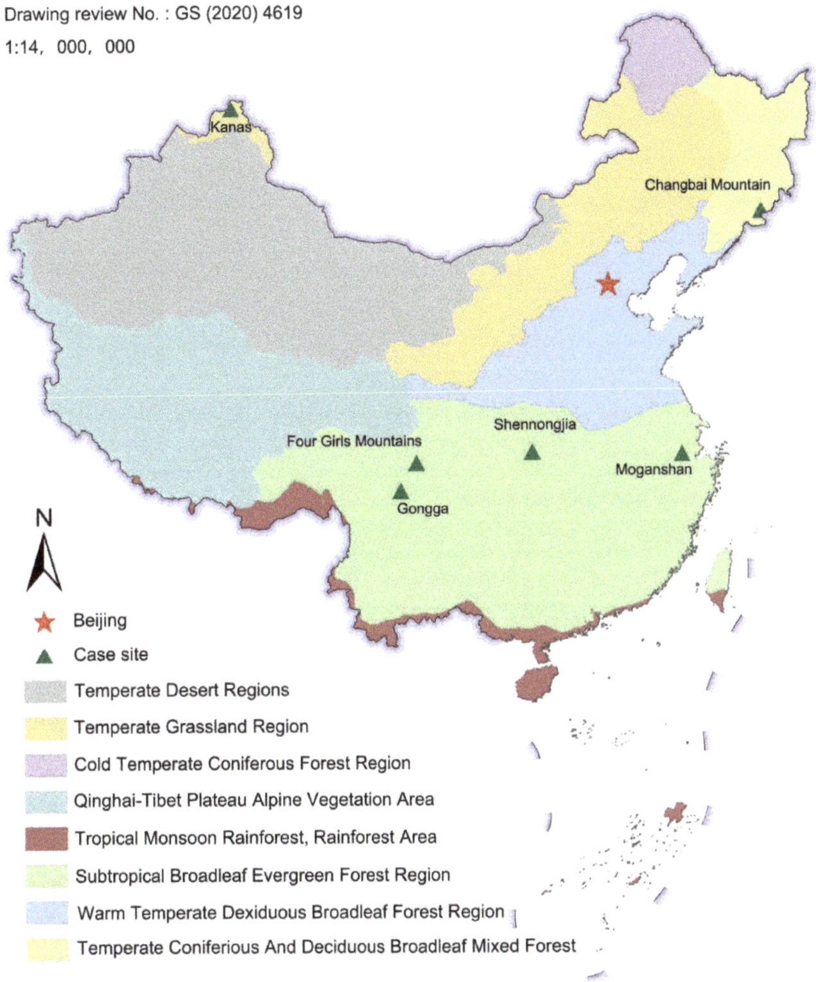

Figure 1. Schematic diagram of case site location. Chinese vegetation zoning data from the Resource and Environment Science and Data Center, Chinese Academy of Sciences. Base map according to the standard map of the Ministry of Natural Resources of China.

3. Methods

3.1. Research Flow

Firstly, we crawled the photos on 2BULU and formed a dataset by using "Tencent API screening port + manual" to filter them. Secondly, the MLP-Mixer model was trained by randomly selecting photos in the dataset to form a training set and finally a classification model was formed. At the same time, sentiment analysis was performed on the dataset to obtain adjectives and determine the sentiment status. The results of the analyses complement each other to provide a more comprehensive assessment of the Chinese forest recreationists' preferences for forest landscapes (See Figure 2).

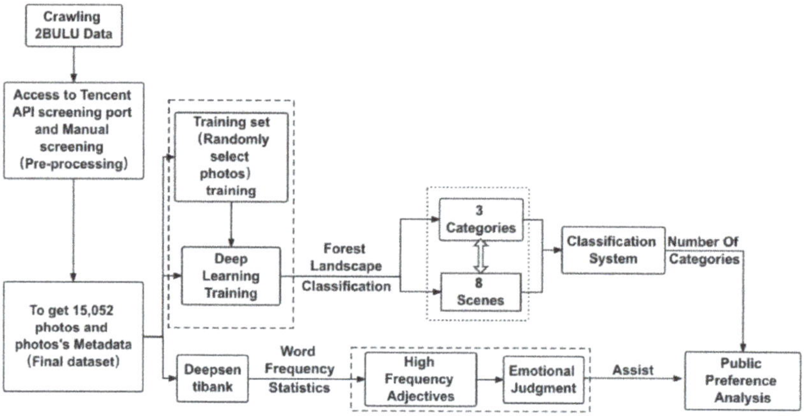

Figure 2. Research Flowchart.

3.2. Forest Landscape Photo Classification

The proportion of forest landscape in the photos taken by people will have a relatively large impact on people's preferences, and the analysis of the content of the photos can also restore, in a better and more detailed manner, what people appreciate, and can also more accurately describe the landscape [50,51]. In order to classify the forest landscape photographs taken by tourists, we developed the forest landscape photograph classification for this study based on the characteristics of the dataset and with reference to the forest landscape classification proposed by Chen et al. in 2001 based on forest beauty. Chen et al.'s classification combines both distance and aesthetic object scales to classify forest (plant) landscapes into seven levels—Single-tree Composed Landscapes, Intra-forest Landscape, Forest Trail Landscape, Detailed Landscape, Near, Medium and Far Landscape [52]. However, during the pre-experiment, we found that it was difficult for machine learning to recognize and judge the near, medium and far views based on the distance between the observer and the aesthetic object, so we then replaced it with look down, look forward and look up, classified according to the vertical foot of the observer and the aesthetic object; the overall view was added to the scale level of the aesthetic object. Therefore, we finally determined 8 forest landscape scenes including: Look Down Landscape, Look Forward Landscape, Look Up Landscape, Single-tree Composed Landscape, Detailed Landscape, Overall Landscape, Forest Trail Landscape and Intra-forest Landscape—8 forest landscape scenes; 3 forest landscape categories, including Spatial Hierarchy, Forest Hierarchy, and Scale Level (See Figure 3). Spatial hierarchy contains Look Down Landscape, Look Forward Landscape and Look Up Landscape. Scale Level includes Single-tree composed Landscape, Detailed Landscape and Overall Landscape. Forest Trail Landscape is a landscape composed of roads and forest stands along the roads; Intra-forest Landscapes are landscapes composed of forest plants in groups rather than single, multiple components rather than single parts, within the forest without the Forest Trail. There are differences between the two. Forest Trail Landscapes and Intra-forest Landscapes are included in the forest hierarchy because they are both non-confined but extensive photographs of forest recreation in the forest interior, and contain more components that distinguish their types from other types.

Figure 3. Examples of forest landscape classification systems.

3.3. MLP-Mixer Model

We decided to try the MLP-Mixer model proposed by Google in 2021 for image classification task scene recognition in order to perform a more objective analysis of forest landscape preferences. This is an architecture based only on multilayer perceptron (MLP), which is optimized and has a simpler structure than models such as the convolutional neural network (CNN). The model proposes a mixer structure that first splits the input images into patches, converts each patch into a feature embedding as per-patch fully connected, sends it to N mixer layers, and finally classifies it as fully connected. It uses spatial-mixing MLPs and channel-mixing MLPs to transfer information between different channels and spatial locations (tokens), respectively [35,53]. These two types of layers are alternately stacked to facilitate the exchange of two input dimensions. Each MLP consists of two fully connected and one GELU. Thus, for Top-1 accuracy on the ImageNet validation set, Mixer achieves a slightly better performance than ResNet and basically the same performance as ViT transformer, and the training speed (img/sec/core) Mixer will be faster than the other two, demonstrating the potential of a simple structure, such as MLP [54]. Thus, on large-scale datasets, MLP-Mixer achieves a very promising performance that can effectively help us in forest landscape classification. The specific operation example is shown in the Figures 4 and 5.

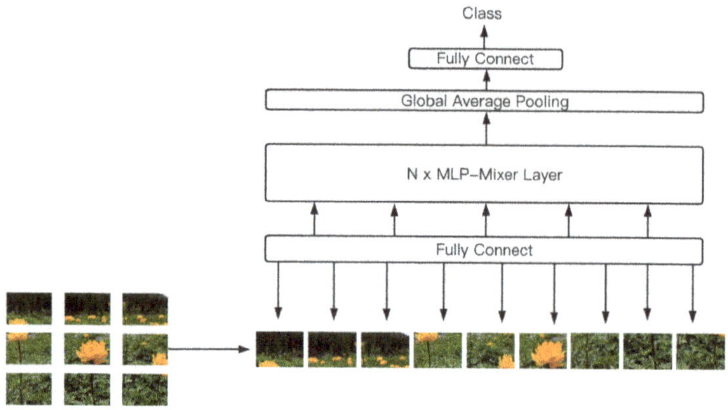

Figure 4. MLP-Mixer Flow Chart.

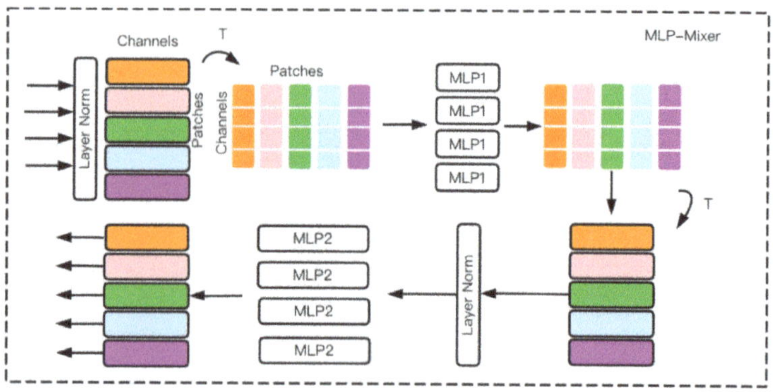

Figure 5. MLP-Mixer Model Diagram.

We use the MLP-Mixer model to form a classifier to classify the obtained forest landscape data, which is a new attempt. There are no ready-to-use models and training sets, so the data set is divided into 2 parts—50% of the data are divided 8:2 into a training set and a test set; 50% of the data will be classified by the model to determine the forest recreationists' forest landscape preferences. Since there may be two categories of images, machine recognition cannot classify an image twice at the same time. We divided the classification categories into two groups: Single-tree composed Landscape, Detailed Landscape, Overall Landscape, Forest Trail Landscape, and Intra-forest Landscapes as the first group, with 69% accuracy after repeated training; look down, look forward, and look up are the second group with 70% accuracy. This indicates that the model can classify forest landscape images more stably, and initially reaches our desired expectation.

3.4. Deepsentibank

In the era of rapid development of information, people's impressions of their surroundings will appear more and more on the Internet, and some studies believe that this is a mental conceptualization of people's relationship with their surroundings [55]. We study what people capture and their emotions during their forest landscape experiences, which can provide a basis and indicators for forest planning and management [28,56]. In this study, to explore tourists' perception of forest landscape images, Deepsentibank, a visual emotion concept classification tool developed by Chen et al. at Columbia University based on deep learning of images (CNN), is utilized. Its development principle is to obtain text tags from web photos, establish the relationship between "adjectives + nouns" used as a basis to identify the content of the images to form emotional keywords, and to transform the image information into textual information [57]. His method uses over 1 million geotagged photos to train the classifier, composing a total of 2089 APNs (Adjective Noun Paris)—231 adjectives as well as 424 nouns—and thus has a high accuracy. This conceptual-level sentiment analysis can extract the implicit sentiment from the ontology and give us a basis for analyzing the emotional attitude of forest landscape images. The content analysis process is shown in Figure 6 (only the first 7 items are listed in the figure). In this case, the program can parse a forest landscape photo into a set of JSON files possessing ANP sorting, with the top sorted words having high weight and greater relevance to the image.

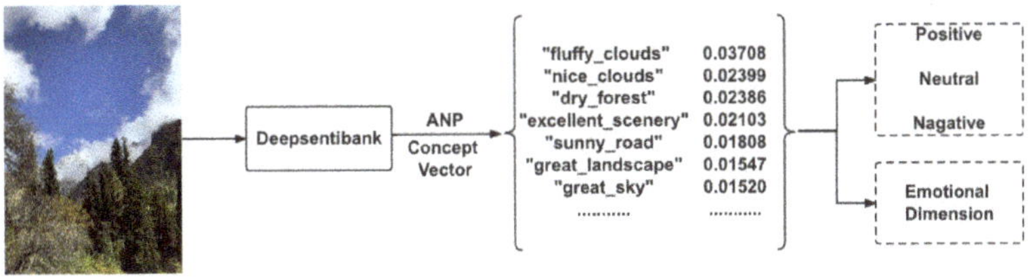

Figure 6. The process of photos analyzed by Deepsentibank.

The results of the study show that Deepsentibank performs well or more significantly in retrieval and annotation compared to support vector analysis models. With its greater use in tourism destination images, these studies also prove that Deepsentibank performs well in sentiment analysis. Since our study is mainly for understanding the emotional state of images, we are interested in adjectives only.

4. Results

4.1. MLP-Mixer Model Classification Results

From the pictures of the classification results of the MLP-Mixer model, it can be seen that there is a significant difference in people's preference for different categories of forest landscapes in both the first and second groups. From the aesthetic spatial level, people will prefer to look at a flatter forest landscape, accounting for 78.32% of the first group classification, followed by looking down at the forest landscape, accounting for 13.8% of the group, and finally, looking up at the forest landscape, accounting for 7.81% of the group. From the forest level, people will take more photos on the forest road and occupy a higher percentage of the second group, with 64.99%; the in-forest landscape will also be noticed by people, with 21.55%. From the scale level, people's attention is lower, Single-tree composed landscape and overall landscape have similar attention, but open overall landscape will be more popular, with 5.65%; single-tree composed landscape in the forest is also relatively easy to be found and recorded by people, with 5.05%; detailed landscape is lower, with only 2.76%. The results are shown in Figure 7.

4.2. Deepsentibank Sentiment Analysis

4.2.1. Emotional High-Frequency Word Analysis

We extracted the first 10 items of the adjective part of the ANP of each forest landscape picture taken by travelers in the six case sites as the salient picture sentiment [39,58,59]. Sentiment lexicon, such as Hownet, was used to compare and analyze with the photo adjectives parsed by Deepsentibank. In general, the emotions embedded in the photographs taken by the Chinese forest recreationists are dominated by positive words and are much higher than negative words. From the results in Table 1, it can be seen that "Classic", "Cute", "Sweet", "Colorful" were the most frequent emotion words. Due to the large volume of word data, words with a frequency of more than 1000 uses were extracted. We found in Table 2, the percentages of the extracted high-frequency words holding positive, neutral and negative sentiments toward the forest landscape were 74.06%, 11.54% and 14.41%, respectively.

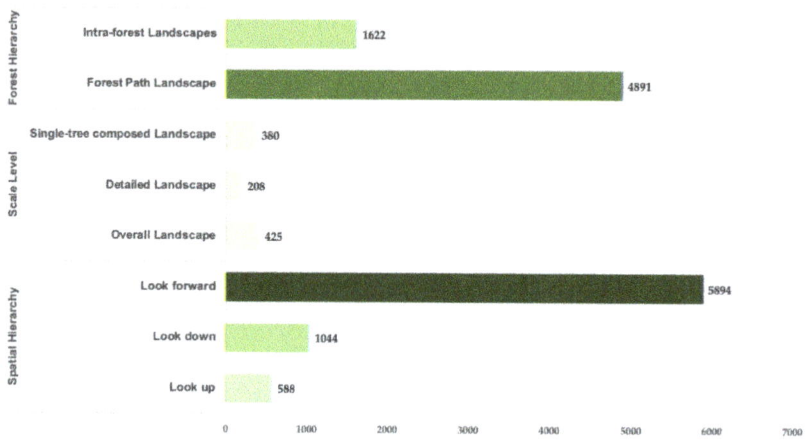

Figure 7. Deep Learning Result Graph.

Table 1. High-frequency adjective of related images (more than one thousand).

Adj	Frequency	Adj	Frequency
classic	9337	broken	2326
cute	8312	delicious	2196
sweet	7441	warm	2186
colorful	7102	hot	2122
sexy	6482	gorgeous	2033
funny	6022	young	1698
amazing	5138	christian	1481
empty	4524	pretty	1407
super	4328	happy	1395
tiny	3837	evil	1342
awesome	3459	upset	1305
yummy	3128	shiny	1215
fresh	2867	energetic	1209
ugly	2657	icy	1173
strong	2500	tired	1142
dirty	2486	clean	1098
adorable	2484	nasty	1027
traditional	2424	favorite	1022

Table 2. Emotional Tendency Gravity Scale.

Emotional Tendency	Frequency	Proportion
Positive	82,872	74.06%
Neutral	12,911	11.54%
Negative	16,122	14.41%

4.2.2. Emotional Dimension Analysis

We classified the high-frequency words (word frequency over one thousand) extracted by Deepsentibank's analysis with the sentiment dimension, based on the Circumplex Model of Affect by Russell in 1980 [60]. The proportions of adjectives in each dimension were obtained and the results are shown in Table 3. "Pleased" has the highest proportion, 45.81%, representing the main emotional tendency of the forest recreationists towards the forest landscape; "Excited" has the second highest response in the Chinese forest recreationists emotional preference (14.45%). The emotional preferences of "Happy", "Delighted" and "Relaxed" are also relatively high (10.11%, 6.65% and 5.02%, respectively). It is worth noting that although people's emotional tendency toward the forest landscape is dominated by positive emotions, there are still negative emotions and feelings toward the forest landscape. The recreationists showed other emotions, such as "Sad" and "Miserable" from the photos (5.47 and 2.37% respectively), while a small number of photos showed the emotions of "Bored" and "Annoyed".

Table 3. Emotional dimension scale.

Type	Frequency	Proportion
Pleased	51,262	45.81%
Excited	16,167	14.45%
Happy	11,310	10.11%
Delightful	7441	6.65%
Sad	6117	5.47%
Relaxed	5622	5.02%
Depressing	3837	3.43%
Miserable	2657	2.37%
Satisfied	2500	2.23%
Calm	1481	1.32%
Afraid	1342	1.20%
Bored	1142	1.02%
Annoying	1027	0.92%

5. Discussion and Conclusions

5.1. Discussion

5.1.1. Preference Characteristics of Different Forest Landscape Categories

The classification results of the MLP-Mixer model show that there are significant differences in different forest landscape scenes as well as categories (See Figure 7), while people have different preferences for different forest landscapes [61].

From the results, it is easy to find that people prefer forest trails, which is consistent with the findings of several studies. Forest trails are important landscape corridors of the forest and have the role of organizing the landscape space, so good trails allow people to enjoy the forest landscape more, and whether or not they have access to plants and other elements in the process will also affect people's preferences [62]. Further, most forest recreationists still enter the forest primarily for walking activities and nature experiences [63]. Forest trail photography is higher than related types of landscapes, because people are more involved in forest activities by walking, and the "2BULU" site we chose is also a photo documentation in this tone, demonstrating the results of You et al., that forest recreationists will prefer forest landscapes along forest trails [64].We corroborate the findings obtained by Gao et al. in 2021, who found that forest tourism participants were prone to interact with the forest on forest trails, for example, by taking photographs [65]. It also corroborates that forest trail coverage has a direct impact on visual quality, with an increase in the proportion of forested landscapes perceived as safe, and a preference for forests with multiple trails or distinct hiking trails [66–69]. Therefore, it is crucial to provide satisfactory forest managers and planners, a well-planned forest trail can increase the attractiveness of the forest landscape from an aesthetic point of view [16,70–73]. In-forest landscapes are less preferred than forest trail landscapes, but in-forest landscapes still have a particularly important position in forest landscape evaluation and management, and can affect the perceived natural beauty and the beauty of the forest stand [52].

In terms of aesthetic spatial entrainment, in the same space, people's preference for landscape will show spatial segmentation differences [74,75]. The results show that people prefer look forward landscapes (see Figure 7), i.e., forest landscapes that can be seen without looking up or down with only horizontal rotation, and look down to see ground cover, fine details or individual fractional landscapes, and look up to view forest landscapes less often. The lower preference for overlooked landscapes can laterally support that ground cover plants have an effect on people's preferences and are not attractive or less attractive in forest landscape preferences [76]. This also demonstrates that visual preferences actually vary and have a significant impact on forest recreationists' preferences in terms of distance area space [77,78]. The overall landscape and the detailed landscape, in line with Gill and Ryan's view that people prefer a relatively open forest landscape, with good visual access to generate points of interest [79,80]. In terms of individual views, people are actually attracted to unique trees or old trees in the forest and stop to take pictures. Helman 2021, for example, demonstrates that Family Forest Owners in Michigan's Upper Peninsula prefer single trees; You confirms that people are interested in old-growth trees [64,81]. From the fine view, it does not match with Nielsen's results [82], as flowers, roots, mosses, etc., did not gain more preference from people, but on the contrary, were the lowest among all types of landscapes. This also shows that people do not objectively prefer flowers, fruits, etc., as much as they subjectively do, and that the perceived facts can differ from the actual true preferences [83], or perhaps the overall level of basic botanical knowledge of the Chinese forest recreationists has something to do with it, which deserves to be studied in depth. The preference for detailed landscapes, overall landscapes and individual landscapes is also low, which we believe may be due to the fact that people pay more attention to the plants in their field of vision during forest tourism, and only when a plant is very special or magnificent in its overall view does it attract attention and generate preference.

5.1.2. Emotional Characteristics Contained in Forest Landscape Photographs

The forest landscape space perceived by humans is a cognitive process from unknown curiosity to the whole, and there will be an interaction of visual behavior and psychological perception [84]. We answered the question "Are people really positive in a forest landscape setting?".

The results show (see Table 2) that recreationists feel different emotions during forest visits and that reactions to the surrounding environment usually involve positive and negative emotions or two bipolar orthogonal dimensions [60,85,86]; however, the participation in forest recreation is still dominated by a positive and pleasant emotional state. Therefore, the heightened perception of the forest is a manifestation of vegetation, which will have a strong psychological impact on what people see, and confirms that visual stimulation is useful as a communication channel in the landscape; "Classic", "Cute", "Sweet", "Colorful" and other emotional words reflect people's positive emotions in the forest (see Table 1), which are directly related to people's innate behavior of pursuing happiness [87–89]. Some studies have shown that people feel comfortable and peaceful in forests and that the main emotions generated are positive [65]. It is also similar to Nielsen's suggestion that people develop emotional and cognitive structures in response to forest landscapes and experience, including "Cosy/uncosy", "Safety", "Serenity", "Care", "Mystery" and "Coherence" [82]. People experience emotions, such as "Pleased", "Excited" and "Happy" (see Table 3), because they gain pleasure through the dynamic function of the landscape and a deeper understanding of the ecological state, evoking a mental response through direct sensory processes and interventions in cognitive structures [90].

The presence of the forest has always proved to be a positive criterion, but this is actually the result of expectations in people's minds. Current spatial analysis based on photographs, objective methods, such as eye-tracking technology, or subjective analysis, such as questionnaires, all consider the forest landscape to be merely attractive, without tapping into the negative emotions that actually exist behind it [8,19,51,74,91–93]. However, it is noteworthy that we found that people, in fact, also have negative emotions towards forest landscapes, such as "Sad", "Miserable", "Bored" and so on (see Table 3), and the negative sentiment even exceeds the neutral sentiment. This corroborates Foltête's suggestion that forest landscapes cannot always be interpreted in a positive way in terms of preferences, which in his view, can be influenced by forest characteristics and cover to produce negative or positive perceptions [74]. In fact, the study by Deng et al. also demonstrated that British tourists suffer from sleepiness when traveling to Beijing [39]. In the course of our analysis, we also found that photographs of forest landscapes with a high degree of grey clutter are more likely to produce negative or more negative emotional states, so the state of the surrounding forest when people travel to the forest can strongly influence the participants' preference for the landscape [77].

We can think of preferences as encompassing emotional factors, with a strong correlation between the two [49,94]; therefore, emotion can also be considered as a predictor of preference, with preference scores being higher when the environment evokes positive and relaxing emotions, and lower when it does not [86,95]. When positive emotions are amplified, people's visual attention is increased and enhanced, and it is the visual appeal combined with the emotional response evoked by the content that influences people to develop preferences [90,96,97]. From the side, we can also see that there is a positive preference and a negative preference for people to take photos.

5.1.3. Shortcomings and Outlook

Social media data still have certain shortcomings in the process of landscape preference research, and the sample selection of forest recreationists photos for forest landscape preference research will have a certain bias [98]; it is also difficult to link to the context of the participants, so we have no way to explore in depth the reasons for the preferences [99,100]. There are many studies that use deep learning for the classification of travel photos, mainly focusing on the classification of large travel segments, such as architecture, plants, food, people, etc. [99,101]; however, our initial attempt to visually discriminate forest landscape type preferences using the computational set has some room for improvement in the accuracy and feasibility of using the dataset [43]. Further, landscape perception is not only visual, it is composed of multiple senses, including hearing and smell, and even the objective physical environment will affect the visual behavior of participants, such as temperature, negative air ion concentration, etc. [102,103]. Thus, the analysis of forest landscape preferences from a photographic perspective is still somewhat inadequate, and a combination of questionnaires and interviews can be conducted afterwards to explore the reasons for preference behavior and whether the perceptions brought about by different sensory combinations have an impact.

5.2. Conclusions

In this study, we analyzed Chinese forest recreationists' preferences for forest landscape classification as well as sentiment, focusing on UGC photos from the "2BULU" website. We novelly used the MLP-Mixer model and Deepsentibank deep learning tool to conduct an objective study. From the categories, we found that people prefer forest trail landscape and in-forest landscape over detailed landscape (e.g., flowers, fruits, etc.), and in terms of aesthetic spatial entrainment, a flat view is more favored and welcomed by people, followed by overhead view, which will focus on ground cover plants or flowers, while elevated view will be ignored by more people; from the perspective of people's emotions in participating in forest tourism, there are positive and negative emotional states, and not all emotions are positive for forest landscapes, so we need to look at people's preferences dialectically. This study is intended to provide a basis for forest planning and design, and to help managers balance resources and needs. China has planned 12 national forest trails since 2017, passing through 20 provinces along the route and covering more than 22,000 km, and is gradually implementing the specific route selection and construction. Our research results are of reference value for the planning and design of nearly 3000 forest parks nationwide and can provide a basis and help for trail construction; for example: constructing high-quality trails to allow recreationists to get in touch with nature on foot; constructing traceless trails that can pass through diverse forest landscape resource areas or maintain the original appearance along the route; planning more landscapes within the forest to highlight features; adopting more tree species to enrich the colors of the forest in different seasons to enhance people's positive emotions, etc.

Author Contributions: Conceptualization, X.Z.; Data curation, X.Z. and X.T.; Funding acquisition, Y.Z. and L.Y.; Methodology, X.Z.; Project administration, Y.Z.; Supervision, Y.Z. and J.W.; Validation, Y.Z., J.W. and X.T.; Visualization, X.Z. and J.W.; Writing—original draft, X.Z.; Writing—review and editing, Y.Z. and L.Y. All authors have read and agreed to the published version of the manuscript.

Funding: This research was funded by the Ministry of Science and Technology of The People's Republic of China, grants 2019YFD1100400. This research was funded by the science and technology innovation Program of Hunan Province: Research and application of expressway landscape and regional culture integration technology based on Huxiang Culture, grant number 2021SK2050. This research was funded by the Central South University of Forestry and Technology, grants CX202102098.

Institutional Review Board Statement: Not applicable.

Informed Consent Statement: Not applicable.

Data Availability Statement: The data presented in this study are available on request from the author.

Acknowledgments: Thanks to the students especially Ding Ningning and Cui Shuaihu and teachers who gave support during the research process and greatly helped with the writing of the article.

Conflicts of Interest: The authors declare no conflict of interest.

References

1. Salisch, H.V. *Forstästhetik*; Springer: Berlin/Heidelberg, Germany, 1911.
2. Baskent, E.Z. Controlling spatial structure of forested landscapes: A case study towards landscape management. *Landsc Ecol* **1999**, *14*, 83–97. [CrossRef]
3. Deng, S.; Yan, J.; Guan, Q.; Katoh, M. Short-term effects of thinning intensity on scenic beauty values of different stands. *J. For. Res.* **2012**, *18*, 209–219. [CrossRef]
4. Gong, L.; Zhang, Z.; Xu, C. Developing a Quality Assessment Index System for Scenic Forest Management: A Case Study from Xishan Mountain, Suburban Beijing. *Forests* **2015**, *6*, 225–243. [CrossRef]
5. Hunziker, M.; Kienast, F. Potential impacts of changing agricultural activities on scenic beauty—A prototypical technique for automated rapid assessment. *Landsc. Ecol.* **2004**, *14*, 161–176. [CrossRef]
6. Nilsson, K.; Sangster, M.; Gallis, C.; Hartig, T.; Vries, S.D.; Seeland, K.; Schipperijn, J. Forests, trees and human health. *J. Adven. Edu. Outdoor Learn.* **2012**, *12*, 95–97. [CrossRef]
7. Nordh, H.; Hagerhall, C.M.P.; Holmqvist, K. Tracking Restorative Components: Patterns in Eye Movements as a Consequence of a Restorative Rating Task. *Landsc. Res.* **2013**, *38*, 101–116. [CrossRef]
8. Zhou, S.; Gao, Y.; Zhang, Z.; Zhang, W.; Meng, H.; Zhang, T. Visual Behaviour and Cognitive Preferences of Users for Constituent Elements in Forest Landscape Spaces. *Forests* **2022**, *13*, 47. [CrossRef]
9. Robinette, G.O. *Plants, People, and Environmental Quality: A Study of Plants and Their Environmental functions*; Department of the Interior, National Park Service: Washington, WA, USA, 1972; p. 51.
10. Jianwen, W. Forests—A vValuable Resource for Wild Play (Foreign Forest Play Use). *Guangdong Landsc. Archit.* **1981**, 16–21. (In Chinese). Available online: https://www.cnki.com.cn/Article/CJFDTOTAL-GDYL198102003.htm (accessed on 12 May 2022).
11. Lundmark, L.; Müller, D.K. *Vilka är ute i Naturen? Delresultat från en Nationell enkät om Friluftsliv och Naturturism i Sverige (Who Visits Nature? Results from a Nationwide Questionnaire Study of Outdoor Life and Nature Tourism in Sweden)*; Fredman, P., Karlsson, S.-E., Romild, U., Sandell, K., Eds.; Report No. 1; Forskningsprogrammet friluftsliv i förändring: Östersund, Sweden, 2008.
12. The 13th Five-Year Plan National Forest Tourism Tourist Volume Reached 7.5 Billion_Data Highlights_Chinese Government Website. Available online: http://www.gov.cn/shuju/2021-01/19/content_5580910.htm (accessed on 27 February 2022).
13. Misgav, A.; Amir, S. Integration of Visual Quality Considerations in Development of Israeli Vegetation Management Policy. *Environ. Manag.* **2001**, *27*, 845–857. [CrossRef]
14. Edwards, D.; Jay, M.; Jensen, F.S.; Lucas, B.; Marzano, M.; Montagné, C.; Peace, A.; Weiss, G. Public preferences for structural attributes of forests: Towards a pan-European perspective. *For. Policy Econ.* **2012**, *19*, 12–19. [CrossRef]
15. Edwards, D.; Jay, M.; Jensen, F.; Lucas, B.; Marzano, M. *Public Preferences for Silvicultural Attributes of European Forests*; Technical Report from the EU FP6 Integrated Project EFORWOOD–Tools for Sustainability Impact Assessment of the Forestry-Wood Chain; European Forest Institute: Joensuu, Finland, 2014.
16. Oku, H.; Fukamachi, K. The differences in scenic perception of forest visitors through their attributes and recreational activity. *Landsc. Urban Plan.* **2006**, *75*, 34–42. [CrossRef]
17. Wilkes-Allemann, J.; Hanewinkel, M.; Pütz, M. Forest recreation as a governance problem: Four case studies from Switzerland. *Eur. J. For. Res.* **2017**, *136*, 511–526. [CrossRef]
18. Filyushkina, A.; Agimass, F.; Lundhede, T.; Strange, N.; Jacobsen, J.B. Preferences for variation in forest characteristics: Does diversity between stands matter? *Ecol. Econ.* **2017**, *140*, 22–29. [CrossRef]
19. Pastorella, F.; Giacovelli, G.; De Meo, I.; Paletto, A. People's preferences for Alpine forest landscapes: Results of an internet-based survey. *J. For. Res.* **2017**, *22*, 36–43. [CrossRef]

20. Weller, P.; Elsasser, P. Preferences for forest structural attributes in Germany—Evidence from a choice experiment. *For. Policy Econ.* **2018**, *93*, 1–9. [CrossRef]
21. Christou, E. Examining Hotel Managers' Acceptance of Web 2.0 in Website Development: A Case Study of Hotels in Hong Kong. *Business* **2016**, 75–88. [CrossRef]
22. Yang, L. Why do we take pictures?—A study on clustering museum visitors based on their motivation for taking photographs. *Chinese Mus.* **2018**, *7*. (In Chinese). Available online: https://www.cnki.com.cn/Article/CJFDTotal-GBWG201803014.htm (accessed on 12 May 2022).
23. Balomenou, N.; Garrod, B. Photographs in tourism research: Prejudice, power, performance and participant-generated images. *Tour. Manag.* **2019**, *70*, 201–217. [CrossRef]
24. Chalfen, R.M. Photograph's role in tourism Some unexplored relationships. *Ann. Tour. Res.* **1979**, *6*, 435–447. [CrossRef]
25. James, A. Travel photography: A methodological approach. *Ann. Tour. Res.* **1988**, *15*, 134–158. [CrossRef]
26. Urry, J.; Larsen, J. *The Tourist Gaze 3.0*, 3rd ed.; Sage Publications: New York, NY, USA, 2011; ISBN 978-1-84920-376-0.
27. Kim, S.O.; Lee, C.H.; Shelby, B.O. Utilization of Photographs for Determining Impact Indicators for Trail Management. *Environ. Manag.* **2003**, *32*, 282–289. [CrossRef] [PubMed]
28. Wilkins, E.J.; Wood, S.A.; Smith, J.W. Uses and Limitations of Social Media to Inform Visitor Use Management in Parks and Protected Areas: A Systematic Review. *Environ. Manag.* **2021**, *67*, 120–132. [CrossRef] [PubMed]
29. Conti, E.; Lexhagen, M. Instagramming nature-based tourism experiences: A netnographic study of online photography and value creation. *Tour Manag. Perspect* **2020**, *34*, 100650. [CrossRef]
30. Jia, Y.Q.; Zhang, Y.J. Preliminary Study of Photographic Behavior of the Tourist—Based on the contrast of the photographic tourist and the mass tourist. *J. Cent. S. Univ. For. Technol.* **2012**, *6*, 5. (In Chinese) [CrossRef]
31. Angradi, T.R.; Launspach, J.J.; Debbout, R. Determining preferences for ecosystem benefits in Great Lakes Areas of Concern from photographs posted to social media. *J. Great Lakes Res.* **2018**, *44*, 340–351. [CrossRef]
32. Lee, H.; Seo, B.; Koellner, T.; Lautenbach, S. Mapping cultural ecosystem services 2.0—Potential and shortcomings from unlabeled crowd sourced images. *Ecol. Indic.* **2019**, *96*, 505–515. [CrossRef]
33. Oteros-Rozas, E.; Martín-López, B.; Fagerholm, N.; Bieling, C.; Plieninger, T. Using social media photos to explore the relation between cultural ecosystem services and landscape features across five European sites. *Ecol. Indic.* **2018**, *94*, 74–86. [CrossRef]
34. Pastur, G.J.M.; Peri, P.L.; Lencinas, M.V.; García-Llorente, M.; Martín-López, B. Spatial patterns of cultural ecosystem services provision in Southern Patagonia. *Landsc. Ecol.* **2015**, *31*, 383–399. [CrossRef]
35. Zhao, Y.; Wang, G.; Tang, C.; Luo, C.; Zeng, W.; Zha, Z.J. A Battle of Network Structures: An Empirical Study of CNN, Transformer, and MLP. *arXiv* **2021**, arXiv:2108.13002.
36. Zhang, K.; Chen, Y.; Li, C. Discovering the tourists' behaviors and perceptions in a tourism destination by analyzing photos' visual content with a computer deep learning model: The case of Beijing—ScienceDirect. *Tour. Manag.* **2019**, *75*, 595–608. [CrossRef]
37. Kim, J. Transfer Learning of a Deep Learning Model for Exploring Tourists' Urban Image Using Geotagged Photos. *ISPRS Int. J. Geo. Inf.* **2021**, *10*, 137. [CrossRef]
38. Deng, N.; Liu, J.; Niu, Y.; Ji, W.X. Different perceptions of Beijing's destination images from tourists: An analysis of Flickr photos based on deep learning method. *Resour. Sci.* **2019**, *41*, 416–429. (In Chinese) [CrossRef]
39. Deng, N.; Liu, J. Where did you take those photos? Tourists' preference clustering based on facial and background recognition. *J. Destin. Mark. Manag.* **2021**, *21*, 100632. [CrossRef]
40. Payntar, N.D.; Hsiao, W.L.; Covey, R.A.; Grauman, K. Learning Patterns of Tourist Movement and Photography from Geotagged Photos at Archaeological Heritage Sites in Cuzco, Peru. *Sociology* **2020**, *82*, 104165. [CrossRef]
41. Cho, N.; Kang, Y.; Yoon, J.; Park, S. Classification of Tourist Photo for Intelligent Tourism Service. *Comput. Sci.* **2019**, *19*, 84–101. [CrossRef]
42. Kim, D.; Kang, Y.; Park, Y.; Kim, N.; Lee, J. Understanding tourists' urban images with geotagged photos using convolutional neural networks. *Spat. Inf. Res.* **2019**, *28*, 241–255. [CrossRef]
43. Wilkins, E.J.; Van Berkel, D.; Zhang, H.; Dorning, M.A.; Beck, S.M.; Smith, J.W. Promises and pitfalls of using computer vision to make inferences about landscape preferences: Evidence from an urban-proximate park system. *Landsc. Urban Plan.* **2022**, *219*, 104315. [CrossRef]
44. Runge, C.A.; Hausner, V.H.; Daigle, R.M.; Monz, C.A. Pan-Arctic analysis of cultural ecosystem services using social media and automated content analysis. *Environ. Res. Commun.* **2020**, *2*, 075001. [CrossRef]
45. China's Forest Coverage Rate 22.96%_Forestry and Grassland Resources_State Forestry and Grassland Administration Government Website. Available online: https://www.forestry.gov.cn/main/65/20190620/103419043834596.html (accessed on 30 April 2022).
46. Zhou, H.; Wang, F. Research on structure characteristics of the inter-provincial tourist flow spatial network in China based on the modified gravity model. *Geogr. Res.* **2020**, *39*, 669–681. (In Chinese) [CrossRef]
47. About "2BULU"—"2BULU" Outdoor Network. Available online: https://www.2bulu.com/about/about_us.htm (accessed on 21 April 2022).
48. Wei, X.; Zhou, G.; Yu, S. A preliminary study of forest landscape classification systems. *South Cent. For. Surv. Plann.* **1997**, *16*, 5. (In Chinese)

49. White, M.P.; Smith, A.T.; Humphryes, K.; Pahl, S.; Snelling, D.; Depledge, M.H. Blue space: The importance of water for preference, affect, and restorativeness ratings of natural and built scenes. *J. Environ. Psychol.* **2010**, *30*, 482–493. [CrossRef]
50. Schirpke, U.; Tappeiner, G.; Tasser, E.; Tappeiner, U. Using conjoint analysis to gain deeper insights into aesthetic landscape preferences. *Ecol. Indic.* **2019**, *96*, 202–212. [CrossRef]
51. Tieskens, K.; Zanten, B.T.; van Schulp, C.J.E.; Verburg, P.H. Aesthetic appreciation of the cultural landscape through social media: An analysis of revealed preference in the Dutch river landscape. *Landsc. Urban Plan.* **2018**, *177*, 128–137. [CrossRef]
52. Chen, X.; Wang, Y. An analytic study on forest beauty—mostly on form beauty of forest plants. *Sci. Silvae Sin.* **2001**, *37*, 122–130. (In Chinese)
53. Sun, C.; Shrivastava, A.; Singh, S.; Gupta, A. Revisiting Unreasonable Effectiveness of Data in Deep Learning Era. In Proceedings of the 16th IEEE International Conference on Computer Vision (ICCV), Venice, Italy, 22–29 October 2017. [CrossRef]
54. Tolstikhin, I.O.; Houlsby, N.; Kolesnikov, A.; Beyer, L.; Zhai, X.; Unterthiner, T.; Yung, J.; Steiner, A.; Keysers, D.; Uszkoreit, J.; et al. MLP-Mixer: An all-MLP Architecture for Vision. In *Proceedings of the Advances in Neural Information Processing Systems, Virtual, 6–14 December 2021*; Ranzato, M., Beygelzimer, A., Dauphin, Y., Liang, P.S., Vaughan, J.W., Eds.; Curran Associates, Inc.: New York, NY, USA, 2021; Volume 34, pp. 24261–24272.
55. Richter, K.-F.; Winter, S. Citizens as Database: Conscious Ubiquity in Data Collection. *Comput. Sci.* **2011**, *6849*, 445–448. [CrossRef]
56. Hull, R.B.; Stewart, W. The Landscape Encountered and Experienced While Hiking. *Environ. Behav.* **1995**, *27*, 404–426. [CrossRef]
57. Chen, T.; Borth, D.; Darrell, T.; Chang, S.F. DeepSentiBank: Visual Sentiment Concept Classification with Deep Convolutional Neural Networks. *arXiv* **2014**, arXiv:1410.8586.
58. Stepchenkova, S.; Zhan, F. Visual destination images of Peru: Comparative content analysis of DMO and user-generated photography. *Tour. Manag.* **2013**, *36*, 590–601. [CrossRef]
59. Fan, M.; Zhang, H.; Chen, Y. Spatiotemporal analysis of visual tourism images in Inner Mongolia from the perspective of tourists. *J. Arid. Land Resour. Environ.* **2020**, *34*, 194–200. (In Chinese)
60. Russell, J.A. A circumplex model of affect. *J. Pers. Soc. Psychol.* **1980**, *39*, 1161–1178. [CrossRef]
61. Petrova, E.G.; Mironov, Y.V.; Aoki, Y.; Matsushima, H.; Ebine, S.; Furuya, K.; Petrova, A.; Takayama, N.; Ueda, H. Comparing the visual perception and aesthetic evaluation of natural landscapes in Russia and Japan: Cultural and environmental factors. *Prog. Earth Planet. Sci.* **2015**, *2*, 6. [CrossRef]
62. Zhang, T.; Zhang, W.; Meng, H.; Zhang, Z. Analyzing Visitors' Preferences and Evaluation of Satisfaction Based on Different Attributes, with Forest Trails in the Akasawa National Recreational Forest, Central Japan. *Forests* **2019**, *10*, 431. [CrossRef]
63. Agimass, F.; Lundhede, T.; Panduro, T.E.; Jacobsen, J.B. The choice of forest site for recreation: A revealed preference analysis using spatial data. *Ecosyst. Serv.* **2018**, *31*, 445–454. [CrossRef]
64. You, S.; Zheng, Q.; Chen, B.; Xu, Z.; Lin, Y.; Gan, M.; Zhu, C.; Deng, J.; Wang, K. Identifying the spatiotemporal dynamics of forest ecotourism values with remotely sensed images and social media data: A perspective of public preferences. *J. Clean. Prod.* **2022**, *341*, 130715. [CrossRef]
65. Gao, Y.; Zhang, T.; Sasaki, K.; Uehara, M.; Jin, Y.; Qin, L. The spatial cognition of a forest landscape and its relationship with tourist viewing intention in different walking passage stages. *Urban For. Urban Green.* **2021**, *58*, 126975. [CrossRef]
66. Abildtrup, J.; Garcia, S.; Olsen, S.B.; Stenger, A. Spatial preference heterogeneity in forest recreation. *Ecol. Econ.* **2013**, *92*, 67–77. [CrossRef]
67. Akay, A.E.; Pak, M.; Yenilmez, N.; Demirbağ, H. Aesthetic Evaluations of Forest Road Templates. *Int. J. Nat. Eng. Sci.* **2007**, *1*, 65–68.
68. Zhang, F.; Zhou, B.; Liu, L.; Liu, Y.; Fung, H.H.; Lin, H.; Ratti, C. Measuring human perceptions of a large-scale urban region using machine learning. *Landsc. Urban Plan.* **2018**, *180*, 148–160. [CrossRef]
69. Clemens, J.; Swaffield, S.R.; Wilson, J. *Landscape and Associated Environmental Values in the Roadside Corridor: A Selected Literature Review*; Lincoln University: Canterbury, New Zealand, 2010.
70. Oku, H.; Fukamachi, K. The Relationships between Experienced Landscape Types and People's Visiting Forms on a Trail in a Forest. *J. Korea Soc. Comput. Inform.* **1999**, *63*, 587–592. [CrossRef]
71. Oku, H.; Fukamachi, K. Occurrence pattern of landscape experience during forest recreation. *J. Jpn. For. Soc.* **2003**, *85*, 63–69. [CrossRef]
72. Kasetani, T.; Okumura, K.; Yoshida, S.; Takayama, N.; Kagawa, T. Differences in the Physiological and Psychological Effects of Walking in Various Satoyama Landscapes. *Environ. Sci.* **2007**, *70*, 569–574. [CrossRef]
73. Silvennoinen, H.; Alho, J.M.; Kolehmainen, O.; Pukkala, T. Prediction models of landscape preferences at the forest stand level. *Landsc. Urban Plan.* **2001**, *56*, 11–20. [CrossRef]
74. Foltête, J.-C.; Ingensand, J.; Blanc, N. Coupling crowd-sourced imagery and visibility modelling to identify landscape preferences at the panorama level. *Landsc. Urban Plan* **2020**, *197*, 103756. [CrossRef]
75. Wu, J.; Yuanyuan, X.; Jianlan, R.; Jie, Z.; Weilong, S.; Mengyu, L.; Bohong, P. Progress and Geographical Response on Landscape Preference. *Sci. Geogr. Sin.* **2021**, *41*, 656–663. [CrossRef]
76. Carvalho-Ribeiro, S.M.; Lovett, A.A. Is an attractive forest also considered well managed? Public preferences for forest cover and stand structure across a rural/urban gradient in northern Portugal. *For. Policy Econ.* **2011**, *13*, 46–54. [CrossRef]

77. Arnberger, A.; Ebenberger, M.; Schneider, I.E.; Cottrell, S.; Schlueter, A.C.; Ruschkowski, E.; von Venette, R.C.; Snyder, S.A.; Gobster, P.H. Visitor Preferences for Visual Changes in Bark Beetle-Impacted Forest Recreation Settings in the United States and Germany. *Environ. Manag.* **2017**, *61*, 209–223. [CrossRef]
78. Buhyoff, G.J.; Wellman, J.D.; Daniel, T.C. Predicting scenic quality for mountain pine beetle and western spruce budworm damaged forest vistas. *For. Sci.* **1982**, *28*, 827–838. [CrossRef]
79. Ryan, R. The influence of landscape preference and environmental education on public attitudes toward wildfire management in the Northeast pine barrens (USA). *Landsc. Urban Plan* **2012**, *107*, 55–68. [CrossRef]
80. Gill, N.; Dun, O.; Brennan-Horley, C.R.; Eriksen, C. Landscape Preferences, Amenity, and Bushfire Risk in New South Wales, Australia. *Env. Manag.* **2015**, *56*, 738–753. [CrossRef]
81. Helman, A.C.; Kelly, M.C.; Rouleau, M.; Dickinson, Y.L. Preferences for Northern Hardwood Silviculture among Family Forest Owners in Michigan's Upper Peninsula. *J. For.* **2020**, *119*, 113–129. [CrossRef]
82. Nielsen, A.B.; Heyman, E.; Richnau, G. Liked, disliked and unseen forest attributes: Relation to modes of viewing and cognitive constructs. *J. Environ. Manag.* **2012**, *113*, 456–466. [CrossRef] [PubMed]
83. Ding, N.; Zhong, Y.; Li, J.; Xiao, Q. Study on selection of native greening plants based on eye-tracking technology. *Sci. Rep.* **2022**, *12*, 1092. [CrossRef] [PubMed]
84. Gao, Y.; Zhang, T.; Zhang, W.; Meng, H.; Zhang, Z. Research on visual behavior characteristics and cognitive evaluation of different types of forest landscape spaces. *Urban For. Urban Green.* **2020**, *54*, 126788. [CrossRef]
85. Russell, J.A.; Pratt, G. A Description of the Affective Quality Attributed to Environments. *J. Pers. Soc. Psychol.* **1980**, *38*, 311–322. [CrossRef]
86. Singh, S.N.; Donavan, D.T.; Mishra, S.K.; Little, T.D. The latent structure of landscape perception: A mean and covariance structure modeling approach. *J. Environ. Psychol.* **2008**, *28*, 339–352. [CrossRef]
87. Gobster, P.H.; Nassauer, J.I.; Daniel, T.C.; Fry, G. The shared landscape: What does aesthetics have to do with ecology? *Landsc. Ecol.* **2007**, *22*, 959–972. [CrossRef]
88. Lôpez-Santiago, A.C.; Oteros-Rozas, E.; Martin-Lopez, B.; Plieninger, T.; Martin, G.; González, J.A. Esther Using visual stimuli to explore the social perceptions of ecosystem services in cultural landscapes: The case of transhumance in Mediterranean Spain. *Ecol. Soc.* **2014**, *19*, 27. [CrossRef]
89. Strauss, E.D.; Schloss, K.B.; Palmer, S.E. Color preferences change after experience with liked/disliked colored objects. *Psychon. Bull. Rev.* **2013**, *20*, 935–943. [CrossRef]
90. Gundersen, V.; Stange, E.; Kaltenborn, B.P.; Vistad, O.I. Public visual preferences for dead wood in natural boreal forests: The effects of added information. *Landsc. Urban Plan* **2017**, *158*, 12–24. [CrossRef]
91. Sahraoui, Y.; Clauzel, C.; Foltête, J.-C. Spatial modelling of landscape aesthetic potential in urban-rural fringes. *J. Environ. Manag.* **2016**, *181*, 623–636. [CrossRef] [PubMed]
92. Yoshimura, N.; Hiura, T. Demand and supply of cultural ecosystem services: Use of geotagged photos to map the aesthetic value of landscapes in Hokkaido. *Ecosyst. Serv.* **2017**, *24*, 68–78. [CrossRef]
93. Youssoufi, S.; Foltête, J.-C. Determining appropriate neighborhood shapes and sizes for modeling landscape satisfaction. *Landsc. Urban Plan* **2013**, *110*, 12–24. [CrossRef]
94. Herzog, T.R.; Bosley, P.J. Tranquility and preference as affective qualities of natural environments. *J. Environ. Psychol.* **1992**, *12*, 115–127. [CrossRef]
95. Real, E.; Arce, C.; Sabucedo, J.-M. Classification of landscapes using quantitative and categorical data, and prediction of their scenic beauty in north-western spain. *J. Environ. Psychol.* **2000**, *20*, 355–373. [CrossRef]
96. Marwijk, R.; van Elands, B.H.M.; Kampen, J.K.; Terlouw, S.; Pitt, D.G.; Opdam, P. Public Perceptions of the Attractiveness of Restored Nature. *Restor. Ecol.* **2011**, *20*, 773–780. [CrossRef]
97. Liang, X.; Fan, L.; Loh, Y.P.; Liu, Y.; Tong, S. Happy Travelers Take Big Pictures: A Psychological Study with Machine Learning and Big Data. *arXiv* **2017**, arXiv:1709.07584.
98. Bubalo, M.; Zanten, B.T.; van Verburg, P.H. Crowdsourcing geo-information on landscape perceptions and preferences: A review. *Landsc. Urban Plan.* **2019**, *184*, 101–111. [CrossRef]
99. Crampton, J.W.; Graham, M.; Poorthuis, A.; Shelton, T.; Stephens, M.; Wilson, M.W.; Zook, M. Beyond the Geotag: Situating "Big Data" and Leveraging the Potential of the Geoweb. *Cartogr. Geogr. Inf. Sci.* **2013**, *40*, 130–139. [CrossRef]
100. Li, L.; Goodchild, M.F.; Xu, B. Spatial, temporal, and socioeconomic patterns in the use of Twitter and Flickr. *Cartogr. Geogr. Inf. Sci.* **2013**, *40*, 61–77. [CrossRef]
101. Kim, J.; Son, Y. Assessing and mapping cultural ecosystem services of an urban forest based on narratives from blog posts. *Ecol. Indic.* **2021**, *129*, 107983. [CrossRef]
102. Liu, Y.; Hu, M.; Zhao, B. Interactions between forest landscape elements and eye movement behavior under audio-visual integrated conditions. *J. For. Res.* **2019**, *25*, 21–30. [CrossRef]
103. Liu, Y.; Hu, M.; Zhao, B. Audio-visual interactive evaluation of the forest landscape based on eye-tracking experiments. *Urban For. Urban Green.* **2019**, *46*, 126476. [CrossRef]

forests

MDPI

Article

Changes in the Abundance of Monoterpenes from Breathable Air of a Mediterranean Conifer Forest: When Is the Best Time for a Human Healthy Leisure Activity?

Domingo Sancho-Knapik [1,2,*], Eustaquio Gil-Pelegrín [1,*], Juan Pedro Ferrio [1,3], David Alonso-Forn [1], Rubén Martín-Sánchez [1], José Víctor dos Santos Silva [1], Junichi Imanishi [4], José Javier Peguero-Pina [1,2] and María Ángeles Sanz [2,5]

1 Departamento de Sistemas Agrícolas, Forestales y Medio Ambiente, Centro de Investigación y Tecnología Agroalimentaria de Aragón (CITA), Avda. Montañana 930, 50059 Zaragoza, Spain; jpferrio@cita-aragon.es (J.P.F.); dalonso@cita-aragon.es (D.A.-F.); rmartin@cita-aragon.es (R.M.-S.); jvdossantos@cita-aragon.es (J.V.d.S.S.); jjpeguero@aragon.es (J.J.P.-P.)
2 Instituto Agroalimentario de Aragón-IA2 (CITA-Universidad de Zaragoza), 50018 Zaragoza, Spain; masanzg@cita-aragon.es
3 Aragón Agency for Research and Development (ARAID), 50018 Zaragoza, Spain
4 Department of Environmental Sciences and Technology, Graduate School of Agriculture, Osaka Metropolitan University, Osaka 599-8531, Japan; junichi.imanishi@omu.ac.jp
5 Área de Laboratorios de Análisis y Asistencia Tecnológica, Centro de Investigación y Tecnología Agroalimentaria de Aragón (CITA), Avda. Montañana 930, 50059 Zaragoza, Spain
* Correspondence: dsancho@cita-aragon.es (D.S.-K.); egilp@cita-aragon.es (E.G.-P.)

Citation: Sancho-Knapik, D.; Gil-Pelegrín, E.; Ferrio, J.P.; Alonso-Forn, D.; Martín-Sánchez, R.; Santos Silva, J.V.d.; Imanishi, J.; Peguero-Pina, J.J.; Sanz, M.Á. Changes in the Abundance of Monoterpenes from Breathable Air of a Mediterranean Conifer Forest: When Is the Best Time for a Human Healthy Leisure Activity? *Forests* **2022**, *13*, 965. https://doi.org/10.3390/f13060965

Academic Editors: Radu-Daniel Pintilii and Diego Varga

Received: 17 May 2022
Accepted: 18 June 2022
Published: 20 June 2022

Abstract: The exposure to monoterpenes emitted by plants to the air might provide human health benefits during forest-based leisure activities. However, forests, especially Mediterranean ones, lack studies to relate forest production and the emission of monoterpenes, considering potential human forest exposure. Thus, the aim of this study was to analyze the variation in the abundance of monoterpenes in the human breathable air under the canopy of a Mediterranean conifer forest, evaluating the influence of different factors. For this purpose, from March to November 2018, we monitored the abundance of monoterpenes in the air at nose height, leaf development, air temperature and soil water potential in a mountain Mediterranean forest of *Pinus pinaster* located in Sierra de Albarracín (Teruel, Spain). We detected six monoterpenes, with α-pinene, β-pinene and limonene being the three most abundant. Temperature was the main environmental factor driving the abundance of monoterpenes in air, with a maxima of abundance found during summer. Leaf development in spring decreased the abundance, while after a drought period, the abundance increased. Thus, people enjoying forest-based activities in Mediterranean conifer areas would be more exposed to air monoterpenes when the temperature increases during the period after leaf development, as long as the trees are not severely water-stressed. If that is the case, the abundance of monoterpenes in the air would increase after the drought period.

Keywords: human health; forest air; monoterpenes; path analysis; *Pinus pinaster*

1. Introduction

Tourists are becoming more and more sensitive to the environment and nature, turning towards rural areas and away from noise and chaos of city life, mainly due to a higher education level and an increased consciousness of a healthier lifestyle [1–3]. Over the last decades, this fact has promoted the development of a great diversity of ecotourism products and leisure activities related to nature and forests, such as ecological farms [3], green hotels [4], and mycotourism [5,6]. The implementation of most of these nature-based activities is mainly influenced by weather variables [7] such as temperature and precipitation, which are the two most common variables used in tourism-based weather impact studies [8–15]. Weather conditions have already been interpreted through relationships with personal characteristics [14] and used to predict thermal sensation [15–17]. Furthermore, the interaction between human and atmosphere has also been used to understand and predict nature-based tourist behavior [18,19] and to develop adaptation strategies [20].

Apart from the weather conditions, there are other environmental factors that could influence forest recreation plans [7], especially those that incorporate forest activities that aim to obtain diverse human health benefits, as in the case of so-called "shinrin-yoku" or forest bathing [21], which is becoming an important forest health practice around the world due to its capacity to provide relaxation and reduce stress [22,23]. Among these other environmental factors, the exposure to atmospheric volatile organic compounds (VOCs), specifically to the monoterpenes group, is one of the aspects most currently being evaluated, as apparently, they can provide human wellness [10,24,25]. Monoterpenes produced by plants, primarily conifers, as a defensive mechanism to pathogens [26–28], are supposed to exert anti-inflammatory, antioxidant, anti-cancer or neuroprotection effects on humans (see references within [24,25]). Thus, exposure to these compounds during forest-based activities may provide additional therapeutic benefits. However, evidence relating forest exposure to monoterpenes and human health benefits remains uncertain, due to the lack of forest type descriptions and the high heterogeneity of approaches and analyses within the studies performed [29]. In this sense, knowing the air abundance of monoterpenes in a particular forest would be a first step to establish further relationships between forest monoterpenes and human health.

The abundance of monoterpenes in forest air may change during the year, as they depend not only on their synthesis by plants, but also on their emission [30]. At the same time, both processes (synthesis and emission) may be subjected to several environmental stimuli [31]. On the one hand, the synthesis of monoterpenes may depend primarily on plant genetics, incident sunlight and air temperature, being influenced by other factors such as plant phenology, soil water availability, wind or physical disturbance that may generate deviations from the general patterns [30]. On the other hand, the emission of monoterpenes may also depend primarily on temperature, being influenced by relative humidity or water availability [30].

The combination and interaction of the different environmental stimuli would result in a particular abundance of monoterpenes for a specific forest type. While temperate and tropical forests from Asia appear to be well studied, Mediterranean forests from southern Europe seem to be scarcely analyzed [29]. As Mediterranean-type forests are characterized by particular climatic conditions such as a summer drought period [32], the variation in the atmospheric abundance of monoterpenes may have a distinct performance in these types of forests compared to others. Therefore, the aim of this study was to analyze the variation in the abundance of monoterpenes in the human breathable air under the canopy of a forest dominated by the Mediterranean conifer *Pinus pinaster* Ait. located in the east of the Iberian Peninsula, evaluating the influence of phenology and environmental factors. This would help to predict, from a few environmental factors, the composition and abundance of monoterpenes at nose height under Mediterranean forest canopies, important in human health-based forest activities.

2. Materials and Methods

2.1. Experimental Site

The study site is composed of a healthy woodland dominated by *Pinus pinaster* located within a protected landscape area of ca. 6800 ha in the east of Spain (40°19′59.00″ N, 1°21′09.49″ W, 1206 m a.s.l., Pinares de Ródeno, Sierra de Albarracín, Teruel, Spain; Figure 1). This protected pine tree landscape is one of the most distinctive natural environments in Spain due to a particular combination of geology, geomorphology, flora and fauna that makes it an attractive area for visitors (http://wildsideholidays.co.uk/pinares-de-rodeno/, accessed on 7 June 2022).

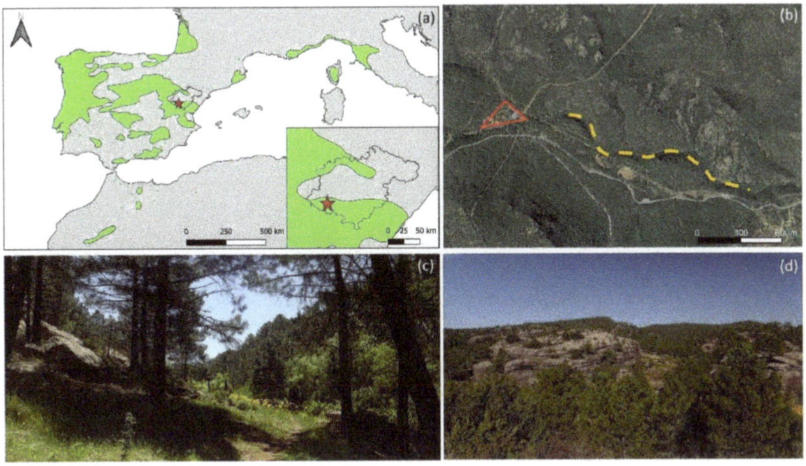

Figure 1. (**a**) Location of the experimental site in Teruel, Spain (red star). Green areas are the natural distribution of the Mediterranean conifer *Pinus pinaster*. (**b**) Aerial photo of the experimental site. The yellow dashed line indicates the sampling path, while the red triangle indicates the tourist information center. (**c**,**d**) provide detail of the sampling path and the landscape, respectively.

The study site comprises naturally-regenerated pine forests with different age classes (from ca. 20 to ca. 60 years), with some interspersed individuals of *Quercus pyrenaica* Willd. and *Cistus laurifolius* L. The mean ($\pm$ se) height and diameter at 1.30 m of pines were 10.5 ± 0.8 m and 30 ± 3 cm, respectively. The trees did not show any symptoms of either abiotic or biotic stress prior to or during the experiment. The area is characterized by a mountain Mediterranean climate with a mean annual temperature of 8.0 °C and annual precipitation of 567 mm (data obtained from the WorldClim database: http://www.worldclim.org/, accessed on 1 March 2022). The forest is mainly utilized for leisure activities, with a tourist information center, marked walking tracks and wild forest gyms. The environmental quality also adds value to the vast cultural heritage preserved for centuries by human hands. The best example is the Cave Paintings, a World Heritage Site by UNESCO, hidden in many rock shelters.

For the measurements of this study, we selected a flat path 1200 m in length, part of a forest walking track close to the information center (Figure 1). The measurements were performed from March to November 2018, a period that included a whole vegetative season and variability in the environmental conditions.

2.2. Environmental Factors and Phenology

Air temperature (T, °C) and relative humidity (RH, %) were recorded throughout the year using a Hobo Pro RH/Temp data logger (Onset Computer Bourne, MA, USA) installed under the canopy 3 m above the ground. To measure soil water availability, soil water potential (SWP, MPa) was recorded using two soil water potential sensors, connected to a data logger (Teros 21/Em50, METER Group, Inc., Pullman, WA, USA), and installed 40 cm below ground, representative of the root system. Both sensors for the air temperature and soil water potential were located halfway along the sampling path. Additionally, precipitation data was obtained from the closest meteorological station (40°31′00.55″ N, 1°16′06.32″ W, 1000 m a.s.l.).

Bud bursting and needle development were periodically observed every month during 2018. Based on the phenology observed, we defined the variable "leaf stage" to refer to the development of leaves. For the input of "leaf stage" in the statistical analysis, we assigned a value between 0 and 11 to each month, so that 0 was assigned to the first month observing leaf development, 1 to the next month, and so on consecutively up to 11, which was assigned to the month prior to the observation of leaf development.

2.3. Monoterpenes Sampling

Air was sampled at nose height (ca. 1.50 m above ground) periodically from March to November 2018, around once per month, for a total of 9 days. For each sampling day, two samples of air per hour were collected throughout the pathway (Figure 1) at three different hours: 11:00 a.m., 12:00 a.m. and 13:00 p.m. h (local time), for 50 min for each sample (i.e., two samples per time x three times). For air sampling, we used adsorbent tubes (Tenax TA, Gerstel GmbH & Co. KG, Mülheim an der Ruhr, Germany) attached with 1 m plastic tubing (Tygon S3 E-3603, Tubes International Sp., Poznań, Poland) to air sampling pumps (flow rate 0.1 L/min, SKC AirChek TOUCH Model 220-5000TC, SKC Inc., Eighty Four, PA, USA). After each air sampling, the adsorbent tubes were hermetically enclosed and kept inside a portable fridge. At the end of the sampling day, a total of six samples (two samples per time × three times) were obtained and carried to the laboratory. The sampling days were selected to avoid weekends (i.e., peak visitor days), rain, or wind force higher than 5 km h^{-1}. The air temperature was also recorded during each sampling time using another Hobo Pro RH/Temp data logger.

Once in the lab, the adsorbent tubes were thermally desorbed and volatile compounds were detected using a gas chromatography tandem mass spectrometer (model 5973N, Agilent Technologies, Santa Clara, CA, USA) [33]. The resulting chromatograms were analyzed to calculate the abundance of each monoterpene (given in area units) as the integrated area below each peak curve [34]. Moreover, we calculated the abundance of total monoterpenes as the sum of the individual areas of the monoterpenes detected.

2.4. Statistical Analysis

For each air-sampling day, we considered the environmental conditions during the previous 15 days (ca. half of the air-sampling period), as well as during the sampling time, as likely to affect, respectively, the synthesis and emission of monoterpenes. In particular, we selected the following environmental variables influencing the synthesis of monoterpenes: the mean maximum daily temperature of the previous 15 days to the air-sampling day (T_{15}), and the minimum soil water potential of the previous 15 days to the air-sampling day (SWP_{15}). As variables influencing monoterpene emissions, we selected the mean temperature of the current air-sampling time (T_c), and the mean soil water potential of the current air-sampling time (SWP_c). For T_c, we calculated three mean values according to the three sampled periods during the day: from 11:00 a.m. to 11:50 a.m., 12:00 a.m. to 12:50 a.m. and from 13:00 p.m. to 13:50 p.m., respectively. For SWP_c, we considered a single value, as it remained constant between 11:00 a.m. and 13:50 p.m.

To characterize the connections among the phenology and environmental variables, and determine their relative importance in influencing the monoterpenes in the air, we conducted a path analysis [35]. This analysis uses a path structure that consists of variables of interest with potential connectivity among the variables, where the connections are based on previously well-established knowledge. Based on the variables and the connections, a set of multiple linear regression models is constructed, and the partial regression coefficients are defined as path values that indicate the causative power of each connection [36]. Structural equation models for the path analysis were fitted using the function 'sem' from package Lavaan 0.6–10 [37]. The model included "leaf stage", T_{15} and SWP_{15} as exogenous variables and T_c (as a function of T_{15}), SWP_c (as a function of SWP_{15}) and air total monoterpenes (as a function of T_{15} SWP_{15}, T_c and SWP_c) as endogenous variables.

3. Results

The evolution of environmental factors in the study area during 2018 showed a general increase in temperature and a decrease in relative humidity from March to the end of July. From then, the temperature started to decrease and the relative humidity increased until

November (Figure 2). The soil water potential remained close to zero for most of the studied period, indicating that there was soil water availability most of the time. However, during mid-October, we recorded a 12-day period with values of soil water potential below −0.5 MPa, indicating soil water scarcity (Figure 2). These general trends influenced the environmental variables selected in this study for each air-sampling day (Table 1). The mean maximum daily temperature of the previous 15 days (T_{15}) and mean temperature of the current air-sampling time (T_c) increased from March to July and decreased towards November (Table 1). The minimum soil water potential of the previous 15 days to the air-sampling day (SWP_{15}) showed values between 0 and −0.3 MPa for most of the sampling days; for DOY = 296, SWP_{15} reached a value of −1.495 MPa. The mean soil water potential of the current air-sampling time (SWPc) showed values close to zero for all sampling days (Table 1).

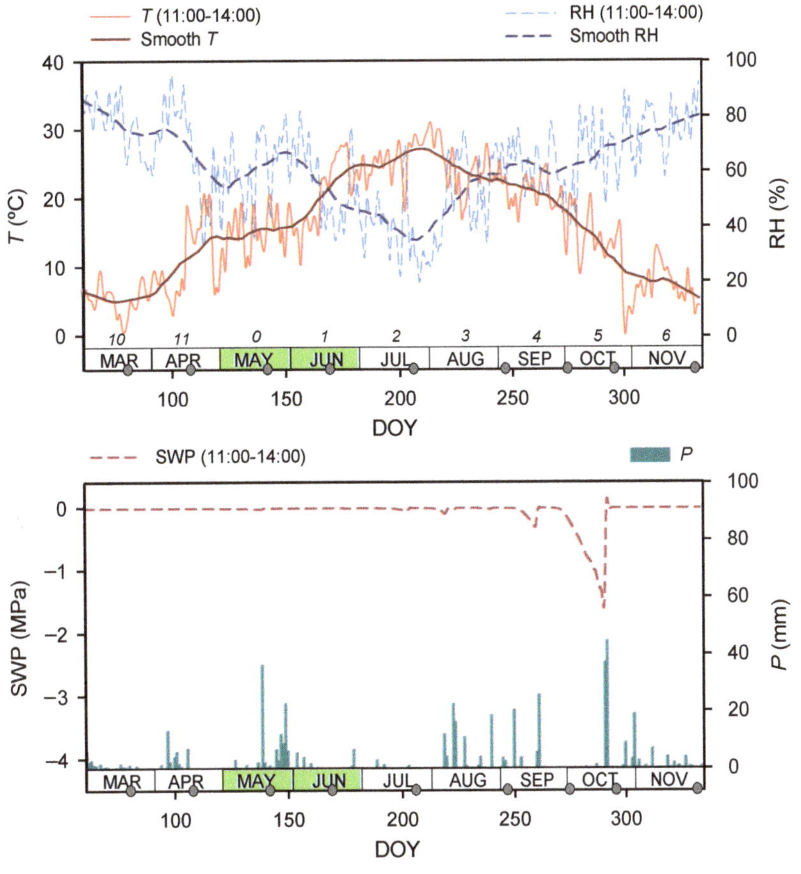

Figure 2. Evolution of environmental factors in the study area during 2018. *T*, temperature; RH, relative humidity; SWP, soil water potential; *P*, precipitation; DOY, day of the year. *T*, RH and SWP are the mean values recorded between 11:00 a.m. and 14:00 p.m. (local time). *P* was obtained from the closest meteorological station. Months colored in green represent the period of leaf development. "Leaf stage" is given by the italic numbers above each month. Grey circles are the volatile air-sampling days.

Table 1. Values of environmental variables for each air-sampling day. DOY, day of year; T_{15} and SWP_{15} are the mean maximum daily temperature and minimum soil water potential of the previous 15 days to the air-sampling day, respectively; T_c and SWP_c are the mean temperature and soil water potential of the current air-sampling time, respectively.

DOY	Month	T_{15} (°C) Mean ± se (min; max)	SWP_{15} (Mpa)	T_c (°C) Mean ± se [1]			SWP_c (Mpa) [2]
				11:00 a.m. h	12:00 a.m. h	13:00 p.m. h	
79	March	9.8 ± 0.7 (5.3; 14.4)	−0.013	1.8 ± 0.3	0.6 ± 0.1	1.1 ± 0.1	−0.013
107	April	12.2 ± 1.2 (2.8; 19.3)	−0.013	18 ± 0.7	18.8 ± 0.4	19.6 ± 0.3	−0.013
142	May	18.8 ± 0.7 (13.8; 23.7)	−0.031	20.3 ± 0.5	18.5 ± 0.2	14.9 ± 0.2	−0.012
170	June	20.4 ± 1.0 (13.2; 27.0)	−0.012	26.2 ± 0.5	27.3 ± 0.3	27 ± 0.3	−0.012
205	July	28.7 ± 0.7 (20.1; 31.3)	−0.063	27.8 ± 0.3	28.9 ± 0.3	31.2 ± 0.5	−0.012
247	September	28.6 ± 0.6 (24.9; 32.7)	−0.034	23.2 ± 0.3	26.8 ± 0.2	28.9 ± 0.8	−0.015
275	October	24.6 ± 0.7 (19.4; 29.6)	−0.288	18.3 ± 0.3	21.5 ± 0.8	22.9 ± 0.2	−0.198
296	October	16.6 ± 0.8 (11.3; 22.6)	−1.495	12.1 ± 0.3	15.3 ± 0.6	15.3 ± 0.4	−0.011
331	November	9.9 ± 0.7 (6.7; 15.1)	−0.011	10.1 ± 1.0	11 ± 0.6	11.5 ± 0.4	−0.011

[1] T_c at 11:00 a.m., 12:00 a.m. and 13:00 p.m. are the mean values of the three sampled periods during a day: from 11:00 a.m. to 11:50 a.m., 12:00 a.m. to 12:50 a.m. and from 13:00 p.m. to 13:50 p.m., respectively. [2] For SWP_c, we considered a single value as it remained constant between 11:00 a.m. and 13:50 p.m.

Bud bursting and needle development occurred during the months of May and June (Figure 2). The needles were fully mature by July. With this phenology, a value of "leaf stage" equal to zero was assigned to May, a value of 1 to June, and so on consecutively, up to 11, which was assigned to April (Figure 2).

Six monoterpenes were detected in the air at nose height below the canopy of the *Pinus pinaster* forest: α-pinene, β-pinene, limonene, *p*-cymene, camphene and 1,8-cineole. The evolution of their total abundance given as total monoterpenes, shows variations throughout the year, with minimums of abundance in March and May and a first peak of abundance in April; an increase in abundance from May to July, where the maximum peak was found; a decrease towards October; and a third peak of abundance at the end of October (Figure 3). The most abundant monoterpene detected was α-pinene, which followed the same pattern of variation as the total monoterpenes (Figures 3 and 4). The next two monoterpenes in abundance (although much lower than α-pinene) were β-pinene and limonene. These two monoterpenes followed a similar general trend of variation throughout the year, but with some peculiarities: β-pinene was not detected until June, whereas limonene had its minimum at the beginning of October, not being detected at the end of the study period (Figures 3 and 4). Concerning the other three monoterpenes detected (*p*-cymene, camphene and 1,8-cineole), the abundance was relatively very low, and variation did not clearly follow the same pattern as the others, being non-detected in some air samples through the study period. The relative maximum abundance of *p*-cymene, camphene and 1,8-cineole were detected in June, October and September, respectively (Figure 3). It should also be noted that the air samples with the most diversity in terms of the number of monoterpenes detected were those collected from June to the beginning of October, while those samples with less diversity were April, the end of October and November (Figure 3).

The model proposed for the path analysis (Akaike information criterion, AIC = 273) showed a good fitting, with both "leaf stage" and environmental factors (temperature and soil water potential) showing a significant effect on the abundance of total monoterpenes in the air at nose height (Figure 5, Table 2). "Leaf stage" showed a positive relation with total monoterpenes in the air, indicating that the older the current-year leaves, the higher the abundance of monoterpenes, or, i.e., the development of new leaves implied a lower abundance of monoterpenes in the air sampled. Concerning temperature, both the mean maximum daily temperature of the previous 15 days (T_{15}) and the mean temperature of the current air-sampling time (T_c) had a similar positive effect on the abundance of monoterpenes in air. That is, an increase in the temperature might favor both the synthesis and emission of monoterpenes to the forest air (Figure 5, Table 2). Regarding soil water availability, the minimum soil water potential of the previous 15 days (SWP_{15}) exerted a negative effect on total monoterpenes in the air (Figure 5, Table 2), indicating that the lower the soil water potential prior to the air sampling, the higher the abundance of monoterpenes in air. By contrast, the mean soil water potential of the current air-sampling time (SWP_c) had a positive effect on total air monoterpenes, which indicates that abundance of monoterpenes in air is favored by well-hydrated soils during the air sampling (Figure 5, Table 2). With regard to the link between the different drivers, variations in T_c were, as expected, largely driven by T_{15}. By contrast, SWP_{15} was not related to either SWP_c or T_{15}. Although both "leaf stage" and T_{15} had a positive effect on total air monoterpenes, the association between these two variables was negative, indicating that their effects were independent of each other.

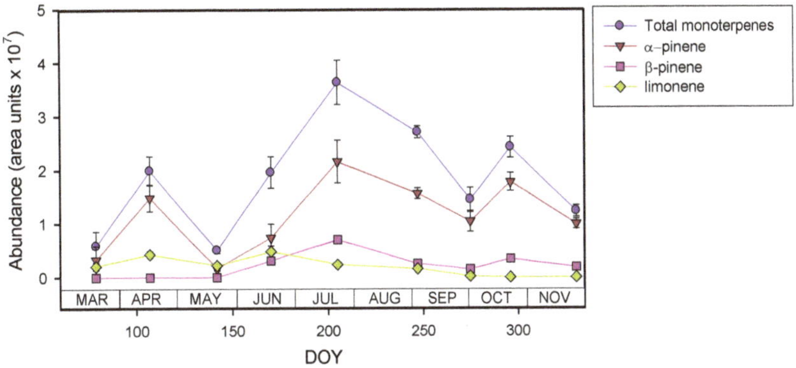

Figure 3. Evolution of the abundance of monoterpenes measured in the air at nose height in the study area during 2018. Figure shows total monoterpenes and the three most abundant monoterpenes measured: α-pinene, β-pinene and limonene. DOY, day of the year. Values are mean ± se (n = 6).

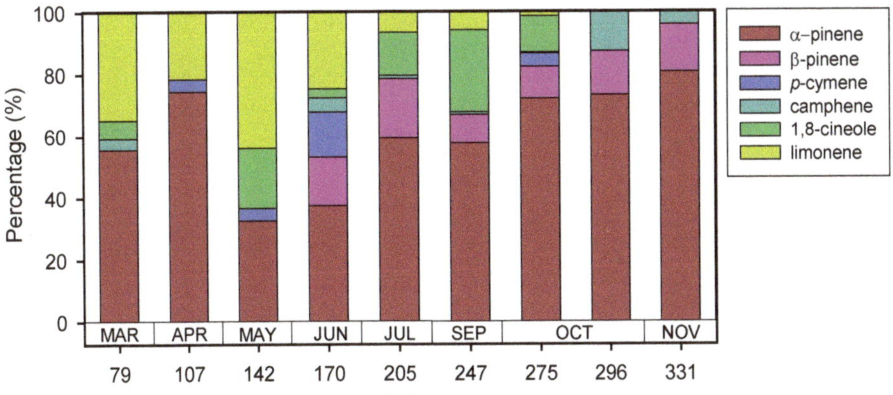

Figure 4. Relative percentage of abundance of each monoterpene measured in the air at nose height in the study area during 2018. DOY, day of the year.

Finally, the relationship between T_c and the abundance of monoterpenes is shown in Figure 6, confirming the result obtained from the path analysis: a higher T_c drives a higher abundance of monoterpenes in the air. α-pinene, the most abundant monoterpene detected, followed the same trend (Figure 6). It should be highlighted that both regression lines in Figure 6 excluded those air samples most affected by leaf development (white triangles below the regression lines) or soil water deficit (black squares above the regression lines).

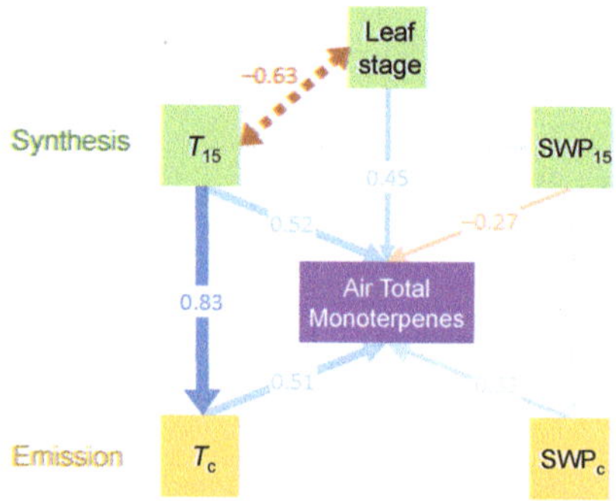

Figure 5. Path analysis relating the abundance of total monoterpenes in air, leaf stage and environmental factors. T_{15}, mean maximum daily temperature of the 15 days prior to the air-sampling day; SWP_{15}, minimum soil water potential of the 15 days prior to the air-sampling day; T_c, mean temperature of the current air-sampling time; SWP_c, mean soil water potential of the current air-sampling time. Blue are positive correlations and red are negative ones. Line thickness is proportional to the standardized coefficients (also shown in numerical labels). Bi-directional dashed lines indicate associations between exogenous variables. Unidirectional solid lines indicate the 'causal' associations included in the structural model.

Table 2. Main statistics of the structural equation model tested in the path analysis (Figure 5). R^2, coefficient of determination for each estimate in the model; UC, unstandardized coefficients and their SE, standard errors; SC, standardized coefficients; *P*, statistical significance. Meaning of environmental variables are in caption of Figure 5.

	R^2	UC	SE	SC	*P*
Monoterpenes~T_c + SWP_c + Leaf stage + SWP_{15} + T_{15}	0.795				<0.001
T_c		0.727	0.221	0.509	0.001
SWP_c		67.507	18.091	0.325	<0.001
Leaf stage		1.55	0.382	0.453	<0.001
SWP_{15}		−7.194	2.303	−0.273	0.002
T_{15}		0.904	0.295	0.521	0.002
T_c~T_{15}	0.683				<0.001
T_{15}		1.004	0.132	0.826	<0.001
SWP_c~SWP_{15}	0.002				0.812
SWP_{15}		0.006	0.024	0.046	0.812

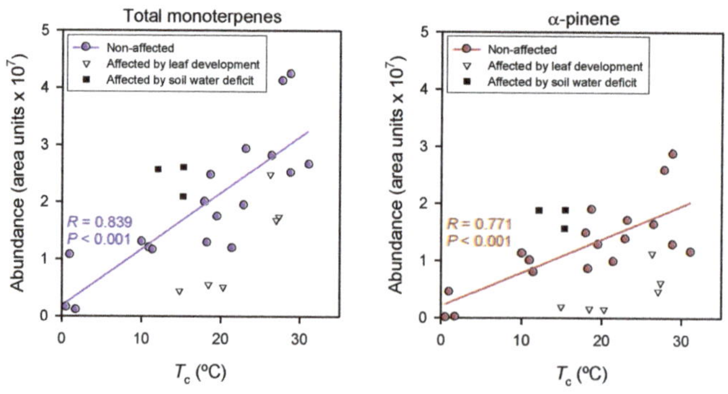

Figure 6. Relationships between the mean temperature of the current air-sampling time (T_c) and the abundance of total monoterpenes and α-pinene. We have excluded from the regression line those air samples affected by leaf development (white triangle) or soil water deficit (black squares). For each linear relationship, the *P*-value is <0.001 and the correlation coefficient (*R*) is shown in the graph.

4. Discussion

Six volatile monoterpenes with apparent benefits to human health [24,25] were detected during 2018 in the air below the canopy of a Mediterranean conifer (*Pinus pinaster*) forest, located in Sierra de Albarracín (Teruel, Spain). The three most abundant monoterpenes found in our study (α-pinene, β-pinene and limonene) coincide with the general abundance trends found in different conifer forests dominated by species from genus *Pinus*, *Abies* or *Picea* [38]. We also detected three additional monoterpenes, with less abundance in the study area (*p*-cymene, camphene and 1,8-cineole), but we did not detect two others (Δ3-carene and myrcene) often mentioned in the literature [31,39–41].

The maximum abundance peak of the monoterpenes in the air during the sampled period was found during the summer, specifically at the end of July, followed by early September, coinciding with the season with the highest resin flow rates found for *P. pinaster* in the northeast of Spain [42]. This maximum peak also coincides with the results found in previous work performed on another Mediterranean forest dominated by *Quercus ilex* L. [43]. Furthermore, the monoterpene emission rates measured at branch or canopy level have also proved to be higher during summer in the Mediterranean *Pinus pinea* [44,45]. This seasonality in the abundance of monoterpenes has been mainly associated with changes in air temperature [43,46,47]. In our study, temperature also seems to be the main driver likely to affect the synthesis and the emission of monoterpenes [30], with higher temperature values resulting in a higher abundance of monoterpenes in the air at nose height. However, we found two other factors—phenology and soil water deficit—that influenced and generated deviations from the main driver.

Concerning phenology, we found that during the months when leaf development occurred, the abundance of monoterpenes in the air was lower than expected, according to the recorded temperature values. This was especially noticed in May, when we detected an abundance of total monoterpenes in air four-fold lower than predicted for a mean temperature of ca. 18 °C (Figure 6). According to previous work [48–50], there is a trade-off between growth and defense, which assumes that plants possess a limited pool of resources that can be invested either in growth or in defense. During leaf development, *Pinus pinaster* seems to allocate most resources to growth, limiting the synthesis of monoterpenes for defense. By contrast, during the rest of the year, when there is hardly any growth, resources might be allocated to the production of defense compounds, such as terpenes [51,52].

Regarding the other factor influencing the general response of monoterpene abundance to changes in temperature, we found that a water deficit period followed by a soil water recovery due to precipitation during October was associated with a slight increase in the abundance of monoterpenes in the air (Figures 5 and 6). In Mediterranean climates, it is common to find a drought period during summer that may affect the functioning of plants [32]. However, in the study site, we registered a drought period during the early autumn of 2018, a displacement likely due to being in a mountain Mediterranean climate [53]. Nevertheless, the literature states that a water deficit period can be associated with increased levels of needle volatile terpenes to increase plant defense [40,54,55], whereas emission may be constrained during water stress due to stomatal closure [56,57]. This suggests that species able to store terpenes, such as *P. pinaster*, would increase the synthesis and storage of monoterpenes during drought, while simultaneously restricting their emission [31,58] as a defensive mechanism to reduce palatability and as herbivore repellents [59,60]. Once water stress conditions disappear, the stomata can open again, and the leaves can release the volatile terpenes accumulated during the drought period into the air.

The highest peak of monoterpenes in the air found during summer in this study also coincides with the season of highest demand in the Mediterranean tourist sector, which is concerned about the future negative impacts of rising temperatures due to climate change [11,61,62]. Summer days with higher temperatures could bring significantly lower visitor numbers due to uncomfortable temperatures [63,64] and, therefore, may change the behavior of tourists [9,13,65]. However, leisure activities under the forest canopy during Mediterranean summers may not only provide greater benefits to human health due to the abundance of monoterpenes, but also may provide thermal comfort [66,67]. Thus, forest-based leisure activities during summer, such as shinrin-yoku [68–70], can be promising adaptive ways to minimize the impacts of climate change on the Mediterranean tourist sector.

5. Conclusions

Temperature was found to be the main environmental factor driving the abundance of monoterpenes in the air of a mountain Mediterranean forest of *Pinus pinaster*, with the maxima of abundance found during summer (July and August, according to this study). Leaf development in spring decreased the abundance, while during post-drought periods, the abundance may increase (end of October in this study). Therefore, people enjoying forest-based leisure activities would be more exposed to air monoterpenes, and apparently obtain more health benefits, when the temperature increases during the period after leaf development, as long as there is soil water available and the trees are not suffering from water stress at the time of forest exposure. If there is a summer drought period and the trees are water stressed, the abundance of monoterpenes in air will increase when the drought period ends.

Author Contributions: Conceptualization, D.S.-K.; methodology, D.S.-K. and M.Á.S.; validation, E.G.-P. and J.J.P.-P.; formal analysis, D.S.-K., J.P.F. and M.Á.S.; investigation, D.S.-K.; data curation, D.S.-K. and M.Á.S.; writing—original draft preparation, D.S.-K.; writing—review and editing, J.J.P.-P., J.P.F., R.M.-S., J.V.d.S.S., J.I. and M.Á.S.; visualization, D.A.-F. and R.M.-S.; supervision, E.G.-P. and M.Á.S.; project administration, D.S.-K.; funding acquisition, D.S.-K., E.G.-P., J.J.P.-P. and M.Á.S. All authors have read and agreed to the published version of the manuscript.

Funding: This research was funded by Gobierno de Aragón—Fondo de Inversiones de Teruel (FITE) and Gobierno de España, grant project FITE-2017-THERAPYFOREST and by Gobierno de Aragón research group H09_20R. The work of D.A-F. was supported by FPI-INIA, grant number BES-2017-081208; the work of J.V.dos S.S. was supported by MCIN/AEI, grant number PRE2020-094944; and the work of R.M-S. was supported by Gobierno de Aragón.

Acknowledgments: We thank Mercedes Vico for her valuable help during the air forest sampling.

Conflicts of Interest: The authors declare no conflict of interest.

References

1. Barreiro-Hurle, J.; Gracia, A.; de-Magistris, T. The effects of multiple health and nutrition labels on consumer food choices. *J. Agric. Econ.* **2010**, *61*, 426–443. [CrossRef]
2. Leelapattana, W.; Hsu, S.Y.; Thongma, W.; Chen, C.; Chiang, F.M. Understanding the impact of environmental education on tourists' future visit intentions to leisure farms in mountain regions. *Sustainability* **2019**, *11*, 1567. [CrossRef]
3. Ceylan, E.N.; Tülbentçi, T. Example of an ecotourism farm in the context of sustainability: Pastoral valley ecological life farm. *Int. J. Adv. Appl.* **2020**, *7*, 116–132. [CrossRef]
4. Wang, S.Y.; Wang, J.; Wang, Y.; Yan, J.; Li, J. Environmental knowledge and consumers' intentions to visit green hotels: The mediating role of consumption values. *J. Travel Tour. Mark.* **2018**, *35*, 1261–1271. [CrossRef]
5. Büntgen, U.; Latorre, J.; Egli, S.; Martinez-Peña, F. Socio-economic, scientific, and political benefits of mycotourism. *Ecosphere* **2017**, *8*, e01870. [CrossRef]
6. De Frutos, P.; Rodriguez-Prado, B.; Latorre, J.; Martinez-Peña, F. Environmental valuation and management of wild edible mushroom picking in Spain. *For. Policy Econ.* **2019**, *100*, 177–187. [CrossRef]
7. Verbos, R.I.; Altschuler, B.; Brownlee, M.T.J. Weather studies in outdoor recreation and nature-based tourism: A research synthesis and gap analysis. *Leis. Sci.* **2018**, *40*, 533–556. [CrossRef]
8. Becken, S. Measuring the effect of weather on tourism: A destination- and activity-based analysis. *J. Travel Res.* **2012**, *52*, 156–167. [CrossRef]
9. Lise, W.; Tol, R.S.J. Impact of climate on tourist demand. *Clim. Chang.* **2002**, *55*, 429–449. [CrossRef]
10. Matzarakis, A.; Hämmerle, M.; Endler, C.; Muthers, S.; Koch, E. Assessment of tourism and recreation destinations under climate change conditions in Austria. *Meteorol. Z.* **2012**, *21*, 157–165. [CrossRef]
11. Brosy, C.; Zaninovic, K.; Matzarakis, A. Quantification of climate tourism potential of Croatia based on measured data and regional modeling. *Int. J. Biometeorol.* **2013**, *58*, 1369–1381. [CrossRef] [PubMed]
12. Gómez-Martín, M. Weather, climate, and tourism: A geographical perspective. *Ann. Tour. Res.* **2005**, *32*, 571–591. [CrossRef]

13. Bigano, A.; Hamilton, J.M.; Tol, R.S.J. The impact of climate on holiday destination choice. *Clim. Chang.* **2006**, *76*, 389–406. [CrossRef]
14. Andrade, H.; Alcoforado, M.J.; Oliveira, S. Perception of temperature and wind by users of public outdoor spaces: Relationships with weather parameters and personal characteristics. *Int. J. Biometeorol.* **2011**, *55*, 665–680. [CrossRef] [PubMed]
15. Pantavou, K.; Lykoudis, S. Modeling thermal sensation in a Mediterranean climate—A comparison of linear and ordinal models. *Int. J. Biometeorol.* **2014**, *58*, 1355–1368. [CrossRef]
16. Rutty, M.; Scott, D. Bioclimatic comfort and the thermal perceptions and preferences of beach tourists. *Int. J. Biometerol.* **2014**, *59*, 37–45. [CrossRef]
17. Rutty, M.; Scott, D. Thermal range of coastal tourism resort microclimates. *Tour. Geogr.* **2014**, *16*, 346–363. [CrossRef]
18. Gómez-Martín, M.; Martínez-Ibarra, E. Tourism demand and atmospheric parameters: Nonintrusive observation techniques. *Clim. Res.* **2012**, *51*, 135–145. [CrossRef]
19. Lindner-Cendrowska, K. Assessment of bioclimatic conditions in cities for tourism and recreational purposes (a Warsaw case study). *Geogr. Pol.* **2013**, *86*, 55–66. [CrossRef]
20. Matzarakis, A.; Hämmerle, M.; Koch, E.; Rudel, E. The climate tourism potential of Alpine destinations using the example of Sonnblick, Rauris and Salzburg. *Theor. Appl. Climatol.* **2012**, *110*, 645–658. [CrossRef]
21. Hansen, M.M.; Jones, R.; Tocchini, K. Shinrin-Yoku (Forest Bathing) and Nature Therapy: A State-of-the-Art Review. *Int. J. Environ. Res. Public Health* **2017**, *14*, 851. [CrossRef]
22. Tsunetsugu, Y.; Park, B.J.; Miyazaki, Y. Trends in research related to "Shinrin-yoku" (taking in the forest atmosphere or forest bathing) in Japan. *Environ. Health Prev. Med.* **2010**, *15*, 27–37. [CrossRef]
23. Wen, Y.; Yan, Q.; Pan, Y.; Gu, X.; Liu, Y. Medical empirical research on forest bathing (Shinrin-yoku): A systematic review. *Environ. Health Prev. Med.* **2019**, *24*, 70. [CrossRef]
24. Cho, K.S.; Lim, Y.-R.; Lee, K.; Lee, J.; Lee, J.H.; Lee, I.-S. Terpenes from forests and human health. *Toxicol. Res. Vol.* **2017**, *33*, 97–106. [CrossRef]
25. Antonelli, M.; Donelli, D.; Barbieri, G.; Valussi, M.; Maggini, V.; Firenzuoli, F. Forest volatile organic compounds and their effects on human health: A state-of-the-art review. *Int. J. Environ. Res. Public Health* **2020**, *17*, 6506. [CrossRef]
26. Gershenzon, J.; Dudareva, N. The function of terpene natural products in the natural world. *Nat. Chem. Biol.* **2007**, *3*, 408–414. [CrossRef]
27. Mewalal, R.; Rai, D.K.; Kainer, D.; Chen, F.; Külheim, C.; Peter, G.F.; Tuskan, G.A. Plant-derived terpenes: A feedstock for specialty biofuels. *Trends Biotechnol.* **2017**, *35*, 227–240. [CrossRef]
28. Martin, D.M.; Gershenzon, J.; Bohlmann, J. Induction of volatile terpene biosynthesis and diurnal emission by methyl jasmonate in foliage of Norway spruce. *Plant Physiol.* **2003**, *132*, 1586–1599. [CrossRef]
29. Bach, A.; Peñuelas, J.; Clarà, J.; Llusià, J.; Campillo, I.; López, F.; Maneja, R. How should forests be characterized in regard to human health? Evidence from existing literature. *Int. J. Environ. Res. Public Health* **2020**, *17*, 1027. [CrossRef]
30. Peñuelas, J.; Lluisá, J. The complexity of factors driving volatile organic compound emissions by plants. *Biol. Plant.* **2001**, *44*, 481–487. [CrossRef]
31. Kopaczyk, J.M.; Warguła, J.; Jelonek, T. The variability of terpenes in conifers under developmental and environmental stimuli. *J. Exp. Bot.* **2020**, *180*, 104197. [CrossRef]
32. Gil-Pelegrín, E.; Saz, M.A.; Cuadrat, J.M.; Peguero-Pina, J.J.; Sancho-Knapik, D. Oaks under mediterranean-type climates: Functional response to summer aridity. In *Oaks Physiological Ecology. Exploring the Functional Diversity of Genus Quercus L.*; Gil-Pelegrín, E., Saz, M.A., Cuadrat, J.M., Peguero-Pina, J.J., Sancho-Knapik, D., Eds.; Springer International Publishing: Cham, Switzerland, 2017; pp. 137–193.
33. Kännaste, A.; Copolovici, L.; Niinemets, Ü. Gas chromatography-mass spectrometry method for determination of biogenic volatile organic compounds emitted by plants. *Methods Mol. Biol.* **2014**, *1153*, 161–169. [CrossRef] [PubMed]
34. Mucciarelli, M.; Camusso, W.; Maffei, M.; Panicco, P.; Bicchi, C. Volatile terpenoids of endophyte-free and infected peppermint (*Mentha piperita* L.): Chemical partitioning of a symbiosis. *Microb. Ecol.* **2007**, *54*, 685–696. [CrossRef]
35. Li, C.C. *Path Analysis—A Primer*; The Boxwood Press: Pacific Grove, CA, USA, 1975.
36. Kimm, H.; Guan, K.; Gentine, P.; Wu, J.; Bernacchi, C.J.; Sulman, B.N.; Griffis, T.J.; Link, C. Redefining droughts for the U.S. Corn Belt: The dominant role of atmospheric vapor pressure deficit over soil moisture in regulating stomatal behavior of maize and soybean. *Agric. Forest Meteorol.* **2020**, *287*, 107930. [CrossRef]
37. Rosseel, Y. Lavaan: An R package for structural equation modeling. *J. Stat. Softw.* **2012**, *48*, 1–36. [CrossRef]
38. Pokorska, O.; Dewulf, J.; Amelynck, C.; Schoon, N.; Šimpraga, M.; Steppe, K.; Van Langenhove, H. Isoprene and terpenoid emissions from *Abies alba*: Identification and emission rates under ambient conditions. *Atmos. Environ.* **2012**, *59*, 501–508. [CrossRef]
39. Simon, V.; Clement, B.; Riba, M.L.; Torres, L. The Landes experiment: Monoterpenes emitted from the maritime pine. *J. Geophys. Res. Atmos.* **1994**, *99*, 16501–16510. [CrossRef]
40. Blanch, J.S.; Peñuelas, J.; Sardans, J.; Llusià, J. Drought, warming and soil fertilization effects on leaf volatile terpene concentrations in *Pinus halepensis* and *Quercus ilex*. *Acta Physiol. Plant.* **2009**, *31*, 207–218. [CrossRef]
41. Lee, D.H.; Kim, M.H.; Park, O.H.; Park, K.S.; An, S.S.; Seo, H.J.; Jin, S.H.; Jeong, W.S.; Kang, Y.J.; An, K.W.; et al. A study on the distribution characteristics of terpene at the main trails of Mt. Mudeung. *J. Environ. Health Sci.* **2013**, *39*, 211–222. [CrossRef]

42. Zas, R.; Quiroga, R.; Touza, R.; Vázquez-González, C.; Sampedro, L.; Lema, M. Resin tapping potential of atlantic maritime pine forests depends on tree age and timing of tapping. *Ind. Crops Prod.* **2020**, *157*, 112940. [CrossRef]
43. Bach, A.; Yáñez-Serrano, A.M.; Llusià, J.; Filella, I.; Maneja, R.; Penuelas, J. Human breathable air in a mediterranean forest: Characterization of monoterpene concentrations under the canopy. *Int. J. Environ. Res. Public Health* **2020**, *17*, 4391. [CrossRef]
44. Staudt, M.; Bertin, N.; Hansen, U.; Seufert, G.; Ciccioli, P.; Foster, P.; Frenzel, B.; Fugit, J.L. Seasonal and diurnal patterns of monoterpene emissions from *Pinus pinea* (L.) under field conditions. *Atmos. Environ.* **1997**, *31*, 145–156. [CrossRef]
45. Staudt, M.; Bertin, N.; Frenzel, B.; Seufert, G. Seasonal variation in amount and composition of monoterpenes emitted by young *Pinus pinea* trees–implications for emission modeling. *J. Atmos. Chem.* **2000**, *35*, 77–99. [CrossRef]
46. Loreto, F.; Forster, A.; Durr, M.; Csiky, O.; Seufert, G. On the monoterpene emission under heat stress and on the increased thermotolerance of leaves of *Quercus ilex* L. fumigated with selected monoterpenes. *Plant Cell Environ.* **1998**, *21*, 101–107. [CrossRef]
47. Loreto, F.; Pinelli, P.; Manes, F.; Kollist, H. Impact of ozone on monoterpene emissions and evidence for an isoprene-like antioxidant action of monoterpenes emitted by *Quercus ilex* leaves. *Tree Physiol.* **2004**, *24*, 361–367. [CrossRef]
48. Coley, P.D.; Bryant, J.P.; Chapin, F.S. Resource availability and plant antiherbivore defense. *Science* **1985**, *230*, 895–899. [CrossRef]
49. Simms, E.L.; Rausher, M.D. Costs and benefits of plant resistance to herbivory. *Am. Nat.* **1987**, *130*, 570–581. [CrossRef]
50. Herms, D.A.; Mattson, W.J. The dilemma of plants: To grow or defend. *Q. Rev. Biol.* **1992**, *67*, 283–335. [CrossRef]
51. Lerdau, M.; Matson, P.; Fall, R.; Monson, R. Ecological controls over monoterpene emissions from douglas-fir (*Pseudotsuga menziesii*). *Ecology* **1995**, *76*, 2640–2647. [CrossRef]
52. Burney, O.T.; Davis, A.S.; Jacobs, D.F. Phenology of foliar and volatile terpenoid production for *Thuja plicata* families under differential nutrient availability. *Environ. Exp. Bot.* **2012**, *77*, 44–52. [CrossRef]
53. Breckle, S.W. *Walter's Vegetation of the Earth. The Ecological Systems of the Geo-Biosphere*, 4th ed.; Springer: Berlin/Heidelberg, Germany, 2002; p. 527.
54. Turtola, S.; Manninen, A.M.; Rikala, R.; Kainulainen, P. Drought stress alters the concentration of wood terpenoids in Scots pine and Norway spruce seedlings. *J. Chem. Ecol.* **2003**, *29*, 1981–1995. [CrossRef]
55. Sancho-Knapik, D.; Sanz, M.Á.; Peguero-Pina, J.J.; Niinemets, Ü.; Gil-Pelegrín, E. Changes of secondary metabolites in *Pinus sylvestris* L. Needles under increasing soil water deficit. *Ann. For. Sci.* **2017**, *74*, 24. [CrossRef]
56. Llusià, J.; Peñuelas, J. Changes in terpene content and emission in potted Mediterranean woody plants under severe drought. *Can. J. Bot.* **1998**, *76*, 1366–1373. [CrossRef]
57. Niinemets, Ü.; Reichstein, M.; Staudt, M.; Seufert, G.; Tenhunen, J.D. Stomatal constraints may affect emission of oxygenated monoterpenoids from the foliage of *Pinus pinea*. *Plant Physiol.* **2002**, *130*, 1371–1385. [CrossRef] [PubMed]
58. Copolovici, L.; Niinemets, U. Environmental impacts on plant volatile emission. In *Deciphering Chemical Language of Plant Communication*; Blande, J.D., Glinwood, R., Eds.; Springer International Publishing: Cham, Switzerland, 2016; pp. 175–210.
59. Hodges, J.D.; Lorio, P.L. Moisture stress and composition of xylem oleoresin in loblolly pine. *For. Sci.* **1975**, *21*, 283–290. [CrossRef]
60. Wainhouse, D. Hylobius abietis—Host utilisation and resistance. In *Bark and Wood Boring Insects in Living Trees in Europe, a Synthesis*; Lieutier, F., Day, K.R., Battisti, A., Gregoire, J.C., Evans, H.F., Eds.; Springer: Dordrecht, The Netherlands, 2004; pp. 365–379.
61. Amelung, B.; Viner, D. Mediterranean tourism: Exploring the future with the Tourism Climatic Index. *J. Sustain. Tour.* **2006**, *14*, 349–366. [CrossRef]
62. Bafaluy, D.; Amengual, A.; Romero, R.; Homar, V. Present and future climate resources for various types of tourism in the Bay of Palma, Spain. *Reg. Environ. Chang.* **2014**, *14*, 1995–2006. [CrossRef]
63. Paudyal, R.; Stein, T.V.; Kc, B.; Adams, D.C. Effects of weather factors on recreation participation in a humid subtropical region. *Int. J. Biometeorol.* **2019**, *63*, 1025–1038. [CrossRef]
64. Wilkins, E.J.; Chikamoto, Y.; Miller, A.B.; Smith, J.W. Climate change and the demand for recreational ecosystem services on public lands in the continental United States. *Glob. Environ. Chang.* **2021**, *70*, 102365. [CrossRef]
65. Buckley, L.B.; Foushee, M.S. Footprints of climate change in US national park visitation. *Int. J. Biometeorol.* **2012**, *56*, 1173–1177. [CrossRef]
66. Schiller, G. Biometeorology and recreation in east Mediterranean forests. *Landsc. Urban Plan.* **2001**, *57*, 1–12. [CrossRef]
67. Takayama, N.; Kagawa, T.; Kasetani, T.; Park, B.; Tsunetsugu, Y.; Oishi, Y.; Hirano, H.; Miyazaki, Y. The comfortableness of the light/thermal environment for bathing in the forest atmosphere. *J. Jpn. Inst. Landsc. Arch.* **2005**, *68*, 819–824. [CrossRef]
68. Richardson, M.; McEwan, K.; Maratos, F.; Sheffield, D. Joy and calm: How an evolutionary functional model of affect regulation informs positive emotions in nature. *Evol. Psychol. Sci.* **2016**, *2*, 308–320. [CrossRef]
69. Farrow, M.R.; Washburn, K. A review of field experiments on the effect of forest bathing on anxiety and heart rate variability. *Glob. Adv. Health Med.* **2019**, *8*, 1–7. [CrossRef]
70. Kotera, Y.; Richardson, M.; Sheffield, D. Effects of Shinrin-yoku (forest bathing) and nature therapy on mental health: A systematic review and meta-analysis. *Int. J. Ment. Health Addict.* **2022**, *20*, 337–361. [CrossRef]

forests

MDPI

Article

Getting Flow: The Place of Production Forests in the Rise of Mountain Biking

Karen M. Bayne [1,*], Matthew B. Scott [1] and Richard T. Yao [2]

[1] New Zealand Forest Research Institute Ltd. (Scion), 10 Kyle Street, Christchurch 8011, New Zealand
[2] New Zealand Forest Research Institute Ltd. (Scion), Tītokorangi Drive, Private Bag 3020, Rotorua 3046, New Zealand
* Correspondence: karen.bayne@scionresearch.com; Tel.: +64-3-363-0911

Abstract: Mountain biking has increased in popularity in many countries over recent decades. Production forests provide not only an outstanding space for mountain bike riding, but also specific characteristics conducive to purpose-built track building. With recreational access to production forests, managers must balance commercial operations and industry environmental obligations with the interests and actions of riders. Production forests in New Zealand have accommodated mountain biking activities for decades. The trend from 1991 to 2022 showed overall increasing opportunities for mountain biking in production forests, particularly in smaller commercial peri-urban plantations and forest parks maintained as dedicated mountain bike parks. Over the same period, public mountain biking access to larger (>1500 ha) forests peaked in 2008. These recreational changes within forests, and the impact these changes have pressed onto forest managers, have not been well documented. This paper explores the rise in popularity of mountain biking in New Zealand's production plantation forests, and the response of forest owners and managers to increased mountain biking activities in their commercial forest estates. The paper discusses implications for forest planning to accommodate active recreational sports such as mountain biking in production forests, suggesting policies and procedures to help protect commercial interests and forest ecology, while allowing for a contemporary mix of recreational activities.

Keywords: mountain biking; recreation; trade-off; use conflict; ecosystem services; forest management; commercial forestry; plantation forest; *Pinus radiata*; New Zealand

Citation: Bayne, K.M.; Scott, M.B.; Yao, R.T. Getting Flow: The Place of Production Forests in the Rise of Mountain Biking. *Forests* **2022**, *13*, 1326. https://doi.org/10.3390/f13081326

Academic Editors: Radu-Daniel Pintilii and Diego Varga

Received: 27 July 2022
Accepted: 16 August 2022
Published: 19 August 2022

1. Introduction

Mountain biking is an increasingly popular form of outdoor recreation, competitive sport and adventure tourism. Since its experimental beginnings in the U.S. in the late 1970s [1–5], the fringe activity has steadily evolved into an established sport, occupying an important public interest and economic niche worldwide [6–9]. In recent decades, the popularity of mountain biking has increased in Australia [9–13], New Zealand [14–20], North America [21–24] and Europe [5,25–28]. Participation in the U.S. and Australia is estimated at 2.5% and 1.3%, respectively [9]; in the UK, an estimated 11.8 million people own a mountain bike with 1.3 million people (or around 2.1% of the population) using them regularly to ride off-road [29]. Mountain bike organisations, clubs [30] and competitions [31] are increasingly widespread internationally and the International Mountain Biking Association (IMBA), a body dedicated to growing mountain bike trail communities, is represented by more than 40 countries [7]. Economically, mountain biking contributed around USD 7 billion to the global market in employment, sales in retail, tourism, and competitions in 2020 [32–36]—just in terms of sales, a staggering 47,670 bikes are being sold each day [8]. Annual growth rates are rising and are expected to continue to rise with further advances in electronic technologies, expanding markets, and growing track networks [7,8,14,32,37–39].

With the expanding interest and improved technologies [40,41], mountain biking has developed to encompass several specialised disciplines, including cross-country, trail, enduro, freeride/downhill, all mountain, slopestyle, 4x, pump track and randonneuring/audax [42]. Mountain biking has been incorporated in multi-discipline races such as adventure racing, mountain bike orienteering and off-road multi-sport events. The sport has also expanded widely into adventure tourism with recreational riders going on multi-day adventures (bikepacking) [43] or touring combinations of cycle trails, forestry tracks and re-purposed disused railways (cycle touring) [44]. There is also a growing number of electric mountain bikes (e-bikes), particularly among riders who might otherwise be physically excluded [7,19,37,45,46]. With multiple riding styles and interest groups, a wide demographic with disparate motivations, vastly different track design requirements and wide-ranging environmental preferences has emerged [47–49].

Most mountain bikers—across disciplines of mountain biking—prefer riding on narrow trails (known as single track) in natural areas [47,50,51]; however, publicly owned natural land in many countries is tightly managed [52–54]. In many places, mountain biking has only recently been included within recreation permitted on public lands [40,41,54–58], largely as a consequence of persisting notions that mountain bikers cause disproportionate environmental damage [4,41,59–63] and promote use conflict from existing users [49,64]. The emergence of mountain biking—among a class of thrill seeking and adventurous outdoor sports—has changed local regional economies [14,16,23,32,40,41,65–69] and the way many natural areas are viewed and used for outdoor pursuit [13,24,25,41,47,70,71]. Mountain bikers are motivated by an appreciation of nature and the natural environment, as well as for the opportunities the environment provides for exercise and fun [25,71,72]. Mountain bikers show similar knowledge and perception of forest biodiversity to other forest users [25], and express commitment to the sustainability of the natural environment [47,71]. However, mountain biking and mountain bike trail building is accompanied by variable levels of concomitant modifications to the environment [13,41,62,63] that some see as a step level change [4,41,49,60,61,64]. Indeed, for many mountain bikers, enjoying the surrounding nature for the nature itself is a secondary interest [13,41,73]. With differences of view, challenging trade-off situations persist [47,49,74].

Production forests (including plantation forests) have emerged as an important alternative space for recreational mountain biking [50,63,72,74,75], particularly where access to public land has been difficult to attain or prohibited [63,76–78]. Notwithstanding, production forests provide outstanding environments with distinctive qualities conducive to track development, riding and competitions. For example, in production forests, a higher tolerance of environmental damage in comparison to natural forests enables more scope for purpose-built track building [41] and fewer environmental concerns for riders [60,61,79,80]. Mountain biking in production forests and the associated track development are seen to pose less of a risk to the ecosystem values, such as native biodiversity, habitat quality and water quality, than biking in natural forests [41]. Additionally, production forests typically have an already established network of roads and tracks to allow for management and harvesting that offers ready-built access across the forest. The importance of commercial production forests to the mountain bike community and the development of the sport raises the questions of how and why access to production forests progressed, and what deciding factors led forest managers to allow or promote mountain biking access. Though some forest managers monitor use of their forests (through permit data, etc.), there is a paucity of collated data available on mountain biking as a recreational activity within New Zealand's commercial production forests—(i) what cycling activities are being undertaken, (ii) by whom, and more importantly, (iii) why users are seeking access to recreate within plantation forests. While past studies have investigated the sport of mountain biking from a user preference perspective [2,41,49,71], as a social phenomenon [48] and the environmental consequences [4,41,60,79], very few have investigated the use of a commercially managed forest for mountain biking from a forest management perspective. Where this has been considered, the focus has been on either use conflict management between forest recreational users [49,75,81], or on the impacts of the activity on the physical forest environment [4,25,79,82,83], (although, see Hrůza et al. 2021 [63]).

This paper considers the rationale and motivations behind mountain bike use of productive plantation forest environments, the response of forest management to societal recreational preferences, and the benefits and risks posed to forest managers with the introduction of mountain biking to their commercial estates. Using New Zealand's experience as a case study, we examine the growth in the sport of mountain biking, and implications for commercial forest managers due to increasing pressures to accommodate recreational mountain biking access and infrastructure within commercially established forestry estates.

2. Background

2.1. Mountain biking in New Zealand

Mountain biking in New Zealand began in 1984 with the importation of 15 off-road "mountain" bikes [15,77]. In these early years, a small cohort of hardy enthusiasts quickly re-defined bicycle touring in New Zealand [84,85], exploring (and reporting on) rural metalled roads, farms, bush tracks and forestry roads. These early riders were also among the first to start to build and improve tracks specifically for improving mountain bike access and rideability; while many of the earliest tracks were unsanctioned, an increasing number of tracks were being developed as a result of negotiations and partnerships between the fledgling mountain bike community and forest managers, both public and commercial [84].

As popularity of recreational mountain biking grew, so did competitive mountain biking. The first mountain bike races were held in 1985 [86]. The first national championship, "New Zealand Off Road Bicycle Race," was arranged by Paul Kennett in 1986 on a rugged course through native forests now known as the "Karapoti Classic" [59,77,86]. With just 45 competitors at the start line, the "championship" could easily have been underestimated with competitors sporting bush shirts and canvas backpacks and riding modified road bikes; however, the following year there was such interest in the competition that the field had to be limited to 1000 competitors [76]. By 1993, popularity had swollen such that New Zealand had laid claim to recreational forest with the "highest use" in the Southern Hemisphere at Woodhill Forest [76]. From latest national participation data (2018), an estimated 7.7% of the adult New Zealand population (>15 yrs) were regularly mountain biking—nearly 300,000 persons—incorporating both competitive level cyclists and recreational riders [14], 14% having participated at least once in the sport during 2018, and 8% wanting to give mountain biking a go [87].

2.2. Production Forests for Mountain Biking Trail Development

In the late 19th century, New Zealand was facing a dwindling supply of timber resources as a result of unsustainable harvest practices and slow growth rates in the merchantable New Zealand indigenous tree flora [88,89]. In response, the New Zealand Forest Service in 1925 led a programme to plant fast-growing, exotic, chiefly radiata pine (*Pinus radiata* D. Don), forests across 300,000 acres of State forest [90]. The success of the afforestation led, over the next decade, to increasingly large-scale exotic plantings across state and private forests, particularly on the North Island [90].

Although primarily managed for timber, planted exotic State forests were important for public recreation throughout the 20th century. By the 1970s, exotic State forests had become popular areas for recreation, as these were more accessible by motor vehicle than many indigenous National Parks and reserves [91–93]. With the closure of the New Zealand Forest Service in 1987, exotic production forest management was reverted to State-run Forestry Corporation, and then on-sold to private companies [89]. With the privatisation of the exotic forest estate, commercial timber extraction became of utmost priority while recreational right to access was left to individual forest companies [89,91,94]. In contrast, indigenous forests had a range of management objectives, both continued selective harvesting as well as the more predominant biodiversity protection and conservation objectives [89,91], including non-vehicular recreation. In the post-privatisation era, most exotic planted forest management objectives included some forms of recreation [90]. In 2018, a concerned group of forestry professionals drafted a national forest policy to guide decisions on legislation and regulation across the sector [95]; in it, the authors acknowledge the importance of New Zealand's productive forests in providing non-productive public services, such as recreation.

Currently, of the 38% of New Zealand land covered in forests (10.1×106 ha), nearly 16.8% of these forests (1.8×106 ha) are occupied by productive exotic, primarily *Pinus radiata* D. Don, plantation forest [88,96]. Most of these planted forests are in the Central North Island region, although exotic planted forests extend throughout the country. In the drier regions, productive pine forests usually support a sparse understorey, whereas in wetter regions, a dense understorey of native ferns and shrubs, as well as exotic weeds such as blackberry (*Rubus fruticosus* agg.) and gorse (*Ulex europaeus*) persist (Figure 1). Production plantation forests provide important surrogate habitats for the conservation of threatened New Zealand biodiversity, including bats, beetles, kārearea (falcon) and kiwi [97–101].

Figure 1. Mountain bike single track in planted production forest in Rotorua, New Zealand.

Despite having expansive natural landscapes, New Zealand does not have the same public accessibility to private land as other nations (e.g., Scotland [73] and Sweden [102]). As mountain biking developed in New Zealand, the Department of Conservation (DOC), the government department managing public conservation lands, considered mountain biking to be in conflict with protecting the interests of conservation, biodiversity and other recreational users [2,4,40,41,77,92]. As with many other national land management agencies (e.g., U.S. National Park Service), DOC did not allow "road" development in National Parks, which included mountain bike tracks—mountain bikes being classed as a vehicle. Mountain bikes throughout the 1990s were legally excluded from most public tracks and restricted to formed roads in public conservation land [2,51]. Consequently, recreational use of public conservation land, which accounts for over 30% of NZ's land area and 79% of all forests [88,103], was largely devoid of opportunities for mountain biking [2]. In addition, most local governments held similar views on mountain bikes, disallowing access to publicly managed exotic production forests until the late 1990s [78,104]. Electronic bikes (ebikes) are currently at the centre of a similar debate on policy around access [105,106].

Access limitations to public conservation land led enthusiasts in the 1980s and 1990s to look elsewhere to ride, build tracks and hold competitions. Exotic production (mainly pine) forests provided a range of conditions appealing to mountain biking and conducive to the development of purpose-built mountain bike track networks. Production forests typically had an established road and track network to connect regions of the forest; forest managers were tolerant of understorey vegetation clearance and earthwork required for track building; and many forest managers were already dealing with access for other recreational user groups (e.g., hunters and horse riders), so some existing offroad trails allowed for further development [76–78]. Some New Zealand production forests were established in peri-urban areas that offered easy access from population centres, whereas other forests were remote and completely undeveloped in terms of purpose-built track networks and with ideal opportunities for experiencing solitude and interacting with nature. Many planted forests were established on steep hill country or in unique landforms (e.g., historic mine sluicing), conducive to building interesting and challenging single-tracks [13,41,51,107].

Enthusiasts partnered with forest companies and local councils to build dedicated track networks within productive planted forest from the early 1990s [108]. The initial forested areas which became important recreational hubs for biking were located near larger population centres—these forests were some of the first to give access and permit track development. The proximity of some production forests in New Zealand to population centres (e.g., Bottle Lake; Hanmer Forest; Golden Downs Forest; Whakarewarewa Forest, and Flagstaff Forest, Figure 2) encouraged the development of track networks within planted production forests [76–78,109].

Figure 2. Production plantation forest areas in New Zealand.

Bottle Lake Forest, on the edge of New Zealand's second largest city, received over 200,000 visits per year from mountain bikers in 2002 [86] accounting for half of all visitors to the forest [3]. As the sport grew, forest managers were under increasing pressure to either restrict mountain bike rider forest access, or to accept the inevitable and address illegal track development and access through community engagement and recreational use policy. These forest managers shouldered the responsibility to safely manage an increasing population of people recreating in the forest, while continuing production operations.

Planted forests are used extensively for other social and recreational activities (hunting, angling access, firewood collection) [110] though these more traditional activities have neither experienced the same growth as mountain biking, nor involved such a large cross section of society participating.

2.3. Contribution of Mountain Biking to the NZ Economy and Society

The primary purpose of New Zealand's 1.8 million hectare planted forests is the extraction of wood products that contribute approximately NZD 10 billion in revenues annually to the New Zealand economy [111–113]. In addition to the above revenues, production forests provide revenue streams and non-market values from the space and infrastructure they provide recreation such as mountain biking [14,72,114]. Increasingly, tourists come to New Zealand to mountain bike [2,15]. In 1995, 22,000 international tourists out of 1.4 million mountain biked during their visit to New Zealand (1.6%), and a further 177,000 New Zealanders mountain biked in landscapes out of their local region [109]. By 2018, 25,000 (4%) of the annual international visitors indicated their prime reason for visiting was to mountain bike [14].

Economic impact studies show mountain biking in production plantation forests contributed to regional economies by generating revenues in retail and hospitality [14,16,18,68]. Mountain biking visits and events across the Nelson-Marlborough region (which include planted forests) contributed approximately NZD 17 million to the economy [14]. Mountain biking activities in the 5700 ha Whakarewarewa Forest in Rotorua have been found to annually provide between NZD 29 and NZD 47 million to the economy, as well as additional employment opportunities of between 210 and 340 full time employment units [16]. The collective value that repeat forest users (i.e., local users) placed on mountain biking access to the forest in 2010 was approximately NZD 5 million per annum, which was greater than the annual timber revenue at the time of the study [115]. Commercial forest owners also benefit from promoting recreational activities, such as mountain biking, as part of achieving industry compliance for international market access [112].

2.4. Mountain Bike Parks

During the late 2000s, dedicated mountain bike parks emerged in New Zealand (e.g., Woodhill Mountain Bike Park; Cougar Mountain Bike Park; Eskdale Mountain Bike Park). Mountain bike parks are dedicated trail centres, mainly in production forest areas (Table 1) set aside and operated as separate enterprises apart from the forest management company operations [116]; often, users pay fees for services although many are free to the public or rely on donations. Their popularity and contribution to increasing mountain bike participation is ascribed to the accessibility to urban centres or popular holiday destinations, the high density of highly engineered, serviced tracks across a range of difficulties, and for some mountain bike parks, the uphill services, such as gondolas, and purpose-built facilities and trail markers [48,109]. The inclusion of beginner level trails for young riders sees mountain bike parks considered safer and less "wild" locations to learn the sport [117]. A 10-fold increase in mountain biking activity occurred in Tangoio forest from the introduction of the Eskdale Mountain Bike Park, from 200 visits in 2001 to 2100 visits in 2009 [109]. Providing gondola lift access during 2011 to the Ben Lomond area also saw mountain bike club numbers in Queenstown grow from 60 in 2010 to 600 in 2011 [118]. Local mountain biking clubs that ride in nearby commercial planted forests have grown in membership, such as Hawkes' Bay Mountain Biking Club, which has 2200 members [1], and the Nelson Mountain Biking Club, which has the largest membership, at over 2500 members [14].

Table 1. Mountain bike parks in production plantation forests (Source: trailforks.com (accessed on 28 June 2022)).

Riding Area	Fees	Track Network (km)	Adjacent Production Forest
Whakarewarewa Redwoods, Rotorua	none	319	Whakarewarewa Forest
Riverhead, Auckland	none	222	Riverhead Forest
Pan Pac Eskdale, Hawkes Bay	use fee or club fee	113	Tangoio Forest
Craters of the Moon, Taupo	use fee or club fee	77	Waiarekei Forest
Woodhill, Auckland	use fee	63	Woodhill Forest
Hanmer Forest Park, Hanmer Springs	none	60	Hanmer Forest
Christchurch Adventure Park, Christchurch	gondola fee	56	McVicar Cashmere Estate
Lake Mangamahoe, Taranaki	donations	51	Mangamahoe Forest
Mcleans Island, Christchurch	none	50	West Melton Forest
Whangamata, Whangamata	use fee	48	Tairua/Matariki Forest
Nasby, Central Otago	none	46	Nasby Forest
Cougar Park, Tokoroa	donations	44	Kinlieth Forest
Bottle Lake, Christchurch	none	43	Bottle Lake Forest
Waitangi, Paihia	donations	40	Waitangi Forest
Arapuke, Manawatu	donations	39	Arapuke Forest
Harakake, Whanganui	donations	30	Harakeke Forest
FourForty, Clevedon	use fee	29	Waytemore Forest
Kingsland Forest, Nelson	donations	27	Silvan Forest
Parihaka, Whangarei	none	25	Onerahi Forest
Ben Lomond, Queenstown	gondola fee	24	Ben Lomond Hills
TECT Park, Bay of Plenty	none	23	OTPP Western Bay
Haven, Banks Peninsula	club fee	22	McQueens Forest
Onepu, Rotoma	none	18	Rotoma Forest
Uenuku Pines, National Park	club fee	17	Waikune Forest
Whitehorse, Waimate	donations	15	Waimate Forest
Tōtara Park, Upper Hutt	none	14	Akatarewa Forest
Millmore, Gisborne	none	10	Millmore Forest
Raincliff, Geraldine	none	8	Raincliff Forest
Blue Spur, Hokitika	none	6	Ngai Tahu Kaniere Forest Estate
Whitehills, Kerikeri	none	3.5	Whitehills Forest
Raumai Forest, Bulls	none	3	Santoft Forest
Wither Hills, Blenheim	none	77	
Mt Hutt, Canterbury	none	48	
Sticky Forest, Wanaka	none	34	
Te Miro, Piako-Morrinsville	none	29	
Parihaka, Whangarei	none	24	
Centennial Park, Timaru	none	23	
Sandy Point, Invercargill	none	23	
Codgers, Nelson	donations	21	
Summerhill Farm, Te Puke	donations	19	
Rivenrock, Wairarapa	use fee	16	
Oropi Grove, Tauranga	none	14	
Carterton, Wairarapa	none	8	
Hotoritori, Thames	none	8	
Whataupoko, Gisborne	none	7	
Wairoa, Hawke's Bay	none	7	

3. Materials and Methods

3.1. Determining Where Mountain Biking Has Been Present

Using published literature and online sources, we looked at production plantation forests in New Zealand (>1500 ha), aiming to answer the following questions: (1) How many production forests in New Zealand (>1500 ha) permit mountain biking? (2) How long has mountain biking been permitted? and (3) What are the policies for each forest relating to mountain biking? Using published mountain bike guides, brochures and mountain biking websites [76,77,86,108,116,119–124], we created a database of commercial forests to map the expansion of recreational and competitive mountain biking access in commercial plantation forestry land for 1991–2022. We created a time-series map of all operating commercial forest estates (both >1500 ha and <1500 ha) in New Zealand with access for mountain biking for the same time period using a forest industry database [125–127] and topographical maps series NZ TopoMap [128]. Statistical analysis using SPSS was conducted on yearly access based on the two size classes (>1500 ha and <1500 ha) via Chi Square. We also collated recorded transitions in forest management policies regarding mountain bike access, permission requirements, and user costs outlined within forest industry websites and annual management plans.

3.2. Assessing Impacts of Mountain Biking on Forest Management

To qualify the impacts to forest management, we interviewed four forest managers and surveyed a further 10 of the 22 forestry management companies with >1500 ha of forest. The aim of the survey questionnaire was to identify risks and benefits to forest management from the presence of mountain biking within their forest estate, and changes in current forest management practices because of these risks and benefits. Survey forms were sent by email directly to all 22 forest managers, inviting them to complete the survey and return the form via email. Responses were collated and open-ended questions were analysed thematically. The questions are provided in Supplementary Data (S1).

4. Results

4.1. Changes in Mountain Biking Access to Planted Forests, 1991–2022

In 1991, 27 commercial forest owners in New Zealand provided recreational access for mountain biking activity (Table 2); however, there were also limited numbers of mountain bikers using the forests.

Table 2. Number of planted forest areas being accessed for mountain biking activity, 1991–2022.

		1991	1993	1999	2002	2008	2013	2022
Number of commercial plantation forests with mountain biking	>1500 ha	18	49	34	36	56	38	40
	<1500 ha	9	15	18	21	35	33	75
Total number of forests		27	64	52	57	91	71	115
No of above forests requiring permits or landowner permission		11	25	19	26	31	32	24
No of above forests requiring payment for access		3	1	6	10	7	20	21

Kennett et al. [76,77] encouraged riders to explore (exotic) forests, advocating them as an ideal environment for mountain biking. Routes described in Kennett's guide from both 1991 and 1993 largely follow 4WD and forestry roads, rather than developed single tracks. We found that in these years, only five of the large forest owners had any formal policy concerning access to the estate for recreational purpose, let alone for mountain biking, mostly through a permit system. In contrast, we found access to small forests (<1500 ha) was often regulated, with 67% requiring a permit and 33% with fees. One forest owner even actively excluded mountain bikers on their production forest trails [78]. By 1993, 49 large (>1500 ha) forests were allowing access for mountain biking [76], an almost three-fold increase from 1991 (Table 2). Access to many of these forests was informal. For example, permission for the construction of one of the first bike-specific tracks—originally known as "the BMX track"—in New Zealand was based only on a verbal agreement between the owner, Forestry Corporation, and riding enthusiast Fred Christensen [86]. In 2022, 115 forests allowed mountain bike entry—more than any other year recorded—buoyed by the steadily increasing number of small (<1500 ha) forests. The number of larger (>1500 ha) forests over the same twenty-year period fluctuated, peaking at 56 in 2008 (Table 2). The main trend in access to forests shows that while in 1991, no significant difference was shown between forest access between forest size (X^2 (1, N = 225) = 0.258, p < 0.389), a significantly larger proportion of forests >1500 ha had mountain biking occurring in 1993 than in small forests (X^2 (1, N = 225) = 7.825, p = 0.004). However, by 2022, this trend has reversed, with significantly more small forests <1500 ha now having access (X^2 (1, N = 225) = 73.35, p < 0.000). The extent of mountain biking in exotic forests during the period 1991–2022 is shown in Figure 3 and in supplementary Tables S2 and S3. We acknowledge some inconsistencies in access regulations in the web and grey literature, so all results are presented based on what was judged to be the most reliable sources.

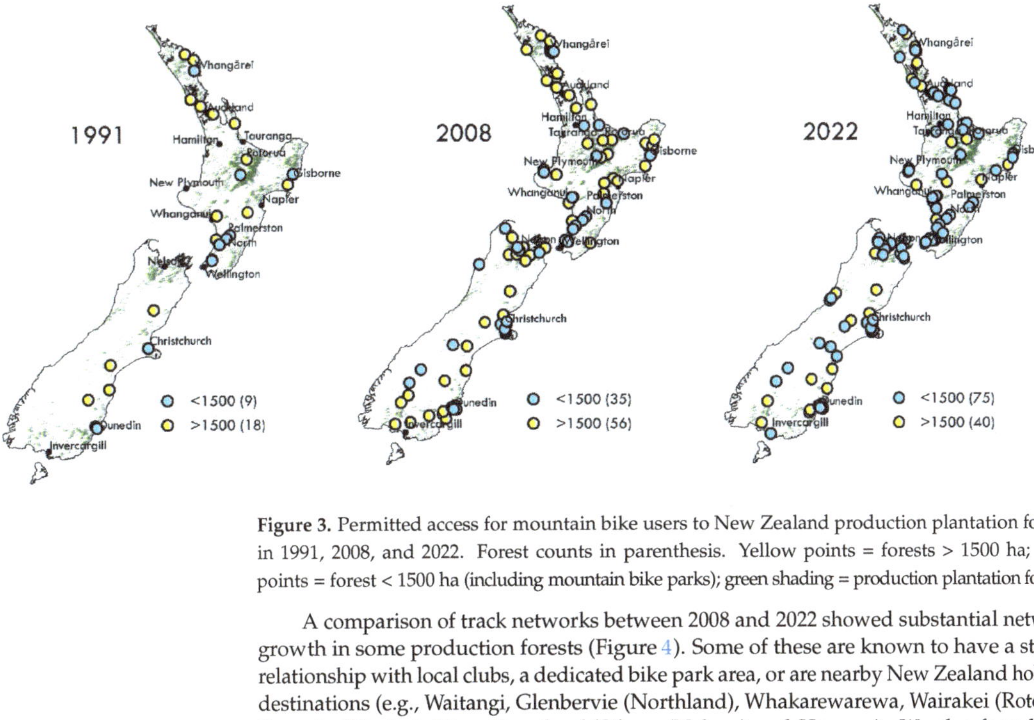

Figure 3. Permitted access for mountain bike users to New Zealand production plantation forests in 1991, 2008, and 2022. Forest counts in parenthesis. Yellow points = forests > 1500 ha; cyan points = forest < 1500 ha (including mountain bike parks); green shading = production plantation forest.

A comparison of track networks between 2008 and 2022 showed substantial network growth in some production forests (Figure 4). Some of these are known to have a strong relationship with local clubs, a dedicated bike park area, or are nearby New Zealand holiday destinations (e.g., Waitangi, Glenbervie (Northland), Whakarewarewa, Wairakei (Rotorua) Tangoio (Napier), Hira, Kingsland/Silvan (Nelson) and Hanmer). We also found that some tracks reported in earlier years by Kennett et al. [108,119,121] do not appear in more recent resources [116,124]. We hypothesise that this is due to the growing development of high-quality mountain bike tracks nationwide, concentrating biking activity on higher-quality tracks. Forests that have purpose-built concentrated track networks have potentially redirected riders from other parts of the forest, or other forests in the region. Examples that seem to support this include the decreased use of Nemona Forest with the development of Blue Spur MTB Park; Rotoma Forest use decreased with the development of Onepu MTB park; and West Dome, Blue Mountains and Hokonui Forest tracks which are no longer listed in Trailforks [124], whereas Bluff and Sandy Point trails have expanded and are regularly referenced in Trailforks in 2022.

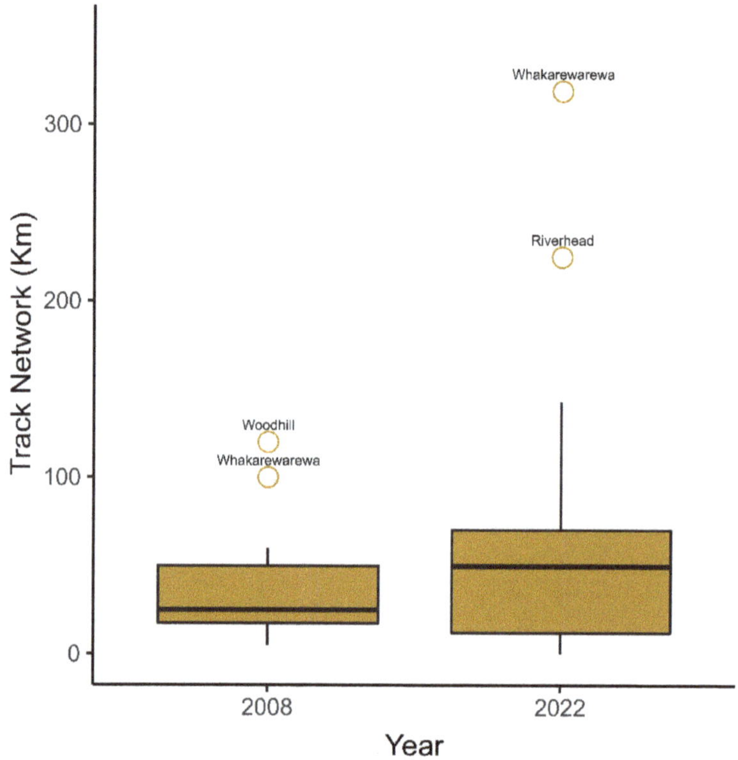

Figure 4. Boxplot showing change in mountain bike accessible track networks in production plantation forests (>1500 ha) between 2008 and 2022. The box bounds the interquartile range (IQR) divided by the median; whiskers extend to a maximum of 1.5 × IQR beyond the box. Open circles are outlier sample data points. Figure based on 23 forests with >5 km track access in 2008, resampled in 2022.

The rise in dedicated mountain bike parks is attributed to the need for greater access to a few, dedicated trails for biking, separated from other recreational users, and for the protection elsewhere of sensitive vegetation. Mountain bike parks concentrate mountain bikers' usage within a defined access area to ensure safe forest operations in adjacent forest areas. Of dedicated mountain bike parks in production forests, nearly 50% were a subset of a larger plantation estate (Table 1).

4.2. Access Conditions for Mountain Biking in Planted Production Forests

From our review of the literature, our survey and online resources, we found that production forest managers have taken four approaches to access: (1) no access permitted, or only by permit for club/private events on a case-by-case basis; (2) only providing access at certain times (e.g., weekends) and require a permit or pass; (3) providing generally open access to mountain bikers, with an agreement with a local club or council of the protocols and procedures for the construction and maintenance of purpose built mountain bike tracks; (4) operating as in (3) but restricting open-access mountain biking and related track building to a portion of the production forest (e.g., mountain bike parks), and often outsourcing recreational management aspects to clubs or partner with a recreation provider.

During the 1990s, mountain bike access to commercial plantation forests was usually without a permit, largely free, but was often restricted to weekends [76–78,120]. We found that relationships between forest owners and cycling enthusiasts or mountain biking clubs helped develop long-term continued access for mountain biking; whereas forest managers in absence of these relationships made access restrictions or destroyed tracks during felling operations to discourage further development [76,77,116,120,121]. Mountain biking enthusiasts worked together with local forest landowners for access and track building, lobbied local authorities, and were willing to put in their own labour to establish infrastructure and develop the sport [20,76,85]. Where strong relationships have been established over time between forest owners and local mountain biking clubs, the access conditions are clear, and they tend to result in an increased level of mountain biking recreation within the forest. For example, Whakarewarewa Forest, situated on the edge of Rotorua, a popular tourist destination, has had a longstanding engagement with the local mountain biking community. The forest management and club agreed, initially as an informal agreement, but later formalised through a "memorandum of understanding" that all tracks should be built in areas away from imminent harvest operations, and that all tracks would be documented, mapped and safely sign posted [85]. A formal track network group was set up from members of Kaingaroa Timberlands forest management, the local mountain bike club, and other interested individuals, to oversee track building. In 2004, Kaingaroa Timberlands further formalised access into the forest with an official document circulated through mountain bike websites and the Redwoods Visitor Centre. This set out the necessity for track mapping and marking and ensured that track construction and maintenance was carried out by sanctioned people under the guidance of the track network group [85]. This forged the way for further track building, and club membership leapt to around 300 members. We noted similar established relationships in the local biking club websites, forest management plans and permits relating to Akatarawa, Hanmer, Lismore, Belmont and Battle Hill, and Tangoio Forests.

4.2.1. Permit Requirements

Our database showed that, while in 1991 around 40% of production forest owners had required permits for mountain biking access, by 2002 this had risen to 46%. Fees have a strong association with permitting; by 2022, 88% of exotic forests requiring a permit to access (24) also required payment (sometimes via club fee registrations). In many cases, the hassles gaining a necessary permit were inefficient and complicated [78], paradoxically leading to uncontrolled entries and clandestine behaviour [76,78,86]. Of the large (>1500 ha) forests we surveyed with mountain biking activities, today 43% of forest companies provide access free of charge, though 62% have mountain bike club affiliations in the provision of access conditions, and trail maintenance. Forty-six percent require permits for access, most through local clubs, though one forest manager provided permits though a local bike shop, and another via a contractor. Fifty-eight percent of the small forests (<1500 ha) are managed by local government. Requirements for a permit to ride in the forest, along with fees for riding, have changed over time, due to changes in forest ownership; affiliations with local clubs; establishment of dedicated mountain bike parks; and forest security requirements. Some forest managers opted to recoup costs through selling maps of the forest trails at local cafes for a nominal fee rather than through permits [78]. Ironically, clandestine behaviour, illegal track building and lobbying worked in some ways to gain long term dedicated access for mountain bikers [129–131], although this behaviour has led to frustration and conflict with forest managers and other lawful users [132,133].

4.2.2. Forest Ownership Change

We found instances of access conditions changing because of a transfer in forest ownership. An example of this occurred in 2004 at forests at Riverhead and Tairua when ownership and management transferred from Carter Holt Harvey (CHH) to Rayonier. CHH had required mountain bikers to obtain a permit to ride in any of the Auckland forests, but by 2010 Rayonier no longer required any permits for passive recreational access on foot (including mountain biking). In a similar example, Baigents required a permit to ride in Hira Forest in 1993; CHH did not require any permit during the latter 1990s, and both Hancock Forest Management and now Tasman Pine Forests require a paid permit (through the Nelson mountain bike club membership) to ride at Hira Forest.

4.2.3. Encouraging Access to Dedicated Areas

In some instances, clubs have actively worked alongside the local forest owners and managers to enable dedicated areas of the forest to be set aside on either long-term lease for mountain biking activity (e.g., Cougar Mountain Bike Park in Tokoroa) or the establishment of a dedicated bike park (e.g., Te Miro Mountain Bike Park and Woodhill Bike Park—the latter being a separate commercial mountain bike park outside the forest company's management). In the case of Eskdale Mountain Bike Park in Tangoio Forest, the relationship with PanPac Forests has seen the forest managed in such a way to accommodate continued mountain bike access safely during harvesting operations. When the Waipunga area of Tangoio forest was due for harvest in 2016, the Eskdale club website noted that new tracks were being established in another section of the park (Pakuratahi Valley) ahead of the closure of access to the area, to ensure continuation of access for mountain biking. Whakarewarewa Forest, managed by Kaingaroa Timberlands, developed a 10-year plan devised with the local mountain bike club to ensure continuous public access to the forest. Other forest owners and managers have subsequently adopted this type of access policy, and today track networks and mountain bike parks are plentiful. Close associations and engagement between forest owners and the local mountain bike clubs means that areas that are most suitable for track development can be identified and managed for recreation, and continued safer forest access for riders can be provided in conjunction with continued harvesting operation.

The last two decades have seen a change in forest company awareness of the need for better mountain biker access, towards at least a social contract for public access (if not a more formalised arrangement with a local biking and track-building club). Most forest management plans now report not only on recreation, but actively use the provision of access for recreational users as a promotional and social benefit. For forests that are Forest Stewardship Council (FSC) certified, the forest can be seen as having "High Conservation Value" if you allow people to recreate in the forest [112].

Today, very few managers of production plantation forests actively discourage recreational mountain bike users, but several management strategies are being employed to ensure safe and enjoyable mountain biking conditions for those recreating within the forest. While only 14 larger companies we surveyed have a formal policy around mountain biking access, most have some form of a public access policy or recreational access via permit (Table 2). The main reason for a permit is to monitor usage, and to ensure the public are aware of access conditions to mitigate hazard risks.

4.3. Risks to Production Forests from Mountain Biking and Associated Management Actions

Our survey of forest managers received responses representing 10 forestry companies. Where mountain biking is occurring within their forests, respondents outlined how they are presently accommodating this activity, the risks and benefits they foresee from the presence of the sport within a production forest, and any policies or procedures in place to mitigate disruption to forest production.

Respondents stated that they had identified risks relating to use conflict, liabilities, and emergency management connected to mountain biking. The following is a summary of identified management issues and risks from survey responses:

1. Fire risk and emergency management: Fire poses significant risk to managers of plantation forests in New Zealand, particularly because of the size and remoteness of many management units, a large cohort of weekend cyclists using the tracks, the difficulty in assessing the number or location of riders at risk, and the intensity of plantation forest fires [134,135]. Forest managers need the ability to evacuate the forest quickly should fire or an emergency event occur. Forest companies manage risk through forest closures in high wind or extreme fire conditions, and by keeping main

fire exit tracks well-cleared. A few companies now have the capability to instantly message club members in emergency or for updates on forest closures. A few forest managers hire security to deter unpermitted access during extreme weather and as part of a wider illicit behaviour management.

2. Illicit track building: Many early mountain biking trails in exotic plantations were developed without permission. Consequently, some tracks exposed riders to active forestry operations (e.g., thinning, felling, hauling, weed control) with serious safety risks. Sixty percent of respondents managed unauthorised track development through partnering with mountain bike clubs or entering formal contract with a trail building enterprise or local government. In some cases, forest managers destroyed clandestine tracks to deter illegal riding. By notifying local clubs of the five-year harvesting plans, this allows the clubs and track builders to schedule new tracks, so that these are available when tracks in compartments being harvested are no longer available. Forest managers want to work with the mountain biking community, but have long term operational plans to meet:

   "We do not let mountain bike activities alter our forest management practices, but instead regularly engage with [Club] so as they can be up to date with 5 year harvesting schedules that may impact their track infrastructure".

Some forest managers stipulated that some high hazard work (i.e., tree felling) was reserved for forestry company personnel. Some forests (e.g., Whakarewarewa Forest) have implemented innovative mechanisms for feedback and suggestions from the local riding community [132].

3. Use conflict with forest operations: Harvesting operations present risks to riders' safety if not managed well. Permits give the forest manager the ability to know how many people, and who, is accessing the forest. The benefit to the owner in clearly outlining the accessible trails on a permit should encourage users to keep to these areas of the forest, rather than providing blanket free access to all forest zones. Some mountain bike parks use counters to provide further details of track usage. This also allows forest managers to monitor track usage for management and planning future development. Managing riders around forest harvesting and thinning operations involves developing concentrated areas of riding to encourage riders to willingly stay on known rides, and using clear messaging (e.g., signage) for track and forest unit closures to avoid riders inadvertently entering operational areas. However, interactions between riders and forest operations are inevitable; one forest manager stated, *"we've had a few near hit reports from log trucks and crew vehicles with mountain bikers,"* in spite of having restricted access to non-operating hours and adequate warning signage at the entrance to forests.

Eighty-two percent of forest companies surveyed allow entities, such as mountain bike clubs, to run mountain biking events at the weekends to avoid the risk that forest crews and cyclists would be on the same roads. Two forest companies indicated that they also give private event organisers special access for hosting of larger events. Forest managers use permit conditions, such as time, area and expected behaviour, to avoid conflict with operations. Nevertheless, *"people may have a valid permit but sometimes see it as their right to use the forest ignoring warning and forest hazard signage."* Several forest managers also reported warning signage were regularly ignored and conflicts between cyclists and forestry staff were not unusual, especially in non-permitted areas: *"We don't allow any other mountain biking (apart from specific events) in our other forests. There is also unpermitted mountain biking in most forests, but these people are asked to leave by whoever finds them".*

4.4. Forest Management Benefits Accruing from Recreational Mountain Biking

Our survey of forest managers also revealed perceived benefits from the presence of mountain bikers in productive forest environments. Some of the benefits seen by forest managers in having mountain biking occurring in the forests include increasing forest security and community engagement. In addition, forest access for mountain biking can lead to increased numbers of people visiting the forest and spending money in the local community, further promoting forestry social licence. This survey response from a forest manager emphasises how giving access helps improve the public perception and understanding of the need for commercial production forestry:

> *"Having stakeholder engagement with the forest also enables them to understand the business model and necessity of harvesting operations for the continued growth of mountain bike tracks in our estate. For if there was no forestry business, there would be no mountain bike tracks there".*

In the context of a commercial harvesting operation, forest managers perceive the construction of mountain bike tracks as having low environmental impact, and negligible economic impact on the forest production and returns. However, they can have positive economic benefits for the forest owners and much greater spinoff economic benefit to the regional economy, for very little effort on the part of the forest company. This is particularly the case where forest companies have entered a partnership arrangement with a local club or mountain bike park operator. Those we interviewed that had such an arrangement in place reported that this arrangement provided multiple benefits to the forest company, such as:

- Helping to prevent illegal entry into the forest. There is a perception that having lawful users in the forest deters other illegal activities: *"Good to have valid forest users which assists to deter unauthorised use of forest"*. Another forest manager stated rider presence "stops some of the dodgy activity". In addition, such partnerships can alleviate the risk of illegal track construction, and ensure long term track building and maintenance that does not compromise commercial forest operations, including harvesting, thinning, pruning.
- Concentrating riders: By actively concentrating mountain bike track networks and permitting tracks to be built to meet a wide range of riding styles, forest companies benefit by having better control of where in the forest riders are recreating. This reduces the risk of interactions between mountain bikers, other recreational users and forestry operations as riders gravitate to purpose-built single tracks. Some forest managers have moved towards establishing dedicated mountain bike parks/zones rather than having recreational use coincide around the harvest plans or developing silvicultural plans that fit around the trail developments. This can be a useful strategy; however, consultation and approval for such long-term lease arrangements when proposals are received for a permanent recreational facility is required from the landowner.
- Public Liability insurance can be managed through the club via membership. Some forest owners partner with local mountain biking clubs and only allow forest access for riders with club member tags on their bikes, allowing access only to affiliated club members. Clubs can also more easily carry liability insurance for the events they hold.

Access via club memberships makes the access conditions straightforward and it is easy for all members to receive updates and know the rules for access and hazards to watch out for. Forest managers need only to engage with one group to inform all legal cyclists of the upcoming changes in forestry operations, track closures, hazards or events. Forest management plans relating to recreation or impacting on the club's activities can also quickly and easily be shared, minimising misinformation and rumours. An example of misinformation is seen in an online forum from 2006 with a rider notifying recreationists that access to Riverhead Forest tracks was being revoked, where the forest manager was simply canvassing a review of users to provide improved access and had to spend time reassuring the user group [89]. In addition, it removes day-to-day management of recreationists from being the concern of the forest manager: *"It is great to have other organisations run the day-to-day management of bike trails and to be the go-between for forest managers to mountain bikers. It simplifies the management of it, while allowing mountain biking to occur"*.

5. Discussion

5.1. Forest Management Responses to the Sport of Mountain Biking

5.1.1. Management and Public Access

Production plantation forests are increasingly part of the fabric of mountain biking in New Zealand. However, we have found that the use of forests is changing. Larger forest areas are becoming less accessible, whereas smaller forests near urban centres and holiday destinations are becoming increasingly important. This trend is likely following people's expectations for higher quality, highly interconnected track networks (e.g., mountain bike parks) in lieu of the overgrown forestry roads once worthy of mention. In 1991, Kennett encouraged riders to explore forest plantations to seek out unknown trails that might exist [77]. Few forests had existing tracks available for mountain biking, and of those that did, most provided access for bikers only to existing operational forestry roads [77]; a lot of riding in plantation forests at the time was not explicitly permitted. In contrast, today's exotic production forest managers are more active in establishing protocols for mountain bike use in their forests than thirty years ago. Forest managers were aware of the potential for issues in having mountain bikers within a forest environment, and two decades ago, Wenita Forests raised similar concerns to those raised through our recent survey: "*Uncontrolled public use of our forests is a concern to us due to the safety issues it raises, the potential risks to the forest and the problems associated with conflicting activities being carried out at the same time.*" (Wenita, 2002 in [78]). Where "illegal" trail building has occurred in the wrong forest locations, this has resulted in increased erosion, damage to ecology, and disturbance to nesting birds [136], along with fines to the track builders.

Whether or not forest owners desire to provide recreation, it is best to manage the recreational activities already occurring in the forest [73,137,138]. There are, however, many benefits for the forest owner in allowing recreational biking access to the forest. Reasons purported for the suitability of large exotic forests for mountain biking activity include the ability to provide economic benefit to neighbouring communities; the sheer size of the resource (provides potential for kilometres of track); and provision of economic returns to the forest owner through payment for ecosystem services (either through rider permits or land lease to clubs for dedicated bike-related enterprise) [41]. Promoting public use has proven to have positive economic, social, and environmental impacts to production forests worldwide. In the case of Coed Llandegla, a 650 ha mostly Sitka spruce (*Picea sitchensis*) forest in north Wales, public use of the forest for mountain biking rose from a handful of competitive events in the late 1990s to over 200,000 visits per annum today, through dedicated trail building. Contractors were hired to create trails, and a visitor centre created 20 local jobs with over £1m per annum turnover from visitors [139]. Similar examples can be seen in New Zealand, most notably in Auckland Forests, Tangoio Forest, the Whakarewarewa Forest, Christchurch Adventure Park, Ben Lomond and Codgers [14,16,68]. All these forests are near to larger urban centres, attracting a pool of local riders and those visiting at weekends.

In 1974, Rennison stated "It has indeed been claimed that if Whakarewarewa were managed for tourism and recreation it would be more profitable than managing it for wood production" ([93] p. 70), a proposition later proven in an economic study from 2012 [115], estimating an annual worth of NZD 10.2 million from mountain biking activity, and predicting biking revenues in the Whakarewarewa Forest could rise to even be five times the value of timber production [72]. A similar economic study from the same period showed indirect benefits of mountain biking to the local Rotorua economy amounted to NZD 33 million (RDC, 2006 in [137]). A related report shows that cycle tourists (mostly mountain bikers) in Rotorua spent about 52% more than average tourists (i.e., NZD 3800/stay and NZD 2500/stay, respectively) [140]. More recently, Nelson's regional mountain biking trails accounted for NZD 17.1 million in revenue in 2018, and mountain biking enterprise employed 211FTEs; however, by 2028, total annual economic impact for the region is forecast to be NZD 39.5 million in GDP and the total employment of 538 FTEs [14]. Planted forests maintained for commercial harvest therefore may offer a complementary tourism economy resource to our national parks, while allowing for more adventurous cycling, but require different management than parks [41] for the safe and enjoyable accommodation of bikes and riders while continuing production forestry.

5.1.2. Managing Risk

A number of known risks exist for the forest manager in relation to permitting mountain biking activities [2]:

(a) Social risks that develop due to use conflict between groups of users;
(b) Physical risks to the trees, soil and water from mountain biking activities;
(c) Safety and security risks due to presence of riders in an operational forest.

The likelihood for use conflict has increased where the number of riders and/or other recreational users has grown [2,17,51]. Recreational use of production plantation forests in the early 1990s was low, so mountain bikers were less likely to encounter use conflict with other recreationists. Still, Kennett et al. [76,77,121] does note incidences of use conflict between mountain bikers and motorised sports, hunters and fishermen [77], particularly where both were using the forest roads and 4WD tracks. Forest managers contributed to the legitimacy of mountain biking in the community through partnering with clubs and leading mountain bike development. The U.S. Forest Service had a similar effect by partnering with the IMBA and local clubs to encourage managed track development [141].

We found that trade-offs occur in shared use areas; for example, the Southland District Council advocated both hikers and mountain bikers avoid forested areas during peak hunting season [118]. The promotion of a mountain biker code [47] that promotes respect for the environment, landowner and other users and club affiliation may lessen use conflict.

Physical risks include soil erosion and compaction, disruption of soil structure, vegetation loss, bark loss and root disturbance, nutrient flow into waterways, and the introduction and spread of weeds and pathogens [41,60]. Several studies [2,4,25,41,50,51] have shown that mountain bikers consider the impact of their activity to be lower than what other users perceive as the impact. While the impact of mountain biking on flat dry trails is similar to walking [51], introducing speed, slopes and wet conditions can exacerbate riders' impacts [41,61–63,142]. Land managers have found the impacts to the landscape are reduced through observing the Mountain Biker's Code and development of dedicated mountain bike tracks [4,121]. However, the environmental impacts vary greatly depending on whether the tracks are wild/natural routes, or hand-dug, or highly mechanised machine built [79]. The sheer number of riders on the trails now makes mountain biking an intensive sporting activity in some production forests (e.g., Esk/Tangoio, Woodhill, Whakarewarewa). While the largest impact on the forest environment from mountain biking occurs during track building, competitive racing is thought to have greater impact than recreational biking, particularly as race events are not postponed due to wet conditions, and the faster speed coupled with tighter cornering can cause ruts and increased soil compaction [2,4]. Pickering et al. [82] suggests that informal trail development has a similar impact to hardened trails, though hardened trails may be effective in lessening damage on sloped or wet areas of track. Manmade jumps and downhill challenges are common in race conditions, and greater damage is likely from the downhill trails, which are often located in steeper exotic planted forest areas. Forest managers can manage mountain biking impacts through legislation and education [17,41].

Increasing recreational visits increases the risk of conflict between forest operations and riders that results in injury, but increasing recreational visits may also mitigate criminal activity. Rennison [93] noted that during the 1950s and 1960s, in response to a large fire event of 1946 forest managers removed all public access to exotic forests, issuing trespass notices at the forest gate and prosecuting those who entered unauthorised. Casual visitors were viewed at that time as a great danger to the planted forests. In contrast, today several forest managers see that providing public access can in fact reduce risks, as there are a greater number of lawful users to report on illicit activity, or able to note emergencies and smoke. Most forest managers control public access to their forested estate for recreational purpose, usually managed via a permitting system, or a security firm. Other forest managers work in league with local clubs to jointly manage access, manage safety plans, and to provide information and support so that local riders are educated about best practice and any owner concerns around the use of their forested resource for mountain biking is discussed and rectified. Some regions have implemented governance and management groups for overseeing the mountain biking trails, as a collaborative community, landowner and sporting club partnership.

Forest managers' response to risks include signage in the forest entrances, and on both forest company and local club websites outlining access and any areas where poison bait (for pest control) or tree felling is occurring or scheduled; making forest maps available to the public, indicating trails that are for mountain biking, and roads or tracks that do not have access for biking; allowing access only by permit, or selling maps (most proceeds are used to maintain tracks); and restricting access during fire seasons or only allowing afterhours access during weekdays when the forest compartment is operational.

5.2. Forest Management Implications

Commercial forests in New Zealand have become an important recreational resource, and recreational needs are now being accommodated alongside production requirements. Forest policies and management must accommodate a growing social desire to recreate in local forests, particularly from mountain bikers. The introduction of mountain biking adds to the responsibilities of managing a commercial forestry operation, placing an impetus on forest managers to provide safe access for an increasing population of recreational users, while continuing production operations. The introduction of the FSC and forestry reporting under the Montreal Process are driving social and environmental responsibilities with respect to production plantation forest management [137,143]. Even if no legal contractual obligation is present, a growing social contract to the surrounding community, coupled with an increasing public desire to access the exotic forests in their locality, may require forest managers (particularly where forests are located close to urban centres) to actively manage and implement policies which accommodate recreational mountain biking access [144,145].

Recreation is compatible with production forestry with potential to enhance forest revenue. Local government bodies throughout New Zealand are increasingly developing bike parks in small-scale plantation forests owned by the local council. In some cases, a small fee for access and the establishment of dedicated track networks have seen exponential growth in recreational bikers entering the forest area. This growth has also given a boost to the local hospitality and retails, as well as mountain bike ventures such as shuttle services and bike repairs [14]. Forest managers now have the view that mountain biking has the potential to be a valued resource. While economic studies have estimated the value of recreation [115,146], determining preferences and use behaviour is important in determining what the public want and what is drawing them to recreate in a specific way. Making this model work economically demands that recreational use must be managed alongside an operational forest. In some cases, partnering with international investors has helped local governments to provide world-class commercial mountain bike parks in their forested estates. Having so much public experience of biking in a commercial plantation setting provides "*a superb shop window, showing how wildlife can thrive and people can enjoy themselves within an economically productive plantation*" [139], p. 2. Cooperating in the design of trails with local volunteer enthusiasts may provide similar public relations benefit to the forest sector [49]. Engagement with mountain bikers is paramount, not only for track building, but in wider aspects of forest management, such as land tenure change, finding ways to incorporate changing rider expectations from tracks, and to identify the most optimal areas of the forest for future track developments to avoid erosion-prone fall zones or note areas that might require additional wildlife and stream protection due to increasing track use [75,147].

In the 1970s, mountain biking was not part of planning and management for production forest plantations. Bignell [55] states that most forest-based recreation in New Zealand was limited to half-day family exploration for short walks and picnics. How things have changed since 1981, when he noted *"there is not a requirement to provide "highly exciting" opportunities for recreational pursuits. Whilst a small segment of the population may enjoy this form of recreation, there is a considerable majority which does not."* (p. 402). Spellerberg [138] stated that there were *"few requirements to provide for amenity or landscape issues"* (p. 21) in the management plans of production forests. The provision of recreation needs to be planned for, and public users educated around both the environment they are using, and best biking practices to adopt while recreating to avoid negatively impacting the forest ecology or potential production return. Determining attitudes to landscapes and recreational use behaviour is important in determining what the public want from the forest environment, what is drawing them to recreate in this landscape over other options. Gathering knowledge of this nature will help to determine the potential of a forested estate for various recreational pursuits and assist in the identification of areas suitable for track design and development.

In incorporating recreational facilities and trails within their forests, it is important that forest managers also look to the future, and account for demographic trends and emerging recreational behaviours as potential drivers of the future recreational activities in forested environments. Forest planners need to plan road and trail access, and construct forest blocks that can accommodate the needs of forest users both now and in the future.

There is also a lack of sufficient data and models concerning recreation and amenity values within forests to assist in investment decisions and forest planning for the incorporation of multiple values into forest management plans. The carrying capacity of production forests should be determined, in terms of balancing ecological and social requirements, while maximising production returns [47].

Experience shows that potential visitation rates of large commercial adventure parks can grow 10-fold within a year [118], compounding the need to plan and manage the resource sustainably. Forest managers we surveyed noted the rising trend of electrically assisted (e-bikes) bikes in their forests. E-bikes provide a new set of challenges for forests managers because of their speed and range [37]. In addition to the anticipated growth from an increased population base and local tourism, these technologies enable a greater number of potential cyclists to enter commercial forests.

Models and research data that increase understanding of how these multiple use complexities can best be incorporated into forest management policies could maximise the returns to forest owners. Incorporating mountain biking (and other) recreational pursuits is highly desirable, especially if best practice learnings can be incorporated into forest standards or management principles.

6. Conclusions

Results from the database of mountain biking presence and access conditions, review of annual forest management plans and analysis of the recent survey of forest managers show key trends in forest management to accommodate the sport of mountain biking. It is most notable that forest management plans have moved towards incorporating a statement regarding recreation and community access, and that mountain biking access is now being separated out in annual reporting from other recreational activities.

Consideration of policies, partnerships and processes for managing active recreational pursuits such as mountain biking has become an essential global aspect of forest management. There has been a long trend in New Zealand production plantation forests towards increased forest access, particularly in smaller forests near urban centres. Most forest management plans already include environmental and ecological management, and a growing number discuss community partnerships. Developing healthy relationships between forest managers and recreational stakeholders will be essential when drafting future forest management plans.

Forest managers are facing a growing recreational presence in their forests. While increased cyclists pose some risks to managers, mitigation strategies exist to ensure production forestry can continue to have a healthy association with mountain biking.

Supplementary Materials: The following supporting information can be downloaded at: https://www.mdpi.com/article/10.3390/f13081326/s1, Supplementary Information S1 Survey Questionnaire; Table S2: Known presence of mountain biking activity within planted production forests (>1500 ha) between 1991–2022. Sources: Kennett Bros "Classic New Zealand Mountain Bike Rides" Vol 1–8 and trailforks.com; Table S3: Known presence of mountain biking activity within planted production plantation forests (<1500 ha) between 1991–2022. Sources: Kennett Bros "Classic New Zealand Mountain Bike Rides" Vol 1–8 and trailforks.com (accessed on 28 June 2022).

Author Contributions: Conceptualization, K.M.B.; methodology, K.M.B.; software, M.B.S.; validation, R.T.Y.; M.B.S.; formal analysis, K.M.B. and M.B.S.; investigation, K.M.B. and M.B.S.; writing—original draft preparation, K.M.B.; writing—review and editing, M.B.S. All authors have read and agreed to the published version of the manuscript.

Funding: This research received no external funding.

Institutional Review Board Statement: The study was conducted in accordance with the organizational Ethics policy and guidelines.

Informed Consent Statement: Informed consent was obtained from all subjects involved in the study.

Data Availability Statement: Data tables used are available in the Supplementary Materials. Survey data is not provided due to commercial confidentiality of the forest enterprises submitting information.

Acknowledgments: The authors extend thanks to Jonathan Dash, Thomas Adams, Tim Payn, Aaron Gunn, Adrian Loo, Colin Maunder and the late Brian Pritchard for early discussions around the topic. We are also grateful to the managers we interviewed and forest companies for managerial insights and survey information. The authors also wish to acknowledge the input of Andrew Dunningham and David Palmer for Geographical Information datasets and mapping advice, and Steven A. Wakelin for programme support.

Conflicts of Interest: The authors declare no conflict of interest.

References

1. Hawkes Bay Mountain Bike Club. 2013. Available online: http://www.hawkesbaymtb.co.nz/ (accessed on 27 January 2014).
2. Cessford, G. Perception and reality of conflict: Walkers and mountain bikes on the Queen Charlotte Track in New Zealand. *J. Nat. Conserv.* **2003**, *11*, 310–316. [CrossRef]
3. Pickles, K. The re-creation of bottle lake: From site of discard to environmental playground? *Environ. Hist.* **2003**, *9*, 419–434. [CrossRef]
4. Burgin, S.; Hardiman, N. Is the evolving sport of mountain biking compatible with fauna conservation in national parks? *Aust. Zool.* **2012**, *36*, 201–208. [CrossRef]
5. Hödl, V.C.; Pröbstl-Haider, U. Mountain biking in forests—Contribution to a critical discussion. *Nat. Landsch.* **2016**, *48*, 122–131.
6. Ernst and Young; Union Cycliste Internationale. *Events Economic Impact Report*; Ernst and Young LLP: London, UK, 2022; p. 24.
7. Moreley, R. The Rise of MTB. Available online: https://bikebiz.com/the-rise-of-mtb/amp/ (accessed on 22 May 2022).
8. Balton, J. Bike Statistics & Facts [Of 2022]. Available online: https://www.bicycle-guider.com/bike-facts-stats/ (accessed on 23 May 2022).
9. Statista. Number of Participants in Mountain/Non-paved Surface Bicycling in the United States from 2011 to 2021. 2022. Available online: https://www.statista.com/statistics/763737/mountain-non-paved-surface-bicycling-articipants-us/ (accessed on 22 May 2022).
10. Burgin, S.; Hardiman, N. Mountain biking: An opportunity to enhance economic development in Australian rural areas? In Proceedings of the Australian Regional Development Conference, Albury NSW, Australia, 15–17 October 2014.
11. Ryan, K. *Mountain Biking in Tasmania: A Summary of Current Trends and Future Opportunities*; Sport and Recreation Tasmania: Hobart, Australia, 2005; p. 17.
12. Newsome, D.; Davies, C. A case study in estimating the area of informal trail development and associated impacts caused by mountain bike activity in John Forrest National Park, Western Australia. *J. Ecotour.* **2009**, *8*, 237–253. [CrossRef]
13. King, K.; Church, A. 'We don't enjoy nature like that': Youth identity and lifestyle in the countryside. *J. Rural. Stud.* **2013**, *31*, 67–76. [CrossRef]
14. Cox, M.; Drew, C.; Dixon, H. *Nelson Mountain Biking Economic Study*; BERL: Wellington, New Zealand, 2018; p. 26.

15. Ritchie, B.W. Bicycle tourism in the South Island of New Zealand: Planning and management issues. *Tour. Manag.* **1998**, *19*, 567–582. [CrossRef]
16. MCa. *Mountain Biking in Whakarewarewa Forest Economic Impact Study*; Report to Rotorua Lakes Council; Michael Connell and Associates: Melbourne, Australia, 2018.
17. Leberman, S.; Mason, P. Mountain biking in the Manawatu Region: Participants, perceptions, and management dimensions. *N. Z. Geogr.* **2000**, *56*, 30–38. [CrossRef]
18. Mason, P.; Leberman, S. Local planning for recreation and tourism: A case study of mountain biking from New Zealand's Manawatu region. *J. Sustain. Tour.* **2000**, *8*, 97–115. [CrossRef]
19. Ridout, A. Powering along Nelson's Mountain Bike Trails on an Ebike. in Stuff online. 2020. Available online: https://www.stuff.co.nz/travel/back-your-backyard/122032268/powering-along-nelsons-mountain-bike-trails-on-an-ebike (accessed on 26 May 2022).
20. Statistics New Zealand. Making Tracks—Strategies to Create New Mountain Biking Terrain. 2014. Available online: https://sportnz.org.nz/resources/making-tracks-strategies-to-create-new-mountain-biking-terrain/ (accessed on 26 May 2022).
21. Needham, M.D.; Wood, C.J.B.; Rollins, R.B. Understanding summer visitors and their experience at the Whistler Mountain ski area, Canada. *Mt. Res. Dev.* **2004**, *24*, 234–242. [CrossRef]
22. Monz, C.; Kulmatiski, A. The emergence of "fat bikes" in the USA: Trends, potential consequences and management implications. *J. Outdoor Recreat. Tour.* **2016**, *15*, 20–25. [CrossRef]
23. Fix, P.; Loomis, J. The economic benefits of mountain biking at one of its meccas: An application of the travel cost method to mountain biking in Moab, Utah. *J. Leis. Res.* **1997**, *29*, 342–352. [CrossRef]
24. Ewert, A.; Attarian, A.; Hollenhorst, S.; Russell, K.; Voight, A. Evolving Adventure Pursuits on Public Lands: Emerging Challenges for Management and Public Policy. *J. Park Recreat. Adm.* **2006**, *24*, 125–140.
25. Heer, C.; Rusterholz, H.P.; Baur, B. Forest perception and knowledge of hikers and mountain bikers in two different areas in northwestern Switzerland. *Environ. Manag.* **2003**, *31*, 709–723. [CrossRef] [PubMed]
26. Pröbstl-Haider, U.; Lund-Durlacher, D.; Antonschmidt, H.; Hödl, C. Mountain bike tourism in Austria and the Alpine region–towards a sustainable model for multi-stakeholder product development. *J. Sustain. Tour.* **2018**, *26*, 567–582. [CrossRef]
27. Christie, M.; Hanley, N.; Hynes, S. Valuing enhancements to forest recreation using choice experiment and contingent behaviour methods. *J. For. Econ.* **2007**, *13*, 75–102. [CrossRef]
28. Schlemmer, P.; Barth, M.; Schnitzer, M. Comparing motivational patterns of e-mountain bike and common mountain bike tourists. *Curr. Issues Tour.* **2020**, *23*, 1186–1190. [CrossRef]
29. Scottish Mountain Bike Development Consortium. *The Sustainable Development of Mountain Biking in Scotland: A National Strategic Framework*; Forestry Commission Scotland: Inverness, Scotland, 2009.
30. Adventure Projects. 2022. Available online: https://www.mtbproject.com/directory/clubs (accessed on 14 July 2022).
31. Union Cycliste Internationale 2022 UCI. Calendar. Available online: https://www.uci.org/calendar/all/2jnxYAuvjgttyHi6YQ94EJ? (accessed on 12 July 2022).
32. Buning, R.J.; Lamont, M. Mountain bike tourism economic impacts: A critical analysis of academic and practitioner studies. *Tour. Econ.* **2021**, *27*, 500–509. [CrossRef]
33. Freeman, R.A.Y.; Thomlinson, E. Mountain bike tourism and community development in British Columbia: Critical success factors for the future. *Tour. Rev. Int.* **2014**, *18*, 9–22. [CrossRef]
34. Taylor, J. *Mountain Bike World Cup 2002-Fort William: Economic Impact Study*; Sportscotland: Edinburgh, UK, 2004.
35. Mander, B. Mountain Bike Getaways Just Got More Enticing. *Wall Street Journal*, 2 August 2019. Available online: https://www.wsj.com/articles/mountain-bike-getaways-just-got-more-enticing-11564753911?mod=mhp(accessed on 4 July 2022).
36. Chakraborty, K.; Keith, J.E. Estimating the recreation demand and economic value of mountain biking in Moab, Utah: An application of count data models. *J. Environ. Plan. Manag.* **2000**, *43*, 461–469. [CrossRef]
37. Mitterwallner, V.; Steinbauer, M.J.; Besold, A.; Dreitz, A.; Karl, M.; Wachsmuth, N.; Zügler, V.; Audorff, V. Electrically assisted mountain biking: Riding faster, higher, farther in natural mountain systems. *J. Outdoor Recreat. Tour.* **2021**, *36*, 100448. [CrossRef]
38. Wengel, Y. The micro-trends of emerging adventure tourism activities in Nepal. *J. Tour. Futures* **2020**, *7*, 209–215. [CrossRef]
39. Newcomb, T. Amid Cycling Surge, Sport of Mountain Biking Is Seeing Increased Sales and Trail Usage. *Forbes Magazine*, 13 July 2020. Available online: https://www.forbes.com/sites/timnewcomb/2020/07/13/amidst-cycling-surge-sport-of-mountain-biking-seeing-increased-sales-trail-usage/?sh=2c02761e3ddf(accessed on 4 July 2022).
40. Cloke, P.; Perkins, H. "Cracking the canyon with the awesome foursome": Representations of adventure tourism in New Zealand. *Environ. Plan. D Soc. Space* **1998**, *16*, 185–218. [CrossRef]
41. Burgin, S.; Hardiman, N. Extreme sports in natural areas: Looming disaster or a catalyst for a paradigm shift in land use planning? *J. Environ. Plan. Manag.* **2012**, *55*, 921–940. [CrossRef]
42. Beginners Guide to Every Types of Mountain Bike 2022. Available online: https://cyclingmagazine.ca/mtb/beginners-guide-mountain-bike-types/ (accessed on 23 May 2022).
43. Demetz, M. Bikepacking: A Growing Niche for Alpine Cycling Tourism—Weblog, in Weblog. 2016. Available online: https://regdev-blog.eurac.edu/bikepacking-alps-new-trend (accessed on 24 May 2022).

44. Atkinson, L. On the rails: Four Australian cycleways built on abandoned train tracks. *The Guardian*, 4 November 2019. Available online: https://www.theguardian.com/travel/2019/nov/04/on-the-rails-four-australian-cycleways-built-on-abandoned-train-tracks(accessed on 24 May 2022).
45. Cherrington, J. Are e-bikes ruining mountain biking? *The Conversation Aug 28 2021 3.03am*. Available online: https://theconversation.com/are-e-bikes-ruining-mountain-biking-166121 (accessed on 24 May 2022).
46. McVicar, J.; Keske, M.A.; Daryabeygi-Khotbehsara, R.; Betik, A.C.; Parker, L.; Maddison, R. Systematic review and meta-analysis evaluating the effects electric bikes have on physiological parameters. *Scand. J. Med. Sci. Sports* **2022**, *32*, 1076–1088. [CrossRef]
47. Symmonds, M.C.; Hammitt, W.E.; Quisenberry, V.L. Managing recreational trail environments for mountain bike user preferences. *Environ. Manag.* **2000**, *25*, 549–564. [CrossRef]
48. Taylor, S. 'Extending the Dream Machine': Understanding people's participation in mountain biking. *Ann. Leis. Res.* **2010**, *13*, 259–281. [CrossRef]
49. Zajc, P.; N Berzelak, N. Riding styles and characteristics of rides among Slovenian mountain bikers and management challenges. *J. Outdoor Recreat. Tour.* **2016**, *15*, 10–19. [CrossRef]
50. Goeft, U.; Alder, J. Sustainable mountain biking: A case study from the southwest of western Australia? *J. Sustain. Tour.* **2001**, *9*, 193–211. [CrossRef]
51. Cessford, G. *Off-Road Mountain Biking: A Profile of Participants and Their Recreation Setting and Experience Preferences*; Science and Research Series; Department of Conservation: Wellington, New Zealand, 1995.
52. Dilsaver, L.M. *America's National Park System: The Critical Documents, 2nd ed*; Rowan & Littlefield: Lanham, MD, USA, 2016.
53. Malchrowicz-Mośko, E.; Botiková, Z.; Poczta, J. "Because we don't want to run in smog": Problems with the sustainable management of sport event tourism in protected areas (A case study of national parks in Poland and Slovakia). *Sustainability* **2019**, *11*, 325. [CrossRef]
54. Hirano, Y. Forests to Revitalize Local Community: Adaptive Contribution Projects for Legitimacy by Mountain Bikers. In *Adaptive Participatory Environmental Governance in Japan: Local Experiences*; Global Lessons; Miyauchi, T., Fukunaga, M., Eds.; Springer Nature Singapore: Singapore, 2022; pp. 33–50.
55. Bignell, A.W. Recreation Behaviour in New Zealand with Special Reference to Forest Recreation. In *Proceedings of the 11th New Zealand Geography Conference, Wellington, New Zealand, August 1981*; NZFS, Ed.; Victoria University of Wellington: Wellington, New Zealand, 1981; NZFS Reprint no 1920.
56. Marquis, A. Wheels of Change. In *National Parks*; National Parks Conservation Authority: Washington DC, USA, 2012.
57. Hurford, M. Mountain Biking in National Parks? Hell Yeah. 2015. Available online: https://www.bicycling.com/rides/a20041577/mountain-biking-in-national-parks-hell-yeah/ (accessed on 20 May 2022).
58. MacBrayne, R. Bikes likely to use tracks in national parks. *NZ Herald*, 14 October 2004. Available online: https://www.nzherald.co.nz/nz/bikes-likely-to-use-tracks-in-national-parks/RFQCIIIF6CCVXSHB44BNGNBHGU/?c_id=1&objectid=3600780(accessed on 20 May 2022).
59. Karapoti Classic. NZ's Original Mountain Bike Race ! Available online: https://karapoti.co.nz/storage/app/media/History/History%20at%202020.pdf (accessed on 14 June 2022).
60. Thurston, E.; Reader, R.J. Impacts of experimentally applied mountain biking and hiking on vegetation and soil of a deciduous forest. *Environ. Manag.* **2001**, *27*, 397–409. [CrossRef] [PubMed]
61. Pickering, C.M.; Hill, W.; Newsome, D.; Leung, Y.F. Comparing hiking, mountain biking and horse riding impacts on vegetation and soils in Australia and the United States of America. *J. Environ. Manag.* **2010**, *91*, 551–562. [CrossRef] [PubMed]
62. Pickering, C.M.; Rossi, S.; Barros, A. Assessing the impacts of mountain biking and hiking on subalpine grassland in australia using an experimental protocol. *J. Environ. Manag.* **2011**, *92*, 3049–3057. [CrossRef] [PubMed]
63. Hrůza, P.; Pelikán, P.; Olišarová, L. Single-Track Bike Trails in the Moravian Karst as Part of Forest Recreation. *Forests* **2021**, *12*, 1601. [CrossRef]
64. Chavez, D.J.; Winter, P.L.; Baas, J.M. Recreational mountain biking: A management perspective. *J. Park Recreat. Adm.* **1993**, *11*, 29–36.
65. Galloway, G. Psychographic segmentation of park visitor markets: Evidence for the utility of sensation seeking. *Tour. Manag.* **2002**, *23*, 581–596. [CrossRef]
66. Boozer, B.B.; Self, M.; Jankoski, M.J. An economic and impact analysis of the Coldwater mountain bike trail. In *Report for the Calhoun County Community Development Corporation*; Center for Economic Development: Jacksonville, AL, USA, 2012; p. 73.
67. Du Preez, M.; Lee, D.E. The economic value of mountain biking in the Baviaanskloof Mega Reserve, Eastern Cape, South Africa: A travel cost analysis using count data models. In *ERSA Working Paper*; E.R.S.A. (ERSA), Ed.; Economic Research Southern Africa (ERSA): Cape Town, South Africa, 2015; p. 546.
68. MCa. *Review of Economic Impact of Queenstown Mountain Bike Trails*; Michael Connell. Assocs, Ed.; Tourism Recreation Conservation: Dunedin, New Zealand, 2017; p. 26.
69. Moran, D.; Tresidder, E.; McVittie, A. Estimating the recreational value of mountain biking sites in Scotland using count data models. *Tour. Econ.* **2006**, *12*, 123–135. [CrossRef]
70. Chaney, R.A.; Hall, P.; Crowder, A.R.; Crookston, B.T.; West, J.H. Mountain biker attitudes and perceptions of eMTBs (electric-mountain bikes). *Sport Sci. Health* **2019**, *15*, 577–583. [CrossRef]

71. Campbell, T.; Kirkwood, L.; McLean, G.; Torsius, M.; Florida-James, G. Trail use, motivations, and environmental attitudes of 3780 European mountain bikers: What is sustainable? *Int. J. Environ. Res. Public Health* **2021**, *18*, 12971. [CrossRef]
72. Turner, J.A.; Dhakal, B.; Yao, R.; Barnard, T.; Maunder, C. Non-timber values from planted forests: Recreation in whakarewarewa forest. *N. Z. J. For.* **2011**, *55*, 24–31.
73. Scott, A.; Carter, C.; Brown, K.; White, V. 'Seeing is not everything': Exploring the landscape experiences of different publics. *Landsc. Res.* **2009**, *34*, 397–424. [CrossRef]
74. Wilkes-Allemann, J.; Ludvig, A.; Gobs, S.; Lieberherr, E.; Hogl, K.; Selter, A. Getting a grip on negotiation processes: Addressing trade-offs in mountain biking in Austria, Germany and Switzerland. *For. Policy Econ.* **2022**, *136*, 102683. [CrossRef]
75. Wilkes-Allemann, J.; Hanewinkel, M.; Pütz, M. Forest recreation as a governance problem: Four case studies from Switzerland. *Eur. J. For. Res.* **2017**, *136*, 511–526. [CrossRef]
76. Kennett, P.; Kennett, S.; Kennett, J. *Classic New Zealand Mountain Bike Rides*, 2nd ed.; The Kennett Bros.: Wellington, New Zealand, 1993; Volume 2, p. 266.
77. Kennett, P.; Morgan, P.; Kennett, J. *Classic New Zealand Mountain Bike Rides*, 1st ed.; GP Publications: Wellington, New Zealand, 1991; p. 193.
78. McDonald, P. *Going out for a Bike Ride: An AOK Diary 2002–3*; Peter Mc Donald: Dunedin, New Zealand, 2003; p. 84.
79. Quinn, M.S.; Chernoff, G.W. *Mountain Biking: A Review of the Ecological Effects: A Literature Review*; Miistakis Institute: Calgary, AB, Canada, 2011; p. 45.
80. Weiss, F.; Brummer, T.J.; Pufal, G. Mountain bikes as seed dispersers and their potential socio-ecological consequences. *J. Environ. Manag.* **2016**, *181*, 326–332. [CrossRef] [PubMed]
81. Neumann, P.; Mason, C.W. Managing land use conflict among recreational trail users: A sustainability study of cross-country skiers and fat bikers. *J. Outdoor Recreat. Tour.* **2019**, *28*, 100220. [CrossRef]
82. Pickering, C.M.; Norman, P. Comparing impacts between formal and informal recreational trails. *J. Environ. Manag.* **2017**, *193*, 270–279. [CrossRef]
83. Marzano, M.; Dandy, N. Recreationist behaviour in forests and the disturbance of wildlife. *Biodivers. Conserv.* **2012**, *21*, 2967–2986. [CrossRef]
84. Grzelewski, D. Ride every mountain. *N. Z. Geogr.* **2000**, *48*. Available online: https://www.nzgeo.com/stories/ride-every-mountain/ (accessed on 22 July 2022).
85. Adams, T. *A History of Public Access and Biking in Whakarewarewa Forest*; Scion: Rotorua, New Zealand, 2009.
86. Kennett, J. *Ride: The Story of Cycling in New Zealand*; Kennett Brothers: Wellington, New Zealand, 2004.
87. Brocklesby, J. *Active NZ—The New Zealand Participation Survey 2018*; Sport New Zealand: Wellington, New Zealand, 2019; p. 44.
88. MPI. About New Zealand's Forests 2022. Available online: https://www.mpi.govt.nz/forestry/new-zealand-forests-forest-industry/about-new-zealands-forests/ (accessed on 25 May 2022).
89. Roche, M. Forest governance and sustainability pathways in the absence of a comprehensive national forest policy—The case of New Zealand. *For. Policy Econ.* **2017**, *77*, 33–43. [CrossRef]
90. Roche, M. *History of New Zealand Forestry*; GP Books: Wellington, New Zealand, 1990.
91. McBean, C.L. *Ranger: The Evolution of the Role of a Protected Natural Resource Manager*; Lincoln University: Lincoln, New Zealand, 1992.
92. Cleland, R.W. Recreational use of forests. *N. Z. J. For.* **1963**, *8*, 743–750.
93. Rennison, G. Forest recreation, with special reference to the proposed Westland Beech scheme. *New Zealand J. For.* **1974**, *19*, 67–74.
94. Williams, F. *Review of 1979 New Zealand Radiata Pine Management Practices*; New Zealand Forest Service: Rotorua, New Zealand, 1982.
95. Cumberland, G. *A Forest Policy for New Zealand*; New Zealand Institute of Foresters: Wellington, New Zealand, 2018.
96. Council, E.W. *The Guide to Forestry*; Eastland Wood Council: Gisborne, New Zealand, 2018.
97. Borkin, K.M.; Parsons, S. The importance of exotic plantation forest for the New Zealand long-tailed bat (*Chalinolobus tuberculatus*). *N. Z. J. Zool.* **2010**, *37*, 35–51. [CrossRef]
98. Pawson, S.M.; Ecroyd, C.E.; Seaton, R.; Shaw, W.B.; Brockerhoff, E.G. New Zealand's exotic plantation forests as habitats for threatened indigenous species. *N. Z. J. Ecol.* **2010**, *34*, 342–355.
99. Brockerhoff, E.G.; Berndt, L.A.; Jactel, H. Role of exotic pine forests in the conservation of the critically endangered New Zealand ground beetle *Holcaspis brevicula* (Coleoptera: Carabidae). *N. Z. J. Ecol.* **2005**, *29*, 37–43.
100. Seaton, R.; Minot, E.O.; Holland, J.D. Nest-site selection of New Zealand Falcons (*Falco novaeseelandiae*) in plantation forests and the implications of this to forestry management. *Emu* **2010**, *110*, 316–323. [CrossRef]
101. Hutcheson, J.; Jones, D. Spatial variability of insect communities in a homogenous system: Measuring biodiversity using Malaise trapped beetles in a *Pinus radiata* plantation in New Zealand. *For. Ecol. Manag.* **1999**, *118*, 93–105. [CrossRef]
102. Hunter, G. Recreational access to privately owned rural land: A case study in Canterbury Hill Country, New Zealand. *New Zealand Geogr.* **2004**, *60*, 48–57. [CrossRef]
103. MfE. *Legally Protected Conservation Land in New Zealand*; Ministry for the Environment Environmental Snapshot; New Zealand Government: Wellington, New Zealand, 2010; p. 7.
104. Kennedy, J.J.; Bignell, A.W. Forest Recreation and Changing Management Attitudes in New Zealand. *N. Z. J. For.* **1978**, *23*, 202–216.
105. Stevens, B. The eMTB Controversy: eBikes Should Be Allowed On Mountain Bike Trails. 2022. Available online: https://www.mountainzone.com/mountain-biking/emtb-controversy-ebikes-allowed-mountain-bike-trails/ (accessed on 1 July 2022).

106. Roberts, J. *Electric Bikes on Public Conservation Land—DOC Guideline*; Department of Conservation, Ed.; Department of Conservation: Wellington, New Zealand, 2017; p. 8.
107. Stavi, I.; Yizhaq, H. Applying geomorphic principles in the design of mountain biking singletracks: Conceptual analysis and mathematical modeling. *Land* **2020**, *9*, 442. [CrossRef]
108. Kennett, P.; Kennett, S.; Kennett, J. *Classic New Zealand Mountain Bike Rides*, 4th ed.; Kennett Brothers: Wellington, New Zealand, 2002; Volume 4.
109. Dignan, A.; Cessford, G. *Outdoor Recreation Participation and Incidents in New Zealand*; Mountain Safety Council: Wellington, New Zealand, 2009.
110. Yao, R.T.; Harrison, D.R.; Harnnett, M. The broader benefits provided by New Zealand's planted forests. *N. Z. J. For.* **2017**, *61*, 7–15.
111. Yao, R.T.; Barry, L.; Wakelin, S.; Harrison, D.; Magnard, L.; Payn, T. Planted forests. In *New Zealand Ecosystem Services: Conditions and Trends*; Dymond, J., Ed.; Manaaki Whenua Press: Palmerston North, New Zealand, 2013; pp. 62–78.
112. Yao, R.T.; Palmer, D.J.; Payn, T.W.; Strang, S.; Maunder, C. Assessing the broader value of planted forests to inform forest management decisions. *Forests* **2021**, *12*, 662. [CrossRef]
113. MPI. *Situation and Outlook for Primary Industries (SOPI)*; Ministry for Primary Industries Economics Unit., Ed.; Ministry for Primary Industries: Wellington, New Zealand, 2021; p. 43.
114. Barry, L.E.; Yao, R.T.; Bayne, K.M. *Estimating Non-Market Values for a New Forest Park in New Zealand*; Report # *48967*; Scion: Rotorua, New Zealand, 2012.
115. Dhakal, B.; Yao, R.T.; Turner, J.A.; Barnard, T. Recreational users' willingness to pay and preferences for changes in planted forest features. *For. Policy Econ.* **2012**, *17*, 34–44. [CrossRef]
116. Kennett, J.; Kennett, S.; Kennett, P. *Classic New Zealand Mountain Bike Rides*; Kennett, J., Ed.; The Kennett Brothers: Wellington, New Zealand, 2011; Volume 8, p. 332.
117. Taylor, S. Understanding mountain bikers' choices of recreational settings. In Proceedings of the 7th International Conference on Monitoring and Management of Visitors in Recreational and Protected Areas, Tallinn, Estonia, 20–23 August 2014; pp. 132–133.
118. Bryant, G. Mountainbike Park Proves Major Tourism Asset. *The Southland Times*. 22 April 2013. Available online: http://www.stuff.co.nz/southland-times/news/8578954/Mountainbike-park-proves-major-tourism-asset (accessed on 22 July 2020).
119. Kennett, J. *Classic New Zealand Mountain Bike Rides*; Kennett, J., Ed.; The Kennett Brothers: Wellington, New Zealand, 2006; Volume 6.
120. Kennett, P.; Kennett, S.; Kennett, J. *Classic New Zealand Mountain Bike Rides*, 2nd ed.; The Kennett Bros: Wellington, New Zealand, 1996; Volume 3, p. 352.
121. Kennett, S.; Kennett, J.; Kennett, P. *Classic New Zealand Mountain Bike Rides*, 8th ed.; Kennett Brothers: Wellington, New Zealand, 2008; p. 416.
122. Connor, R. *The Mad Keen Mountain Biker's Road Trip*; Random House: Auckland, New Zealand, 2007.
123. Mitchell, D. *Mountain Biking South: 41 Great Rides in New Zealand's South Island*; Craig Potton Publishing: Nelson, New Zealand, 2010.
124. Trailforks, Trailforks Database. 2022. Trailforks Mapping. Available online: https://www.trailforks.com/ (accessed on 28 June 2022).
125. Innovatek. *New Zealand Forest Products Industry Map*; New Zealand Forest Products Industry Map series; Innovatek, Ed.; Forest Industry Engineering Assn: Rotorua, New Zealand, 2012.
126. *Wood Availability Forecasts—New Zealand 2014–2050*; Indufor: Auckland, New Zealand, 2016; p. 71.
127. MPI. *Ministry for Primary Industries Forest Ownership Map 2008*; Forest Ownership Map Series; New Zealand Government: Wellington, New Zealand, 2008.
128. NZ Topographical Map. 2022. Available online: https://www.topomap.co.nz/ (accessed on 20 June 2022).
129. Uenuku Pines MTB Park. 2022. Available online: https://www.mountainstosea.nz/uenuku-the-pines-mtb-park/#:~{}:text=The%20new%20park%20takes%20a,to%20intermediate%20to%20advanced%20riders (accessed on 3 July 2022).
130. Anon. Vorb Website n.d. Available online: http://www.vorb.org.nz/thoughts-the-port-hills-tracks-t86242-300.html (accessed on 25 May 2022).
131. Cooke, E.L. *Use and Perceived Impact of Recreation on the Port Hills of Canterbury with Examples from Kennedy's Bush Scenic Reserve and Ahuriri Scenic Reserve*; Lincoln University: Lincoln, New Zealand , 1999.
132. Ride Rotorua, "Illegal" Trails in Whaka Forest. Available online: https://www.riderotorua.com/blog/post/73218/illegal-trails-in-whaka-forest/ (accessed on 12 October 2021).
133. Dangerfield, E. Native Trees Suffer at the Hands of Mountain Biking Vandals. *Stuff.* 13 August 2020. Available online: https://www.stuff.co.nz/environment/122421410/native-trees-suffer-at-the-hands-of-mountain-biking-vandals(accessed on 24 May 2020).
134. Pearce, H.G.; Anderson, S.A.; Clifford, V.R. *A Manual for Predicting Fire Behaviour in New Zealand Fuels*; S.R.F.R. Group, Ed.; Scion: Christchurch, New Zealand, 2008; p. 130.
135. Zald, H.S.J.; Dunn, C.J. Severe fire weather and intensive forest management increase fire severity in a multi-ownership landscape. *Ecol. Appl.* **2018**, *28*, 1068–1080. [CrossRef] [PubMed]

136. Urie, H. Boulder discovers illegal, 'extreme' mountain bike trail on Flagstaff. *Daily Camera*, 18 August 2011. Available online: https://www.dailycamera.com/2011/03/28/boulder-discovers-illegal-extreme-mountain-bike-trail-on-flagstaff/ (accessed on 24 November 2021).
137. Hock, B.; Payn, T.; Clinton, P.; Turner, J. Towards green markets for New Zealand plantations. *N. Z. J. For.* **2009**, *54*, 9–18.
138. Spellerberg, I.; Sawyer, J. Multiple-use, biological diversity and standards. *N. Z. For.* **1995**, *39*, 21–25.
139. Anon. Trailblazers: UPM's Forest Mountain Biking Centre. Available online: https://newgenerationplantations.org/multimedia/file/b4651e2d-77cc-11e3-92fa-005056986313/ (accessed on 26 November 2021).
140. Grow Rotorua (Ed.) *Grow Rotorua Annual Report 2013/2014*; Grow Rotorua: Rotorua, New Zealand, 2014.
141. Tory, S. Illegal bike trails and a Forest Service crackdown divide a town. *High Country News*. 20 July 2015. Available online: https://www.hcn.org/issues/47.12/illegal-mountain-bike-trails-and-a-forest-service-crackdown-divide-an-arizona-town (accessed on 24 November 2021).
142. Evju, M.; Hagen, D.; Jokerud, M.; Olsen, S.L.; Selvaag, S.K.; Vistad, O.I. Effects of mountain biking versus hiking on trails under different environmental conditions. *J. Environ. Manag.* **2021**, *278*, 111554. [CrossRef]
143. Klinger, S.; Bayne, K.M.; Yao, R.T.; Payn, T. Credence Attributes in the Forestry Sector and the Role of Environmental, Social and Governance (ESG) Factors. *Forests* **2022**, *13*, 432. [CrossRef]
144. Egunyu, F.; Reed, M.G.; Sinclair, A.J.; Parkins, J.R.; Robson, J.P. Public engagement in forest governance in canada: Whose values are being represented anyway? *Can. J. For. Res.* **2020**, *50*, 1152–1159. [CrossRef]
145. Harshaw, H.W. Outdoor recreation participation in BC forest-dependent communities. *For. Chron.* **2008**, *84*, 210–220. [CrossRef]
146. Tyrväinen, L. The amenity value of the urban forest: An application of the hedonic pricing method. *Landsc. Urban Plan.* **1997**, *37*, 211–222. [CrossRef]
147. Newsome, D.; Stender, K.; Annear, R.; Smith, A. Park management response to mountain bike trail demand in South Western Australia. *J. Outdoor Recreat. Tour.* **2016**, *15*, 26–34. [CrossRef]

MDPI
St. Alban-Anlage 66
4052 Basel
Switzerland
Tel. +41 61 683 77 34
Fax +41 61 302 89 18
www.mdpi.com

Forests Editorial Office
E-mail: forests@mdpi.com
www.mdpi.com/journal/forests

www.ingramcontent.com/pod-product-compliance
Lightning Source LLC
Chambersburg PA
CBHW061551210926
51850CB00009B/2066